HOW IT HAPPENED

HOW IT HAPPENED

From the Big Bang to Civilization and Beyond

Franklin Robinson

Breezemont Publishing
Charleston, West Virginia

Breezemont Publishing
915 Breezemont Drive
Charleston, WV 25302

Edited by Robert W. Walker, www.robertwalkerbooks.com
Cover Design by Mary Drolet, www.marydroletdesign.com

ISBN: 978-0-578-00709-0

Library of Congress Control Number; 2008944179

Table of Contents

Table of Figures

Index of Tables

Preface

Ernest Hemmingway rewrote the ending of *A Farewell to Arms* twenty-nine times before satisfied. While not comparing myself to Mr. Hemmingway, at least in this one respect, I may have equaled his performance; but in my case, rather than a search for the perfect ending to a great classic, my persistence came at the *front end* of a nonfiction book; attempts to "break the ice" and get things moving. One clichéd beginning is for the author to explain why she or he wrote the book. Having worked on *How IT Happened* for several years before deciding to put it in book form, I struggled to recall the precise motivating circumstances that set me on this path. Not surprisingly, the answer finally came at a New Years Eve party; it had been the unintended result of a New Year's resolution. I wanted to address a nagging concern that had bothered me periodically, usually while watching television or reading about subjects dealing with our origins. These might include segments from Carl Sagan's Cosmos series about the workings of our Universe; or perhaps the discovery of fossilized remains of a distant human ancestor, or creatures that preceded human habitation of the Earth. Subjects that added to our understanding of where, when, and how we—and all that we see around us—came to being.

I wanted to understand the history of the Universe. It would begin with the instant it all came into existence some 14 billion years ago when, according to the theory now accepted by science, the

Universe began with a Big Bang. I would then identify mileposts depicting the major events along this timeline that culminated with civilizations of intelligent beings tucked away in a tiny corner of that Universe.

Most of us with such interests in our past feel comfortable with our understanding of post-civilization history, after our forbearers began documenting human activity for future generations. However, my interest predated recorded history to times we know only through unintended clues left behind by our ancestors or by nature itself. This segment of the Earth's history meant reliance on more indirect information such as fossils and artifacts; and more recently, the discovery of DNA. Instead of historians, scientists had to write this portion of the human drama by applying reasoning and imagination to such clues; and they did so while at the same time overturning centuries of firmly implanted mythology. Mythology that had served as answers prior to these scientist's discoveries.

A professional career in chemistry and engineering, along with a more than casual interest in other sciences, had led me to a spotty knowledge of various stages of life's evolution; and how the Earth, Sun, and the Universe came into being. But knowledge accrued in this manner is too disjointed, and I felt the need to fill in the gaps: to get an appreciation of the continuous flow of events from the beginning until the present day. I wanted answers to the basic questions that have intrigued me, and I believe much of the rest of humankind since modern humans first appeared on Earth some 200,000 years ago. Scientists of the last Century have made such remarkable progress that for the first time in human history, we have access to systematic information describing all of the events leading from the instant of creation to the existence of a species blessed with the awareness to ponder the questions: What am I? And, where did I come from? It further concerned me that in the twilight of my years, I might not be willing to forgive myself, had I spent a lifetime as a part of this wondrous creation without attempting to understand its legacy.

These events would include the transforming of Big Bang energy into the basic sub-atomic particles of matter, their further conversion to hydrogen and helium, followed by the gravitational collapse of immense clouds of these primordial atoms, and their ignition to become the first stars and galaxies. Further, along the timeline, stars themselves would change when the nascent hydrogen fueling them had converted to helium, an event that signals an end to the lives of smaller stars. The massive gravity of larger stars would allow them to continue shining after this stage by the further burning of helium, converting them to heavier atoms; eventually producing a complete periodic table of elements. After spending the last of their usable atomic fuel, the ultimate fate of many of these giant stars was a fiery explosive supernova death that cast its accumulation of elements into the spatial confines of the surrounding galaxy. There, elements would cool and condense, many combining to form chemical compounds. Swirling accumulations of original Big Bang hydrogen and helium, destined to become new stars, would entrap some of these new compounds, while the others would agglomerate to form the planets that orbit them.

The timetable would then follow one of the many trillions of such planets that happened to be composed of a serendipitous mixture of the necessary conditions for life. One of the Edens of the Universe, it was blessed with all of the atomic elements and chemical compounds, a Moon of just the right size, located the correct distance to stabilize its spin while orbiting at a comfortable distance from a properly sized star. It would then record the major steps in Earth's maturation that formed oceans, landmasses, an atmosphere, and all of the ingredients needed to initiate the phenomenon we call life. Next, it would follow the evolution of the first life form into subsequent plant and animal species with special emphasis on a particular hominid. Finally, it would identify the steps that transformed that first naked human who competed for sustenance on a level playing field with the rest of the animal kingdom, into a speaking, writing, worshiping, clothed, civilized creature.

Returning to the inspiration behind this book, I'd have to say it comes from the same urge and curiosity that motivates research into one's own family tree—the need to know our past while discovering unanticipated ancestors, their triumphs and disasters. Those fortunate enough to uncover information stretching several hundred years back in time, will almost certainly find a relative or two who brought credit to the family name, and perhaps an equal number who brought shame. However, regardless of the thoroughness of the search or how renowned the family, it will eventually stall due to ancestral information dead ends. Unfortunately, the truly fascinating part of any family tree actually begins where this conventional tree ends, and it is the portion shared by every person on Earth. If we could continue tracing the unbroken chains further back, we would find that only ten generations or about 200 years ago, we had 1,024 great[8] grandparents (that's a string of 8 greats preceding grand), and the total number of relatives from aunts, uncles, and cousins, is greater still.

The further back we go, the longer our list of relatives becomes, and at a point about 2,000 years ago, we would find ourselves related to most of the famous (as well as infamous) people living on our section of the globe. Since we usually measure a generation as 20 years, this was the state of our ancestry only 100 generations ago. Eventually on this backward journey, we reach a point where we realize that all of us are relatives; we all received a portion of our DNA from one single individual. In fact, as we progress downward toward the larger lower branches of the tree, we come to the unsettling conclusion that we share common ancestors with all living things on Earth—from baboons to bacteria to broccoli.

We all received one particular strand of DNA, called *mitochondrial DNA*, from the same woman—aptly though posthumously named *mitochondrial Eve*—who lived about 150,000 years ago. All other parts of our DNA unique to modern humans came from about 86,000 other individuals living at different points in time. This does not make everyone's DNA identical since the DNA from the various Eves and Adams has been altered to some degree with every succeeding generation. It only means that progressing backwards in time, there is an unbroken chain of the particular strain that terminates with the individual Eve or Adam. A portion of their DNA remains with us while none remains from their forbearers or peers. Therefore, before descending very far down the tree, at least in biological terms, the ancestral heritage of all humans is the same; the only missing information from our individual trees is in their uppermost branches.

4

Of course, our ancestral lineage continues past mitochondrial Eve and our other ancestral DNA donors, since they had parents as well; it is only that none of their DNA survived. There are some appearance differences between the people of mitochondrial Eve's time and us; they were stockier with thicker brow ridges and, except for an occasional individual with a genetic skin pigment mutation, all had dark skin since they lived in equatorial Africa.

However, a man from 148,000 BC brought kicking and screaming into the present Century—given a haircut, manicure, shave, and dressed in a business suit—would have looked enough like us that he would not be overly conspicuous on a New York subway train. Of course, that depends on which variety of "us" we mean. Just as the anatomies of current human populations vary from African Pygmies to Nordic Swedes, similar variation existed then. However, the daily life style of this ancestor of 7,400 generations ago was very different from ours since his generation had not yet developed language, agriculture, and civilizations. Instead of buying meat from the butcher, or for that matter from slaughtering domesticated livestock, his sustenance and survival depended on competition with other animals, many far stronger and quicker. However, this ancient ancestor's upright stance, prehensile thumbs and a level of intelligence not very different from ours, more than offset other inadequacies; this permitted him to survive and even prosper in a hunter/gatherer existence.

As we backtrack from Eve to more than 200,000 years ago, differences that are more important become evident. We find humans different enough that, while they belong to our genus *homo*, they no longer belong to our species. They are nevertheless ancestors; however, one individual of their group had been born with a genetic mutation that made him or her different from the others, thereby spawning the new species of modern humans.

Two million years ago, our closest ancestors belonged to a different genus named *Australopithecus*. This species proves substantially different in appearance with longer arms, sharply sloping foreheads, huge jawbones, and definitely had inferior intelligence. However, australopithecines fashioned crude sharp edged tools by splitting rocks, and used animal skins for warmth just like us. At six million years ago, our relatives included chimpanzees, preceded a million years earlier by gorillas. By 14 million years ago, we could count orangutans among our kin, and by the 18 million year mark, we could have dined with crested gibbons. We share an ancestor who lived 25 million years ago with Old World Monkeys of South America and with New World Monkeys of Africa 40 million years ago. Our last common ancestor with all carnivores, including cats and dogs, lived about 60 million years ago and traveled on all fours. At 65 million years ago, the last common ancestor of humans and other mammals such as horses, cows, whales, and elephants was alive.

During the Jurassic period 65 to 150 million years ago, our mammalian relatives were scarce, and the few we did have served primarily as appetizers at dinosaur banquets. But our direct ancestors of the period, which we share with kangaroos, anteaters, and bats, did have four limbs and bore their young from internal eggs just like us. Continuing back to 350 million years ago, we have the same ancestors as all land-dwelling vertebrates, including birds, reptiles, salamanders, newts, and frogs. At 400 million years ago, our closest ancestors lived in the oceans. They were able to breathe air and had four limbs; however, these appendages took the form of fins rather than arms and legs. They hadn't developed the ability to sustain themselves on dry land, although a few had ventured beyond the ocean into freshwater and even used their fins to propel themselves on land for brief periods to take advantage of feeding opportunities not available at home; but they wore their skeletons on the inside of their bodies just like us.

As we look even further into our genealogical past beyond the 500 million year mark, the variety, size, and number of species decreases. Our nearest relatives were small multi-limbed creatures—vertebrates like us; however, their skeletons took the form of external shells. One billion years ago, jellyfish-like animals were human ancestors, but like us, they were mobile and moved about in search of food. Near microscopic beings constituted our ancestry two billion years ago, but just like us, their cells contained the nuclei that made possible the replication process needed for further evolution of complex life forms. Beginning three billion years ago, Cyanobacteria, one of our most industrious ancestors, exercised their talent for photosynthesis throughout the next billion years to convert carbon dioxide into oxygen thereby preparing Earth's atmosphere for future inhabitants.

We finally come to our ultimate ancestor 3.5 billion years ago. Our humble beginning was a microscopic creature born from a miraculous coincidence of events, perhaps in boiling hot waters adjacent to a volcanic fault somewhere in a deep ocean trench. Our oldest ancestor likely found sustenance in hydrogen sulfide spewed from a crack in the ocean floor that exposed the molten core below. Our quintessential Eve is a strange creature indeed, reproducing simply by splitting in half; however, her body is composed primarily of carbon, hydrogen, nitrogen, and oxygen—and she is alive—just like us.

We have reached the beginning. But have we? As many intriguing questions remain as have been considered. What was the origin of the atoms that made up our most distant ancestor, and the oceans that spawned her? How did the composition of the atmosphere, proper temperatures, and all the other conditions that made the synthesis of life possible come to be?

The search for these answers carries us even further back to a time about five billion years ago. Our Sun had just condensed from the primordial elements formed during the infancy of the Universe. The Earth, Moon, and other planets of our solar system had just begun to form out of debris from stars that had ended their lives in gigantic explosions called *supernovae*. About seven billion years ago, the very first galaxies and stars began forming from hydrogen and helium atoms that had condensed from the various forms of matter and energy that made up the early Universe. Approaching a point only 380,000 years out from the beginning of our 13.7 billion year old Universe, atomic nuclei and electrons were just beginning to combine to form atomic structures. At three seconds after creation, the Universe was a tiny hot place only five million billion miles across with a temperature of ten billion degrees Centigrade. Protons and neutrons had just begun bonding to form hydrogen and helium nuclei. At one ten millionth of a billionth of a billionth of a billionth (10^{-33})

of a second, the Universe was about the size of a baseball with a temperature of one billion billion billion (10^{27}) degrees, and matter was in a more fundamental form, namely quarks and antiquarks. At 10^{-34} seconds, matter did not exist; the Universe consisted only of energy particles with zero mass. The earliest time that can be described or even studied with the current state of scientific knowledge is 10^{-43} seconds. At this point the Universe was indescribably hot and many orders of magnitude smaller than a single atomic nucleus. This was the moment of creation, the birth of the Universe, its galaxies, stars, and planets, and life itself—the Big Bang.

Making sense

Understanding the Big Bang theory of the origin of the Universe requires us to accept certain concepts that are contrary to our senses and to some may even border on the outrageous. It makes the astonishing claim that all that constitutes our entire Universe, the Earth, our solar system, the Milky Way Galaxy, and all the billions of other galaxies, at one time occupied space many, many times smaller than the point of a sharp needle. *Ridiculous! No Way! Can't Be! Who are you kidding? No wonder Hollywood portrays scientists as nut cases! This is the dumbest theory I've ever heard—fortunately, it's only a scientific theory and not a law!* If any or all of these responses come to mind, you belong to a large majority. Some of these concepts are foreign to our senses, and even their discoverers can only visualize them in mathematical terms. Nevertheless, they provide answers to the questions of when and how the Universe evolved, all the way back to the first fraction of a second following the instant of creation. They not only provide explanations, but thousands of proofs have established their validity without a single disproval; and more importantly, they have withstood many challenges by other scientists bent on disproving them.

One problem in dealing with such radical concepts stems from our heavy reliance on our senses; senses that have served us well in solving the problems of every day experience. However, these senses have been hard-wired to the environment occupied by modern humans for the last two hundred thousand years: the pull of gravity generated by a planet of a particular mass; and an atmosphere and temperatures that result from being in orbit at just the right distance from a star of just the right size and age. Except for occasional jet trips near 600 miles per hour, we travel mostly at speeds less than 80 and consider the 25,000-mph journeys of astronauts extreme. Things that "make sense" are those that behave as we expect in our familiar environment. Therefore, it is not terribly surprising that exploring areas where these factors differ dramatically, challenges our senses. However, we need to keep in mind the fact that as helpful an asset that good common sense represents, if we had relied on it exclusively for the last 4,000 years, our picture of the Earth would still be that of a flat rectangle perched on the back of a turtle. When confronted with phenomena outside the experience of our "earthbound senses" we are handicapped much like the cave dwellers of Plato's Republic. Obstructions prevent the cave dwellers from seeing the opening to their cave, limiting their view of events of the outside world to the shadows cast onto the opposing cave wall by light from the cave mouth. Dark, two-dimensional beings that occasionally moved across the cave wall represented their reality with respect to beings other than themselves.

Another problem lies in the public misunderstanding of what actually constitutes a theory. Many believe a theory is nothing more than musings tossed out by scientists for others to either accept or disprove. In truth, in order for science to accept a hypothesis as a theory, it needs to clear a host of major hurdles. It must be consistent with all experimentally verified pre-existing theories, although it may show a pre-existing theory to be wrong in a precise sense. Rather than resting on a single foundation, the originator must provide supporting evidence from a variety of bases. In order to establish the usefulness of a theory, it should make predictions about things not yet observed that are "falsifiable," e.g., they can be disproven or, from the point of view of the theorist, hopefully proven. The theory wins added credibility if such predictions are risky and clash with established scientific thinking; however, the primary purpose behind testing of a theory is to disprove it.

Even though proven, to gain creditability in the scientific world, a theory must also be dynamic, allowing it to change as others discover new information. It should also be parsimonious; make as few assumptions as possible by eliminating irrelevancies. Stephen Hawking, in his best selling book *A Brief History of Time* stated, "…a theory is a good theory if it satisfies two requirements: It must accurately describe a large class of observations on the basis of a model that contains only a few arbitrary elements, and it must make definite predictions about the results of future observations." He goes on to state, "…any physical theory is always provisional, in the sense that it is only a hypothesis; you can never prove it. No matter how many times the results of experiments agree with a theory, you can never be sure that the next time the result will not contradict it. On the other hand, you can disprove a theory by finding even a single repeatable observation that disagrees with predictions."

However, the intent of a scientific theory is not to be the "last word"; rather, it should be subject to amending or incorporation into wider theories. Using the word theory instead of law to define new explanations of scientific phenomena is more of a 20th Century practice brought about by the rapid expansion of science and the recognition that the truths uncovered may only be subsets of a larger truth yet to be found. If scientists in 1800 discovered the Big Bang theory and tested it to the current level of confidence, they would have designated it a law. To appreciate the Big Bang theory and understand the physics that embody it, to some degree, we need to set aside our "common senses" and replace them with a level of confidence in scientific method.

Comprehending subjects like the Big Bang forces us to consider environments so foreign to the familiar arena in which our senses evolved that said senses are overwhelmed. The creation of the Universe involves phenomena from worlds of immensely different sizes and densities, demanding that our normal instincts be set aside in order to consider strange concepts that conflict with "common sense". Noted physicist and author Michio Kaku introduced his highly acclaimed 1994 book *"Hyperspace"* with the comment "If all our common sense notions about the Universe were correct, then science would have solved the secrets of the Universe thousands of years ago". It is a study that involves particles and accumulations of matter at scales that range from near infinitesimally small to mind-bogglingly enormous. These phenomena involve bits of matter so tiny scientists measure them in units as low as 10^{-33} (one thousandth of a billionth of a billionth of a billionth) of a centimeter that travel at speeds approaching the speed of light. They also include objects larger than a billion Suns located at distances we measure in billions of light years (a single light year being the distance light travels in a year at 670 million miles per hour).

So what motivates this fascination and interest in our origins? It's a no brainer. A desire to appreciate that to which the greatest minds of our species have devoted their lives and energies; a quest for the holiest of grails; to answer these simple questions: What am I? And, where did I come from? I believe their answers constitute the greatest story ever told.

Numbers

As you may have guessed by now, the story of "How IT Happened" is an exercise in extreme numbers. Furthermore, this not only holds true for measuring mass, distance, and speeds, but also for understanding the likelihood that events of extremely low-probability can actually occur. Events so improbable in fact, we might conclude they could never happen (like the infinite number of monkeys with typewriters authoring all the books ever written—word for word—no mistakes). We must keep in mind however, that, even though an event has one chance in 10^{36} (a trillion trillion trillion) of happening, it has an even chance if given an equal number of opportunities.

Scientists report measurements that require numbers of these sizes in logarithmic shorthand, which involves raising the number 10 to an exponential power. Simply put, an exponent is the reduced-size digit placed to the right and slightly above the number 10; e.g., the exponent is 5 in the value 10^5. This simplifies the viewing of extremely large numbers by avoiding the need to string long lines of zeros across the page requiring the reader to count them in order to know its value. It is also useful when multiplying or dividing extremely large numbers. When the exponential number is positive, we can convert it to the familiar form by simply following the numeral 1 with the number of zeros equal to the exponent. Therefore, 10^0 equal 1; 10^1 equals 10; 10^2 equal 100; 10^9 equals 1,000,000,000 or 1 billion and so on. To get the value of the product of exponential numbers, you only need add the exponents, e.g., 10^9 x 10^6 = 10^{15} or 1,000,000,000,000,000, or a quadrillion (a million billion). To divide one exponential number by another, change the sign of the exponent in the denominator and add it to the exponent of the numerator: 10^{46} ÷ 10^{32} = 10^{14}. Negative exponential numbers behave similarly but represent fractional quantities of 1, e.g. in the series: 10^0 = 1; 10^{-1} = 0.1 (one-tenth); 10^{-2} = 0.01 (one hundredth); and 10^{-13} = 0.0000000000001 (one ten trillionth).

We multiply and divide numbers with negative exponents in the same manner as positive numbers; to multiply, simply add the two negative exponents; 10^{-4} x 10^{-7} = 10^{-11}. The same holds true for dividing negative exponential numbers; change the sign of the denominator exponent and add it to the numerator; e.g. 10^{-33} ÷ 10^{-15} = 10^{-18}. Mixed exponents function the same way; e.g. 10^{-24} ÷ 10^{13} = 10^{-37} and 10^{24} ÷ 10^{-13} = 10^{37}, or in the case of multiplication, 10^{-4} x 10^7 = 10^3.

Of course, problems are not always kind enough to allow us to express them in even multiples of ten. When this is the case, we precede the exponential number with a multiplying factor that consists of enough digits to express the number at the desired accuracy. For example, we normally express the number 786,400,000,000,000 as $7,864 \times 10^{11}$; however, we can write it just as accurately as 78.64×10^{13}, or 7.864×10^{14}. We often do this when comparing a series of numbers where it is convenient to maintain the same exponent throughout and express variation in the multiplier.

The Ouraborus in figure 1 below illustrates the numerical orders of magnitude encompassed in the physics of the Universe, going from the very small to the very large. The Ouraborus shows how humans fit as intermediates in this scale—with a micro-world below us, and a vast cosmos above. The ancient Egyptians and Greeks used the Ouraborus to represent the unity of all physical and spiritual things. It portrays a giant serpent with its tail in its mouth continually devouring and being reborn from itself. On a height basis, we occupy a size category in the Ouraborus at about 2 times 10^2, which is 200 centimeters or 2 meters. The Ouraborus shows the sizes of particles from the micro-world on the left side, and those of the macro-world consisting of planets stars and galaxies on the right.

Objects within the Ouraborus body depict views of the individual subjects from different distances measured in centimeters, each separated by 5 orders of magnitude, or 100,000. A view from 10 centimeters would have us uncomfortably close to an ancient ancestor for most non-confrontational situations at a distance of 10^1, or 10 centimeters (cm). Backing off 1 order of magnitude places us at the distance illustrated in the figure 1, a more sociable 10^2, or 100 cm. If we back off to 3 orders of magnitude at 10^3, or 1,000 cm, we would see him walking toward a tree. A fourth order would reveal a barren area behind and after 5 orders of magnitude, or 100,000 cm, we see a distant mountain. Continuing outward in 5 orders of magnitude increments to 10^{10} cm, we can view the entire Earth. Zooming another 5 orders of magnitude to 10^{15} brings the inner three planets of our solar system into view, and at 10^{20}, we see the entire Milky Way galaxy with our Sun indistinguishable from a hundred billion other stars.

Following the left side of Ouraborus, one order of magnitude inward toward the very small from our 100 cm vantage point to 10^0, brings us to 1 centimeter or just about the limit of the human eye's ability to distinguish detail. Those with exceptional eyesight however can appreciate differences in the ancient face. At 10^{-3}, details of human tissue are discernable and we can examine individual cells at 10^{-5}. Two more levels inward at about 10^{-7} we see large molecules like DNA, and from there, into the ever-shrinking world of atoms, nuclei, and elementary particles.

In addition to providing a very good example of perspective, the Ouraborus is also fitting icon for the Big Bang. It graphically illustrates the intimate wedding of the very large and the very small at the beginning of the Universe.

Figure 1 The Ouraborus

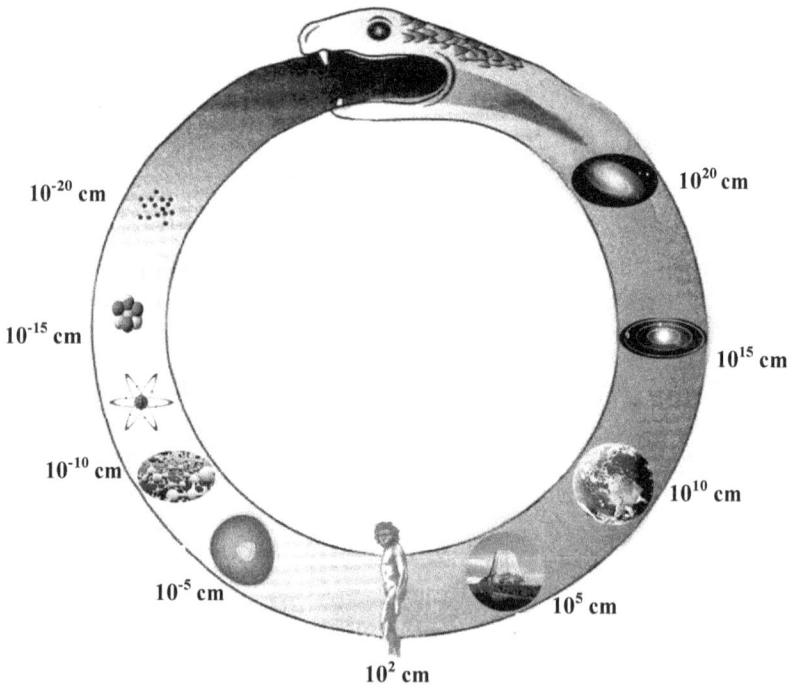

10^{-20} cm

10^{20} cm

10^{-15} cm

10^{15} cm

10^{-10} cm

10^{10} cm

10^{-5} cm

10^5 cm

10^2 cm

1   Realm of Uncertainty

If you aren't confused by quantum physics, then you haven't really understood it.

Niels Bohr, primary contributor to quantum theory

Shoulders on Which We Stand

George Lemaître first proposed the Big Bang Theory in 1927, but this date hardly marks its beginning. It was only one milepost in humankind's long journey toward understanding the Universe in which we live. The Big Bang Theory is about matter and energy and, since they are not quite what they appear, a review of man's progress toward understanding them is in order. Democritus (c400BC) first proposed that all matter must ultimately consist of indivisible particles, which he called *atoms*, a name that has stuck for two and a half millennia. By the early 1800s, knowledge of atoms had advanced substantially, prompting Dimitri Mendeleev to arrange the

15

accumulated information in an orderly fashion. The result, published in 1869 arranged all of the elements according to their atomic weights and repeating chemical properties, into the periodic table of the elements. In the late 1800s Henri Becquerel and the Curies discovered that certain heavy elements would undergo radioactive decay, a process in which large unstable elements such as radium and uranium convert to lighter elements by emitting small particles now known as neutrons.

This discovery provided the first proof that atoms are not indivisible but made of smaller particles. The discovery of the electron in 1900 by J. J. Thompson led Baron Kelvin to propose the Plumb Pudding model, showing the atom composed of positive and negative particles. In 1911, Ernest Rutherford and Hans Geiger conducted experiments by shooting high-energy alpha particles from a radioactive substance at a thin piece of gold foil. They noticed that while most of the particles went through the foil, a few bounced back. From this, they concluded that atoms are mostly empty space with a tiny positive charged nucleus that contains most of the mass, while electrons orbit in the space around them.

Figure 2 Atomic Models

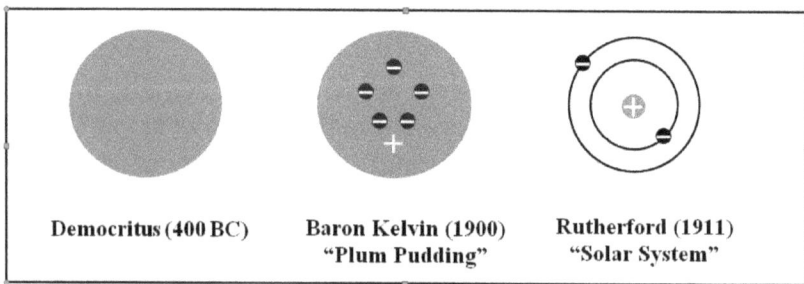

Democritus (400 BC)   Baron Kelvin (1900) "Plum Pudding"   Rutherford (1911) "Solar System"

While the Rutherford atom represented a major step forward in our understanding of matter, due to rapid advances in nuclear physics in the early 20th Century its shortcomings became evident. One of the predictions of Maxwell's equations (see section on Relativity) is that electric charges in accelerated motion should emit electromagnetic waves (light). Therefore, since all light waves are forms of energy, Maxwell's calculations insisted that emission of such waves would represent an energy loss. Radiation would rob the electron of a bit of its energy with each orbit around the nucleus. This would cause the electron to gradually spiral inwards to the nucleus the same way a satellite gradually loses energy due to friction with the upper atmosphere. While the popular conception of accelerated motion is that which accompanies increasing speed, an object traveling at a constant speed also is undergoing accelerated motion anytime it deviates from a straight line. Therefore, an object that follows an arcing path such as the circular orbits of Rutherford's electron is always in accelerated motion and should radiate these waves and in so doing, lose energy.

One could only conclude that atoms should not exist at all since Maxwell's calculations indicated this energy loss would be sufficient to cause the electrons to spiral into the nucleus within a fraction of a second. Another prediction, based on the structure of the Rutherford atom and the prevailing beliefs about light behavior, stated that all atoms should emit full spectra of light; whereas in fact, each atom emits its own unique pattern of discrete wavelengths or colors.

Early 20th Century scientists faced other problems brought about by inconsistencies in the rapidly developing body of knowledge. The *ultraviolet catastrophe* was the most obvious of these. Maxwell's equations predicted that hot glowing objects should emit light due to vibrations of their constituent atoms. Use of the word "ultraviolet" in describing the phenomenon refers to the frequency of light emitted by heated objects, namely the high frequency, short wavelength ultraviolet light. The light bulb and molten volcanic lava are verifications of the prediction, but based on classical physics (a term identifying the state of physics prior to Albert Einstein) the emission should be infinite. Scientists reasoned correctly that heating atoms should increase the vibrations of their orbiting electrons causing them to emit electromagnetic waves. However, in the early 1900s, everyone assumed electron vibrations occur over a continuous range. If true, it would mean availability of an infinite number of frequencies, and that electrons would share all of these equally.

The problem centered on the *equipartition theorem* of classical physics, which states that all modes (degrees of freedom) of a system at equilibrium have some specific average energy. However, classical electromagnetism says the number of these electromagnetic modes per frequency is proportional to the frequency squared; therefore, according to the Rayleigh-Jeans law, the radiated power per unit frequency must be proportional to frequency squared. The Rayleigh–Jeans law was an early twentieth Century construct that attempted to describe electromagnetic radiation from black bodies for all wavelengths at a given temperature. Physicists found the law useful for a period since it agreed with experimental results for long wavelength, low frequency measurements. It broke down however, when testing short wavelengths, predicting an infinite energy output. This suggests the impossible situation where both the power at a given frequency and the total radiated power approach infinity as higher and higher frequencies are considered.

The *photoelectric effect* presented early 20th Century physicists attempting to model atomic structure with another problem. Physicists first noticed the effect in the late 1880s from experiments when they projected light beams onto a metal surface in a vacuum, and noticed an immediate emission of electrons from the surface. Other scientists had recognized the effect, but based on the wave nature of light, classical physics predicted the emission of electrons should happen very slowly. Considerable time should be required for electrons to absorb enough energy from successive waves of light to initiate electron ejection. Classical physics further predicted that increasing the projected light intensity would reduce the waiting time for ejection, and the color of the light (frequency of the wavelength) should make no difference at all.

The experiment, illustrated in figure 3 disproved these beliefs. It shows that light projected on a metal plate causes it to emit a flow of electrons toward the opposite plate, completing a circuit, while a meter measures the electron flow. Using such an apparatus researchers found that ejection of electrons happened immediately regardless of the light intensity; increasing intensity simply produced more electrons but did not alter the time required. Furthermore, the color of the light is important; ultraviolet light emits electrons at higher energy levels, while longer wavelength red light emits none at all.

Figure 3 Photoelectric Effect

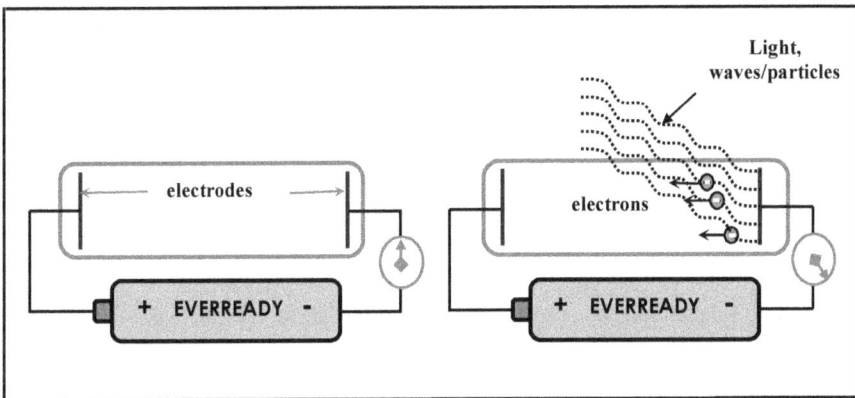

The search for answers to problems such as the ultraviolet catastrophe and the photoelectric effect cast doubt on our understanding of light, motivating early 20th Century physicists to embark on the *quantum revolution*. Quantum theory, the heart of that revolution, is based on the premise that all the "stuff" of the Universe, whether matter or energy is not infinitely sub-dividable but comes in "chunks" or *quanta*. While physicists had long considered matter—atomic nuclei—in these terms, they still looked at energy as strictly a wave phenomenon. The first inkling to the contrary came in 1900 from German physicist Max Planck when he suggested that the ultraviolet catastrophe could be explained if one assumed that atomic vibrations do not occur continuously over a given range but exist only as whole number multiples of some basic amount. Planck's solution stated that all forms of energy are *quantized*, and electrons can only vibrate at frequencies equal to an integer multiple of some basic amount. Light could have frequencies 2 times, 3 times, 10 times or 10 million times the basic amount but could not have fractional multiples such as 0.5, 1.5, or 0.0001. Planck's equation established this basic energy quantity as:

$$E = hf$$

> where: E is the energy of a vibrating particle
> f is its frequency, or the number of vibrations per second
> h is a very small number, 6.626 x 10^{-34} called *Planck's constant* that establishes the proportionality link between the frequency of vibration and the energy.

By stipulating that radiation can only oscillate or be emitted in discrete packets of energy proportional to their frequency, Planck reduced the number of frequency modes from infinite to finite, which effectively limited the average energy at those frequencies. Therefore, the total predicted power is finite, and the high frequency vibrations responsible for the ultraviolet catastrophe do not occur. The vibration frequencies of ultraviolet light are extremely high—ranging from 7 x 10^{14} to 3 x 10^{16} Hertz (cycles per second). Therefore, even when multiplied by the extremely low value of h (6.626 x 10^{-34}) in Planck's equation, E= hf, energy requirement for atoms to emit light in the ultraviolet frequency range is so high that radiation does not occur, thereby eliminating the infinite energy requirement. Planck's explanation did not address the basic cause, only that if quantization proved correct the ultraviolet catastrophe would disappear.

A few years later in 1905, Albert Einstein published results of work unrelated to relativity that provided an explanation for the photoelectric effect, an effort that earned him his only Nobel Prize. Einstein said light energy is not spread evenly over a wave but is quantized in individual bundles he called *photons*, and that quantity of energy can be calculated by Planck's equation $E = hf$. This proposal explains the photoelectric effect because it predicts that a particular amount of energy is required to eject an electron from metal. The low frequency of red light means that it has insufficient energy to eject an electron. Increasing the brightness of red light only increases the number of low frequency photons, each of which have too little energy to eject electrons. However, once the frequency threshold is reached, electron ejection occurs immediately. Further increases in the photon frequency, and therefore the energy, does not increase the rate of electron ejection but simply ejects electrons with higher energy. However, increasing the flow of sufficiently high frequency photons does increase the rate of electron ejection. Furthermore, all ejection incidents happen immediately since the energy to eject each electron comes from a single photon, not from spreading out over a wave, signifying the particle or quantum behavior of light.

While Planck and Einstein had opened the bottle by quantizing energy, it remained to Niels Bohr to put the quantum genie to work. In 1913, Bohr proposed an atomic model that dealt with the problem mentioned earlier regarding energy loss associated with the accelerated motion of electrons (in this case, motion deviating from a straight line) as well as the spectral emissions of atoms. Bohr connected quantization with atomic orbits, stating that the various orbits of atoms (see Chapter 4) require different discrete energy levels, and that the energy of electrons occupying them will match the requirements of the orbit. The orbits quantize according to the angular momentum of the orbiting electron's rotational motion by the following equation.

$$L = h/2\pi$$

where: L is the electron's angular momentum
h is Planck's constant

Bohr did not base his atom on any theoretical concepts but stated simply that electrons in their allowed orbits do not emit electromagnetic waves, and therefore do not crash into the nucleus. However, he did explain the different spectra of atoms. When electrons move from a higher energy orbit to a lower one, they emit the energy difference between the orbits as electromagnetic wave radiation in the form of a single photon. The liberated photon has energy equal to the energy difference between the two orbital's causing it to emit specific colors of light based on the frequencies of the orbits involved.

Light, Particle or Wave

Sir Isaac Newton first suggested that light was made up of particles. Scientists accepted the concept and continued to do so for more than a Century since it explained many of the properties of light including reflection, refraction, and color. However, in 1800, Thomas Young provided what most scientists considered incontrovertible evidence that light is actually a wave. Young reasoned that if light truly is a wave, it must have observable wave characteristics, and one of these should involve wave interference, a phenomenon similar to what one sees when waves from two separate sources intersect on an otherwise quiet pond. Young first passed *coherent*, or "in phase" light through two tiny openings in a dark box in order to create two separate wave fronts of coherent light, that should—assuming light is really a wave—recombine to form wave interference patterns on a screen located at the back of the box. Wave interference patterns result when waves merge due to synchronization of the wave peaks and troughs as they combine. When the peak portions of two waves come together, they are in phase, and reinforce one another, producing a higher amplitude wave, whereas when the peak and trough portions coincide, they are out of phase and effectively cancel each other—essentially killing the wave.

Figure 3 gives an example of in phase and out of phase light. The left side of the figure shows two waves in phase; their amplitudes (peaks) and troughs perfectly coincide. As shown by the combined wave line at the bottom of figure 3, if the two waves have the same amplitude and are perfectly in phase, the combined wave will have twice the amplitude of the individual waves, demonstrating *constructive interference*. The right side of figure 3 shows the two waves 180° out of sync, totally canceling each other out and producing a combined wave of zero amplitude; an example of *destructive interference*.

Figure 3 Wave Phase Relationship

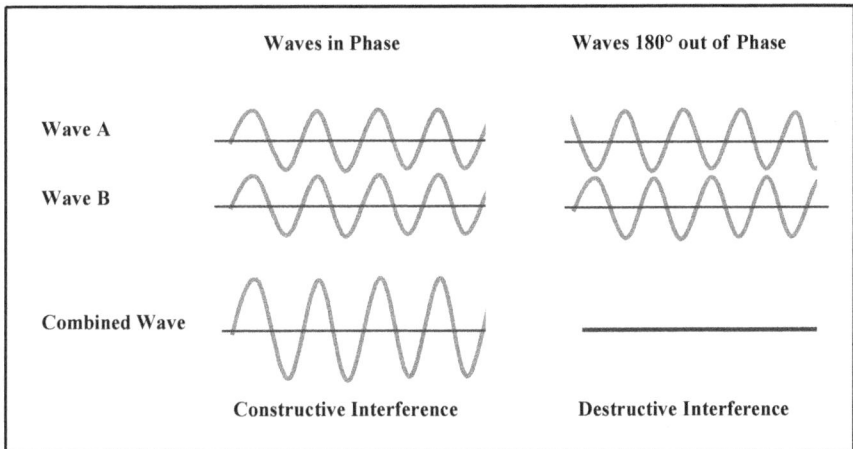

Figure 4 is a modernized version of Thomas Young's double slit experiment (at least an advance from the 18th to the19th Centuries since a photographic plate is added to illustrate the wave interference patterns). Drawing I shows how the experiment would have turned out if light was indeed a particle. Individual light particles would travel in straight lines toward the barrier, and all particles but the ones aligned precisely with the slits would reflect back from the barrier. The ones passing through would strike the screen directly opposite the slits, producing black lines on the negative consistent with the shape of the slits through which the light passed. Drawings II and III show the wave behavior of light with II demonstrating the effect with only a single slit open. As shown in drawing II, with only one slit open, there is no interference pattern and only one pattern produced on the screen.

Figure 4 Light: Wave or Particle

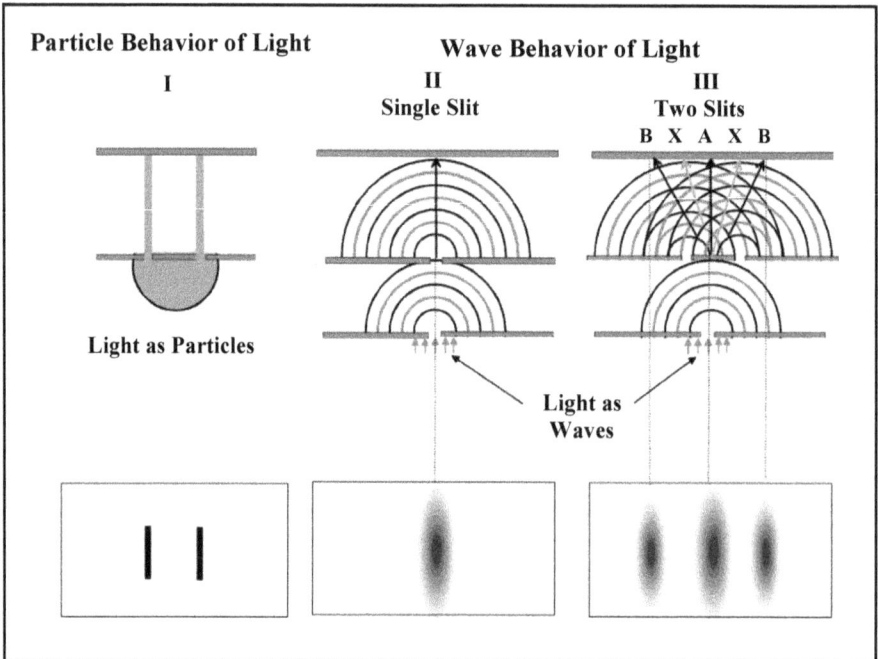

In the two-slit experiment of drawing III, a beam of light first passes through a tiny slit in a barrier, a necessary step to provide a *coherent* light source. The coherent, in phase, light then passes through a second barrier containing two thin slits (actually the slits were razor thin but drawn wider for clarity), and the resulting pattern observed on a screen. The black semicircular lines emerging from the slits denote the peaks of waves while gray lines represent troughs. As light passes through the slits, two new wave fronts, perfectly in sync with one another emerge from the other side. They remain in phase because both slits are situated at equal distance from the light source. If the experimenter moved either of the slits to the left or of right, they would be out of phase since the two waves would not have traveled the same distance.

As the two new wave fronts move toward a photographic plate in the back, they begin to merge, producing black regions on the developed negative that correspond to the intersection of wave peaks from both slits. Light energy in the areas where wave peaks merge with wave troughs is not great enough to darken the film, producing clear regions. Black arrows go through the points on each wave where the peaks intersect to make black areas on the negative and gray arrows through the intersections of their respective peaks and troughs that produce clear areas.

Note that point A on the screen produces the largest black pattern on the photographic negative. This is because point A is equidistant to both slits. This guarantees that both waves will arrive in sync, and as A is the closest point to the slits, point A receives more photons, thus producing the darkest blackening of the negative. As you move to the left and right of central point A, the two slits are no longer equidistant, so light from one slit will need to travel farther to reach the negative. This makes the wave peaks gradually go out of phase and destructive interference kicks in. At point X, they are 180° out of phase, totally canceling each other, and the negative is clear. As you move left or right from the central point A toward the two point Bs, (to the left and right of point A) the wave peaks once more begin to merge and become totally in phase again at 360°. However, at points B, both wave fronts have traveled a greater distance so the pattern on the negative is smaller. Note that the black arrows in the interference pattern, identifying the points of maximum blackening of the negative, run precisely through the points of convergence of the various wave peaks, while the gray arrows that identify clear areas run through the points of convergence of peaks and troughs. Drawing III shows only three darkened areas, but if the experiment is designed properly, more dark areas will appear to the left and right sides of the negative; however, each pair will be smaller than the previous ones.

An even more dramatic example of wave behavior came from experiments run with the light source dimmed until only one photon at a time went through the slits. Researchers found that given enough time, the same interference pattern appeared. This can only mean that a photon approaches the barrier spread out as a wave capable of going through both slits before recombining in a wave interference pattern, and that each photon passing through is aware of both slits.

The wave nature of light received further credibility through work in the 1860s on electromagnetism by James Clerk Maxwell and continued to be the favored concept among scientist throughout the 19th Century. Then in 1905, the pendulum once again swung in favor of particle theorists when Einstein showed he could explain the photoelectric effect if light behaved as a particle with its energy concentrated in particle-like photons. Then in 1923, Arthur H. Compton provided what appeared to be the coup de gras for particle behavior. After bouncing light (actually X-rays) off electrons, he found that light loses energy from the collision in much the same way colliding pool balls transfer energy, with the lost energy manifested by lower frequency light.

These seemingly contradictory aspects of light—that it could reasonably be proved a wave with no particle properties or a particle with no wave behavior—presented a major problem to early 20th Century scientists. The *Principle of Complementarity* answered the riddle by introducing the *wave/particle duality* of light, democratically calling it both wave and particle. The relationship is a statistical one described by an amplitude/frequency curve, which says the probability of finding a photon at a particular point on the wave curve is related to the amplitude (height) of the wave at that point. The probability gradually increases until the wave peak is reached, then diminishes as it nears the trough.

Light continues to behave as a wave until the point of its "capture" or detection, at which time it has the annoying habit of completely changing its behavior. At that time, probability considerations no longer apply, and the photon becomes a particle. Any attempt to detect a photon, such as using it to eject an electron in the photoelectric effect experiment, will cause it to lose its wave properties and bring about its particle nature. Such apparent absurdities permeate quantum physics, prompting many great scientists from first half of the 20th Century to question its fundamental soundness. In fact, Einstein, believing God would not base his physics on such random and undefined principles, spent the last half of his career in unsuccessful attempts to disprove quantum physics. However, the doubters died with Einstein; the Universe must work in a very similar manner since quantum theory has successfully explained the mysterious micro-world through nearly a Century of scientific examination.

In 1923, a young French prince, Louis de Broglie (pronounced de Broy), while pursuing his doctorate, considered this proposition: If light has both wave and particle behavior, perhaps the same property applies to matter as well. He proposed that every form of matter has a wave function (a tool used to describe the possible physical states of a system in mathematical terms). De Broglie suggested that all massive objects from complex molecules to planets do in fact oscillate in the same manner as light; however, the mass of larger objects make oscillations undetectable and insignificant. De Broglie proposed the following equation to describe the relationship:

wavelength $= h/mv$

where: h = Planck's constant, 10^{-33}
v = velocity of particle
m = mass of the particle

While de Broglie's equation applies to all massive objects, effects become negligible for all but the tiniest particles of matter. The small constant value of h in the numerator, combined with the large value for m in the denominator for large objects, produce wavelengths so small relative to the object's size they become insignificant, and therefore unaffected by such quantum weirdness. Some subatomic particles and electrons, however, have masses the same order of magnitude as h and therefore wavelengths similar in size to the objects with which they interact. Furthermore, since wavelength also depends on velocity, larger particles can be cooled to temperatures near absolute zero (the point where molecular motion ceases), which

27

increases their wavelength to the point where it becomes significant in relation to the size of the atom. Experimenters also verified the wave behavior of particles through a modified version of the double-slit experiment in which they used crystals for slits, and replaced photons with electrons. The results showed the same type of wave interference patterns observed with light.

DeBrogli's experiments showed that objects with mass have the same wave properties as light and are therefore subject to the same probabilistic criteria. In 1926, Erwin Schrodinger proposed that every matter particle has a wave equation that tells the probability of finding the particle at a particular place and time. This matter/wave concept provided the explanation for the fact that electrons occupying different atomic orbits have different but discrete energy levels as described in Bohr's atomic theory. The wavelength of an electron must be such that the number of complete waves, or *standing waves*, fitting into the space allotted to the orbit can be expressed as a whole number. Musicians consider this phenomenon when designing the strings of a guitar or other stringed musical instruments. The guitarist obtains each note through a string with a length equal to the particular number of complete wave vibrations that produce the desired note, multiplied by its wavelength. Figure 5 shows the vibration for a note that requires five complete vibrations between the ends of the string. The vibration at the right is not allowed since the fifth wave is not a complete vibration wave.

Figure 5 Standing Waves – Musical Strings

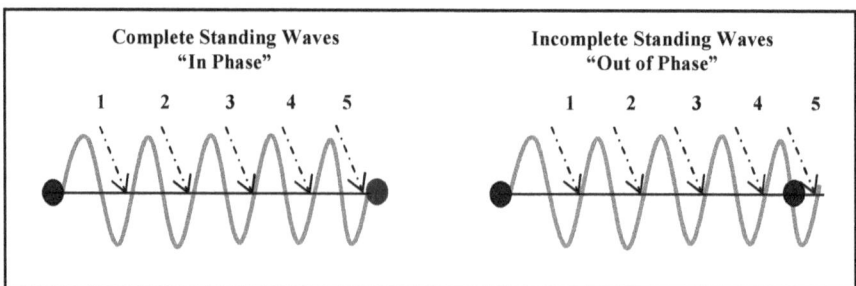

Since wavelength is a function of frequency, which is further related to energy, each orbit has a distinct energy level and therefore is quantized. Figure 6 shows this effect for electrons occupying atomic orbits where the wave function of the electron must fit exactly within the orbit. The wavelength of the electron on the right is too long (i.e., too low energy) and is not permitted in the specific orbit; the one on the left fits precisely indicating it has the correct energy level.

The probabilistic aspect of Schrödinger's equation has very surprising and not readily apparent implications. At first glance, it seems to imply that a particle actually prefers to be in a given spot but will occasionally stray to another location, providing nothing hinders it from doing so. This is not the case however; according to Schrodinger, it will be in other locations some percentage of the time regardless of barriers in its path. This fact is born out in the phenomenon of *quantum tunneling* wherein particles actually pass through barriers that, according to classical physics, require more energy than they possess. This does not mean barriers are irrelevant; they reduce the probability of the particle passing through, but the probability never goes to zero.

Figure 6 Standing Waves - Atomic Orbitals

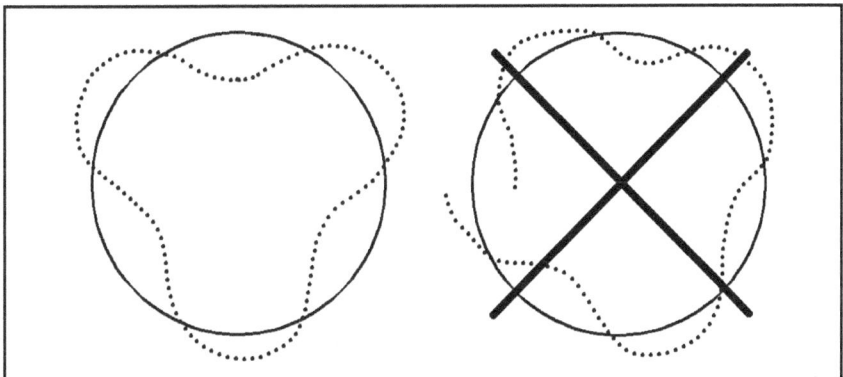

Quantum physics however, states that we can only assess a probability of finding a particle at any given location, and the amplitude of its particular wave function dictates that probability. Most of the time, the particle (or most of the particles in cases where we consider large numbers of particles) will be at the location corresponding to the highest wave amplitude. The particle is less likely to be at locations represented by low wave amplitudes. Figure 7 illustrates this by use of a particle wave that suggests a high probability of the particles locating within the walls of a dense barrier, and a reduced but non-zero probability of straying beyond the barrier. (The high amplitude of the wave inside the vertical barriers indicates high probability, while the low amplitude portions of the wave means low probability outside) As indicated, some particles will be located beyond the barrier walls even though conventional physics dictates that its energy level is too small to overcome the barrier presented by the walls.

Figure 7 Particle Behavior, Newtonian vs Particle Physics

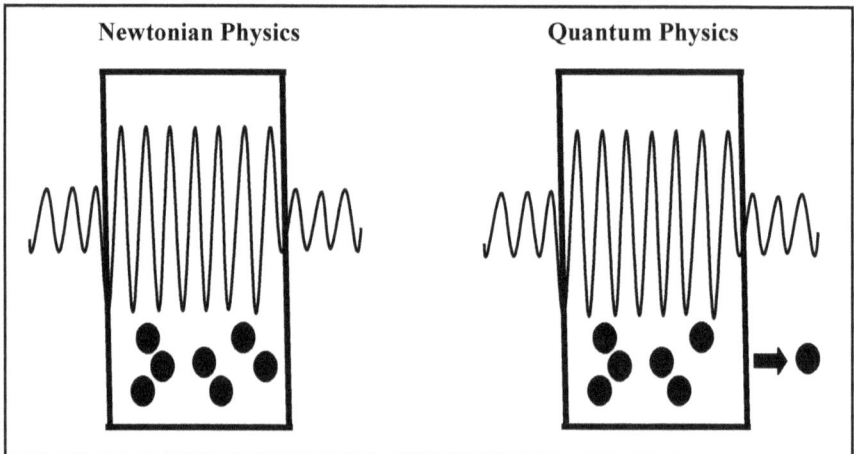

The particle mass component in the wave equation (wavelength = h/m) is in the denominator, so particles much larger than those of the atomic world have wavelengths too small to have meaningful probabilities of quantum tunneling. However, the equation still applies to all objects regardless of size. Therefore, if an infinite number of universes inhabited by humans existed, eventually someone living within one of them would walk through a brick wall unscathed, and in another, a little leaguer would hit a baseball ten miles.

Uncertain World

Planck's quantization theory presented scientists of the 1920s with yet another enigma. Scientists use light to study the behavior of atomic particles by bombarding them with photons, and Planck's theory requires a minimum amount of energy—that of a single photon—in order to make measurements. Therefore, when using light in this way, the energy of photons colliding with small mass particles such as electrons is high enough to affect the velocity and location of the particle under observation, the same way a cue ball affects the velocity and position of the pool ball it contacts. Furthermore, the higher the energy of the observed electron, the higher the energy required of the probing photon, which translates to a loss of information regarding an electron's velocity. When this is countered by using lower energy photons to minimize the change in velocity, lower frequency, longer wavelength light is needed, meaning the photon is less localized and information about the electron's position is lost.

These factors prompted Werner Heisenberg in the mid 1920s to propose the *Uncertainty Principle* which stated that the more accurately one attempts to measure the position of a particle, the less accurately one can measure its speed, and vice versa; therefore, it is impossible to define a particle's position and momentum simultaneously. The very act of measuring the location of an atom creates an uncertainty in its velocity that will cause the atom to move in an unknown and unknowable direction. Heisenberg quantified the principle by showing that the sum of the uncertainty as to its position and the uncertainty in its momentum (velocity x mass) must always be greater than Planck's constant.

Heisenberg's Uncertainty Equation

$$x\,v\,m \;=\; \text{or} > h$$

x = uncertainty of the position of the particle
v = uncertainty of the velocity of the particle
m = mass of the particle
h = Planck's constant (10^{-33})

As in the de Broglie's equation, the very small value of h means that measurements of uncertainty are significant only in the atomic arena. The principle still applies to normal sized objects; however, the uncertainty of their location at levels of 10^{-33} is far beyond our ability to take notice of them. Even objects of bacterial size are still much too large to be significantly affected.

The uncertainty principle brought about a profound change in the way scientists visualized the atomic world. Since one can never simultaneously measure both the velocity and position of an electron or proton, it made no sense to reference them in such terms. They were no longer tiny little balls with even tinier little balls orbiting them. Instead, they became vague cloud-like entities with statistical descriptions of their properties. Again, the "common sense" description of the workings within the micro-world did not easily give-way. Even today, some physicists do not totally accept quantum theory because of its indeterministic aspects. Albert Einstein stated his disdain with the now famous remark, "…that God would choose to play dice with the world is something that I cannot believe for a single moment." An unknown wag from a campus science center somewhere added:

Heisenberg's principle mustn't be forgot

For sometimes its here
And sometimes its   .

Heart of the Matter

While scientists for many years had speculated about the existence of more fundamental forms of matter, proof had to wait until 1968 when Stanford Linear Accelerator Center researchers, using new technology, proved that protons and neutrons contain even smaller particles. The following year, Nobel Prize winning physicist Murray Gell-Mann dubbed them *quarks* (pronounced kworks), taken from a lighthearted passage in James Joyce's *Finnegan's Wake*, "Three quarks for Muster Mark." Quarks come in two main varieties, up quarks and down quarks, which differ by the fractional electric charges they carry. Up quarks carry a charge of +2/3 and down quarks –1/3.

Table 1 Elementary Particles of Matter

Particle	Charge	Mass, Grams
Up Quark	+2/3	5.528×10^{-25}
Down Quark	-1/3	5.528×10^{-25}
Electron	-1	9.109×10^{-34}
Neutrino	0	$<10^{-35}$

All of the matter comprising the Universe comes from these basic particles; however, they are not the only particles housed in an atomic nucleus. In fact, a slew of them exist, many of which seem to serve no function in the present Universe beyond providing physicist the opportunity to dream up clever names like strange quark, charmed quark, bottom quark, and top quark. They did however play an important role in the very early moments of the Universe. Charm quarks and top quarks are heavier analogs of up quarks, identical in every respect except for their masses, which weigh 340 and 40,000 times more respectively. We do not see these particles in nature since their very high masses make them extremely unstable. Quantum theory predicts their existence and nuclear physicists have created them in high-energy particle accelerators; however, they quickly decay to conventional particles. Strange and bottom quarks are similarly high mass relatives of down quarks. Each of the quarks comes in three different colors, red, green, and blue, an interesting name assignment since atomic nuclei are smaller than the smallest wavelength of light and therefore colorless. As we will explain later, the color-coding derives from the manner in which quarks combine to form protons and neutrons. They must always pair in such a manner as to yield a colorless product.

Table 2 Atomic Nuclei Composition

Particle	Up Quark	Down Quark	Charge	Mass Grams x 10^{-24}
Proton	2	1	+1	1.672
Neutron	1	2	0	1.675

Electrons also qualify as matter, although scientists thought them zero mass energy particles until the early 1900s. The electron family also has its unstable overweight counterparts, the *muon* and the *tau*—about 200 times and 3,500 times heavier respectively. *Neutrinos* are extremely low mass particles that permeate our entire solar system. Their existence, first predicted by Wolfgang Pauli in 1930 and verified in experiments by Frederick Reines and Clyde Cowan in the mid 1950s, went unnoticed for so long because their impact on other forms of matter is very mild and infrequent. The Sun bombards us daily with billions of neutrinos, but they pass unnoticed through our bodies. If our solar system consisted entirely of lead, a neutrino could pass through it unaffected.

Quarks combine in groups of three to form protons and neutrons. A proton contains two up quarks that provide a charge of 2 x +2/3, or +4/3, and one down quark that contributes a –1/3 charge; adding +4/3 and –1/3, gives a total charge for the proton of + 3/3, or +1. Neutrons form from 1 up quark and 2 down quarks, which result in a net particle charge of 0, since the –1/3 charges of the two down quarks exactly cancels the +2/3 charge of the single up quark.

Quarks also combine to form a host of other particles, which, along with protons and neutrons, belong to a grouping called *hadrons*. All hadrons other than protons and neutrons are unstable and decay into other particles; this leaves protons, neutrons, and electrons to construct essentially all the visible matter of the Universe. While we do seem to get along quite well without these seemingly superfluous creatures, their discovery has aided our understanding of many mysteries surrounding the history of the Universe.

This does not conclude the particle zoo tour however; in fact, we are only half way through. It turns out that for every particle in the Universe, there is an antiparticle, identical in mass and all other properties except electric charge. The antiparticle of a proton is a negative charged antiproton, quarks have oppositely charged antiquarks, and the electron an antielectron, (or in this case a positron) with a +1 charge. When a particle and its antiparticle meet, they immediately annihilate one another, yielding nothing but energy. The very early Universe contained nearly equal quantities of matter and antimatter that combined in a phenomenal explosion, destroying essentially all of the antimatter. Fortunately, a tiny fraction of excess matter remained, for it is from this excess that our entire Universe evolved.

One of the more speculative universe formation theories suggests there may be other universes that began from a Big Bang where the imbalance favored antimatter. Everything in the antiuniverse would be identical to our Universe, except that it would be composed entirely of antimatter. However, it's my guess that antiphysicists of such a place are still prone to earthly ethnocentricities and call their stuff matter and ours antimatter.

Natural Forces

Identifying the most basic constituents of matter does not explain its existence and behavior as we observe it in our daily lives. Many questions remain, such as: What holds it together? What keeps it apart? Where does it get many of its properties? We find these answers in the four fundamental forces of nature; namely the *strong force*, the *electromagnetic force*, the *weak force*, and the *gravitational force*, listed in descending order of strength. Each natural force is carried by a particle, which in quantum theory is the smallest indivisible unit or bundle of the respective force. Natural force particles are often referred to as *virtual particles*, since unlike particles of matter and energy, we cannot detect them directly. We can only measure their effects. Except for *bosons*, the particle associated with the weak force, matter particles such as quarks and electrons carry the force particles. Matter particles feel the forces between one another by exchanging force particles; each matter particle emits a force particle that collides with the other matter particle that absorbs it.

The *graviton*, the force particle of gravity is the most familiar and surprisingly the weakest force being about 10^{30} times weaker than the boson, which represents the weak nuclear force. The gravity force is always "attractive" and is exerted and felt by every bit of matter or energy in the Universe. However since the impact of gravity is directly proportional to the masses of the objects involved, its effect is only significant when considering large concentrations of matter and/or energy. Massive objects transmit gravitational force by exchanging gravitons emitted from the combined particles that make up the massive bodies. For our Earth, Sun, and Moon, one could describe the cumulated force as benevolent. Fortunately, the particular mass of Earth provides a harmonious relationship between the gravitational and electromagnetic forces favorable to the development of the complex molecular structures necessary for life. If a few times smaller, Earth's gravitational force would be too low to prevent the light components of its atmosphere from escaping into interstellar space. If too large, electromagnetic forces that hold atoms and molecules together would not be able to resist the gravitational pressure, making their formation impossible. For extremely massive accumulations of matter and energy—such as that leading to the formation of black holes—awesome is more descriptive. In black holes, the gravitational force is so strong it overwhelms the other three forces, reducing all matter within its grasp to an infinitesimally small point.

Table 3 The Fundamental Forces

Force	Force/Particle	Relative Strength
Strong Nuclear	Gluon	10^{44}
Electromagnetic	Photon	10^{42}
Weak Nuclear	Weak Gauge Bosons	10^{30}
Gravity	Graviton	1

Another important property of gravity is that we feel its impact over long distances. Its attractive force diminishes rapidly as separation between objects increases (actually it decreases in proportion to the square of the distance; e.g. increasing the separation of two objects by a factor of 10 decreases their mutual gravitational attraction by 100). Still the attractive force of any two objects is always present.

Although gravity is the weakest of the four natural forces, on the very large scale of the Universe it becomes dominant. Gravitons hold the Earth in its orbit around the Sun 90 million miles away and keep all of us from drifting away from the Earth to become astronauts sans rockets. Two protons for example have a gravitational attraction pulling them together; however, their mass is so small (6 x 10^{33}, or 6 million billion billion billion of them weigh only 1 gram) that it is completely insignificant when compared to the electromagnetic force acting on their positive charges to push them apart. Only in the presence of massive nuclei accumulations does the gravitational influence become important. The Sun consists of 10^{57} protons and neutrons (together referred to as *nucleons*), giving it sufficient gravity to maintain the Earth, with 10^{52} nucleons, in its orbit, despite their great distance apart. The planet Jupiter is about 100 times larger than the Earth with a mass equivalent to 10^{54} nucleons causing a gravitational force so great that it crushes, or does not permit the formation of, any solid matter. The surface of Jupiter is essentially a dense sea of hydrogen and helium pulled from the outer part of the disc that surrounded the Sun during the early stages of its formation, and held in place by its extreme gravity. Such factors would limit our exploration to orbital missions; any astronaut unfortunate enough to take a small step for mankind on Jupiter would be taking his last, and would not leave a footprint.

In the presence of mass accumulations similar to Earth, the strong and weak forces of atomic and subatomic matter hold sway. The strong force is so named due to the Herculean task it fulfills while building protons and neutrons from quarks. The strong force overcomes the desperate need of the positive charged quarks to be as far apart as possible, a task you can appreciate if you have ever tried to hold the positive ends of two magnets together. Particles called *gluons* carry the strong force, and unlike gravitons, the distance separating them does not diminish their effectiveness. Gluons also take responsibility for the absence of single quarks in nature due to a property they possess called *confinement.*

This strange property operates through a color-coded representation that assures quarks will always assemble in colorless combinations when they form protons and neutrons. Confinement causes gluons to form protons and neutrons by binding quarks, arranged in combinations of three that always produce a white end product; e.g. one red quark, one blue quark, and one green quark. A single quark or any other combination would result in a colored particle. The strong nuclear force is responsible for the energy released in nuclear fusion reactions that power the Sun, the stars, and the hydrogen bomb.

Photons are force carriers for the electromagnetic force, which is responsible for the structure of matter, and functions through interactions with electrically charged particles like electrons and quarks. The force between like charged particles (+ /+ or −/−) is repulsive while it is attractive for unlike charged (+/−) particles. Electromagnetic forces become unimportant in larger frames of reference since the division between positive and negatively charged species is about equal, cancelling the electromagnetic forces between them. Atomic particles also feel gravity; however, electromagnetic forces, being stronger than gravity by a factor of 10^{42}, make gravity's role insignificant at the atomic level wherein positive and negatively charged particles interact. The electromagnetic attraction transmitted by the exchange of mass-less photons between negative charged electrons and atomic nuclei containing positive charged protons, causes electrons to orbit in much the same manner that gravity induces the orbiting of planets about the Sun. While photons carrying the electromagnetic force are *virtual particles* that cannot be directly detected, as referenced in Bohr's explanation for the quantum nature of orbiting electron energy levels, one can detect the light emitted when electrons move to different energy level orbits.

The weak nuclear force is carried by particles called *bosons*, which come in three varieties, W_+, W_-, and Z_0. The +, −, and 0 signs describe the +1, −1, and neutral, electrical charge carried by the particles. The electrical charges of atoms that emit bosons can become more positive, negative, or be unaffected depending on which of the three are emitted. Bosons are the force particles responsible for radioactive decay of heavy elements such as cobalt and uranium, a process that proceeds by converting neutrons within the nucleus of an atom to protons, accompanied by the emission of one electron for each converted neutron. Since a neutron weighs more than the proton/electron combination (about 0.2% more), the weight differential shows up as radiation energy according to $E=mc^2$ (to be described in the next chapter). This radioactive decay occurs regularly within the Earth's interior and energy emitted serves as one of the main driving forces of volcanic eruptions near the surface.

2   The Einstein Effect

Galilean Relativity

Although the implications of Einstein's *special theory of relativity* are far-reaching and complex the basic theory itself is quite simple and can be stated so. At its most fundamental level, special relativity merely says that everyone in a uniform state of motion experiences the same laws of physics. However, this apparently bland statement has far-reaching implications that force radical alterations of our notions of space and time. It suggests the possibility of travel through time and points out the awesome explosive potential locked up in atoms.

Appreciating relativity, and why and how Einstein discovered it, first requires some understanding of how the physics of motion had evolved prior to Einstein's time—usually called the period of *Newtonian physics*—and the problems relativity solved. The Universe according to Aristotle (c. 349 BC) consisted of celestial and earthly components with different laws governing the behavior of objects within each realm. The Earth resided at the center of Aristotle's Universe and the natural state of motion of earthly objects was at rest in a position as close to the Earth as possible. Overcoming this natural state to set objects in motion and maintain that motion required force. The celestial component consisted of the Sun, stars, and other planets. In this perfect, heavenly realm, the Sun and planets orbited the Earth in precise circles—a prerequisite for divine perfection—forever in motion requiring no input of force. Ptolemy (c. 140 AD) only slightly modified Aristotle's model of the Universe when he changed the circular orbits to circles-on-circles, or *epicycles* in order to explain observations made by astronomers of the day regarding the erratic "loopy" behavior of Mars and other planets. Although the second Century intelligentsia initially resisted Ptolemy's model since it clashed with the heavenly perfection philosophy, his view of the Universe went unchanged for a millennium and a half until 1543 when Copernicus announced the revolutionary concept of a universe with the Sun at its center. Copernicus however, maintained the distinction between earthly and celestial objects and gained the adulation of the church by overthrowing Ptolemy's elliptical orbits for the Earth and other planets and returning them to perfect circles.

In 1610, Johannes Kepler once again upset the heavenly perfection concept when he developed mathematical laws that reinstated planetary orbits as ellipses rather than circles. Shortly after, the heavenly realm received another setback from Galileo (1564-1642). Using his newly invented telescope, Galileo observed that the Sun was not a perfect spherical orb but blemished with sunspots. Furthermore, he noted that Jupiter was not a strange aberrant star but very earthlike, and in fact had four moons. Galileo also conducted the first scientific experiments about the motion of objects on Earth and found that all objects, regardless of their mass, fall with the same acceleration. From these observations, he developed the law of inertia, which states that an object continues in straight-line motion at constant speed unless disturbed by an outside force; and the only thing that prevents moving earth-bound objects from continuing in motion forever are interfering forces like friction or atmospheric resistance. This observation overthrew the Aristotelian concept of natural law that had survived for two millennia and redefined the natural state of motion as straight-line motion at constant speed.

While Galileo is deservedly the father of modern science, the great proliferation of scientific progress did not begin until Isaac Newton, born in 1642, the year Galileo died. If anything, the reference to Isaac Newton as the "genius of Cambridge" grossly understated his contribution to science when we consider the scope of his influence. He quantified Galileo's idea of the natural state of motion and expanded its scope with the development of his three laws of motion. The three laws essentially stated that all motion is completely predictable once we know the forces acting on its objects, making the Universe completely deterministic—a concept dubbed "*the clockwork universe.*"

Newton's Laws of Motion:

> The first law is a restatement of Galileo's discovery that objects move in a uniform manner unless acted upon by outside forces.

> The second law states quantitatively how a given force F acting on an object of mass, m produces a change in its motion, acceleration, a.

> Newton's third law declares that forces always come in pairs; "for every action there is an equal and opposite reaction". A force exerted on object A by object B will be countered by an equal force exerted by object A on object B.

Newton also finally put to rest—at least scientifically and philosophically—the concept that different physical laws applied to the earthly and heavenly realms when he discovered the concept of universal gravitation. More popularly known as "the laws of gravity", universal gravitation showed that the force responsible for an apple to falling to Earth is the same force that holds the Moon in its orbit around the Earth, and the Earth in its orbit around the Sun. He stated that every object in the Universe exerts a force, which he called gravity, on every other object, and the strength of that force depends only on the masses and the distance separating the objects.

Since 17th Century mathematics was incapable of quantifying the newly discovered relationships, Newton invented a new form of mathematics, namely calculus, for this specific purpose. By applying calculus to his laws of motion and gravity, he proved that planets must move in elliptical orbits as Kepler had observed earlier. His understanding of gravity and the mathematics describing it also allowed him to demonstrate the feasibility of artificial satellites. He proved that if one could throw an object from a very high mountain (actually so high that it would be completely above the atmosphere) with enough force that it would fall toward the Earth at the same speed the Earth curved away beneath it, the object would continue falling around the Earth in a circular orbit forever (assuming there is no air resistance or other mitigating factors to impede its progress). Figure 8 below illustrates Newton's satellite "thought experiment":

While scientists did not use the term relativity in conjunction with the ideas of Galileo and Newton until after Einstein's theories became mainstream, the concept is implicit in the context of their laws of motion and has since been named The *Principle of Galilean Relativity*. It states that the laws of motion work the same for everyone as long as they are moving uniformly with respect to the same point of reference. In other words, the laws of physics apply in the same manner to activities such as playing tennis or bowling, whether they take place on land or on a cruise ship moving in a straight line at a constant speed over a calm sea. The player with his back to the direction of the ship does not need to be concerned that the velocity of the approaching ball is the sum of that imparted by his opponent plus the speed of the ship relative to a stationary buoy the ship is just passing. Nor does the player facing forward have a much easier task due to the slow speed of the return relative to the same buoy. For both players, the frame of reference is the moving ship and, as long as the sea is calm and they remain protected from wind generated by the ship's movement, the game will proceed in the same manner as if played on dry land. This concept can also be stated as follows: If a plane is moving in a straight line, on a perfectly calm day, at steady speed of 600 miles per hour, there is no experiment a passenger on that plane can perform that will tell him whether or not he is moving. All questions about motion have meaning only when stated as motion at rest relative to something else, be it the Earth or the fellow passenger he passes while walking down the aisle of the airplane.

Figure 8 Newton's Satellite

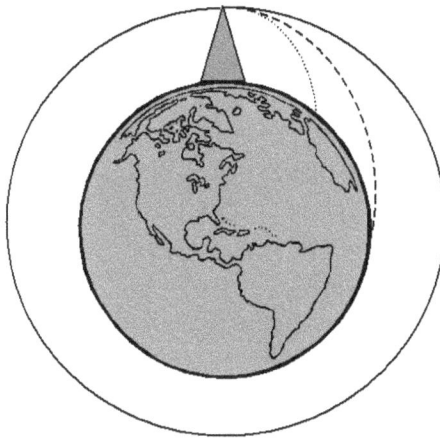

·············· Object thrown with modest force falls toward Earth in a curved path

– – – – Object thrown with greater force falls further

——— Object thrown with the precise amount of force that allows it to fall towards the Earth at the same rate the Earth is curving away beneath it. This results in a permanent circular orbit.

Light Speed

In 1676, Ole Christensen Roemer discovered that light always travels at the same fixed speed. He observed inconsistencies in the times at which the moons of Jupiter pass behind the planet and that their eclipses appeared later proportional to their distance from Earth. The variable distance results from their elliptical orbits as the two planets revolve around the Sun. He concluded that when Jupiter was further away, the light from its moons took longer to reach the Earth meaning it traveled at a fixed speed. Roemer calculated the speed of light as 140,000 miles per second, which agrees surprisingly well with modern calculations of 186,282 mps or 670,616,629 mph for light traveling through a vacuum.

Michael Faraday demonstrated in 1821 that both electricity and magnetism behaved as force fields and that a changing magnetic field (such as one produced by a rotating magnet) would create an electric field. Then in 1865, James Clerk Maxwell realized the reverse must also be true. A changing electric field could produce a magnetic field and result in the creation of electromagnetic waves—waves of linked electric and magnetic fields—and these would move through space like ripples on a pond. Then using a series of mathematical relationships between the two forces, now referred to as Maxwell's equations, he determined that electromagnetic waves travel through space at a fixed speed, the speed of light. He proved that a light beam was just another variation of this back and forth conversion of magnetism and electricity. As a light beam moves forward, it essentially powers up a bit of magnetism, and then as that magnetism

46

moves on, it creates yet another surge of electricity, and thus continues forward in a leapfrogging manner. Other electromagnetic waves include radio waves, which have long wavelengths measuring more than one meter, microwaves a few centimeters, infrared, less than 0.0001 cm, and visible light with wavelengths ranging from 40 to 80 millionths of a centimeter. All however, travel at c, the abbreviation selected for the speed of light to represent its constant nature.

The inevitable outcome of all advances in science must be the generation of new questions; and Maxwell's discoveries proved no exception. The laws of motion in Galilean Relativity required one to state the motion of an object or wave in terms of a frame of reference. In other words, to what is the motion relative? For a wave, this frame of reference was the medium through which the wave traveled. The ground is the medium for waves from an earthquake, water for ocean waves, and air for sound waves. The same question needed answering for the motion of electromagnetic waves. Electromagnetic waves, including light, travel through space at a constant speed c. But at speed c relative to what?

One possible answer: the speed of light is relative to its source. The speed of light emitted from the object would increase or decrease according to its velocity and direction. However, this clashed with many astronomical observations. One of these involved observations of light emitted by stars that rotate in orbits around each other, called *double star systems*. In the course of their rotation, there is a point when both stars are equidistant from the Earth, but one star approaches the Earth while the other moves away from it. If the speed of light depended on (relative to) the motion of its source, light from the approaching star would reach us sooner than light from the star moving away. This would result in a host of unusual effects. For example, light emitted from them at several different locations would reach us at the same time, and we would simultaneously see multiple images of the individual stars instead of observing the continuous rotation of two stars.

In the absence of experimental proof, the theory accepted by 19th Century physicists said that light travels at speed c relative to the *ether*;

a substance they believed permeated the Universe. To the ancient Greeks, the ether personified space and heaven; to alchemists it became the "fifth element" (in addition to earth, water, air, and fire), and to physicists, the medium through which light propagates. Scientific acceptance of the ether had survived several Centuries due to a very logical assumption; since all known vibration phenomena have a medium through which they propagate, such as air for sound, light must also have a medium, the ether. All celestial bodies moved through the ether and created "ether winds" in the same manner that objects on Earth encounter atmospheric winds. Savants assigned qualitative properties to the ether in order to explain their observations in terms of their knowledge of the Universe at the time. The ether had to be a very tenuous medium since the Earth traveled through it with no apparent resistance. They also felt it had to be "stiff" in order for waves to travel through at the incredible speed of light, much like waves travel faster through a tightly stretched length of string than through a loose one.

Scientists assumed that light and other electromagnetic waves would propagate through the ether the same way other waveforms propagated through their media. A sound wave, they assumed, would travel through air that is at rest relative to an observer (or if you prefer, listener) at about 700 miles per hour. However, a tail wind or head wind blowing relative to the observer would slow or speed up the sound. Therefore, if the physics of electromagnetic waves obeys the same relativity principle of motion as other waves and objects, then "ether winds" should affect light in the same manner.

In 1887, Two years after Maxwell's discoveries, Albert Michelson and Edward Morley devised a series of ingenious and very accurate experiments specifically to calculate how much the ether wind increased or decreased the speed of light as it moved through the ether. They designed the experiments to measure differences in the speed of light from one source traveling in the same direction of the Earth's motion as it revolves around the Sun, and another at right angles to it. They reasoned that as the Earth moves through the ether wind, a light beam that is aligned into the ether wind would travel slower than one aligned with or across it. The experiment involved

48

splitting a beam of light into two beams and directing them at right angles to each other toward mirrors spaced equal distances from the splitter. The mirrors then reflected the two beams back to the splitter where they recombined, and a spectrophotometer examined the resulting wave pattern.

If the speed of light was affected by—or relative to—the ether, one or the other of the light beams, depending on its alignment with the ether wind, would need more time to complete its journey from the source to the analyzer. The difference would be manifested by the same interference patterns discussed in Chapter 1, when the two light beams recombined. If the two beams had traveled at exactly the same speeds, the peaks and troughs would reach the detector precisely in phase, and it would mean that the speed of light is constant and not relative to the ether. This would produce constructive interference resulting in a combined wave pattern that would appear identical to the original split beams, except that the wave amplitude would be higher. If the two beams traveled at different speeds, consistent with the speed of light being relative to the ether, then the wave peaks would not arrive in phase, causing the peaks and troughs to at least partially cancel each other, producing a destructive interference pattern.

The Michelson/Morley experiment proved conclusively that the speed of light was not relative to the ether, its supposed medium. In spite of performing a host of experiments, considering different configurations of the test equipment and times of the year that corresponded to different directions of the Earth's orbit, the experiments always produced the same constructive interference pattern signaling that light traveled at the same speed regardless of its motion with respect to the ether. The result shocked scientists at the time, since they had assumed the speed of light followed the same physical laws as other wave phenomena, and that the ether would slow its journey through space.

Special Relativity

Many scientists attempted to explain the Michelson/Morley result as being caused by objects contracting or shrinking in the direction of their motion and by clocks slowing down when they move through the ether (not due to Einstein relativity, but to special properties of the ether). However, these amounted to only ad hoc answers for explaining away the results without any physical basis backing them. In 1905, a clerk in the Swiss patent office named Albert Einstein resolved the contradiction in one of the most important scientific papers of all time titled, *On the Electrodynamics of Moving Bodies*, received by *Annalen der Physik* on June 30. A brash confident young graduate of the Zürich Polytechnic, Einstein, relegated to patent office employment since he had not impressed his professors sufficiently to earn a doctorate in physics, stunned the scientific world by proclaiming the ether and all the statements associated with it irrelevant. Furthermore, the young upstart replaced previously undisputed scientific truths of Galileo and Newton that had survived more than two Centuries by simply stating that all observers in uniform motion experience the same laws of physics. At first glance, the impact of Einstein's statement seemed rather benign since Newton's laws of motion already accepted this for massive objects. Einstein simply extended Newton's laws to include Maxwell's theory as well; specifically, that the same laws also apply to electromagnetism.

The radical nature of Einstein's new theory becomes evident on closer inspection. However, for many people, making the leap to accept its consequences requires changing deeply held beliefs and concepts of space and time. This does not mean that relativity is not real, only that our common senses are not tuned to the greater realities of our Universe. We inhabit a very mild but extremely miniscule and highly non-representative portion of a vast Universe where temperatures range from minus 457 to tens of millions of degrees Fahrenheit, filled with subatomic particles hurling through space at speeds approaching that of light.

One of the predictions of Maxwell's equations is that light and other electromagnetic waves all travel at the same speed, c, which is 186,000 miles per second (the origin of the letter c comes from the Latin word celeritas meaning "swiftness", although normally taken simply as an abbreviation for the word constant). Maxwell did not know however, even though his equations contained the answer, that this constant speed of light applies regardless of the state of motion of the person observing it. Einstein's special relativity revealed what Maxwell had missed; namely that all observers will measure the same value for the speed of light whether motionless, moving toward it on an airplane, or away from it while having hitched a ride on Haley's comet. Imagine standing by a road holding a device that could accurately measure the speed of light, when a car equipped with an identical contrivance passes by at 70 mph. Up the road, a traffic light flashes sending a beam of light toward the approaching car. When the speed of the light beam is measured by the instrument in the moving car and in the stationary one on the roadside, both obtain the same value, c—or 186,000 miles per second.

There is no problem with the quality of the clocks or the speed of the moving car; you would get the same result from a spaceship traveling at half the speed of light. How is this possible? Surely if a driver traveling at 70 mph meets a trooper moving in the opposite direction at 60 mph, the radar gun will show the driver approaching at a speed of 130 mph; and in order for the trooper to know the correct speed, he will need to subtract the patrol car speed of 60. But this is not the case if he is measuring the speed of a beam of light; he would not need to account for motion regardless of his speed.

We can only explain the fact that observers in different states of motion will obtain identical results when measuring the speed of light

51

emitted from a common source, if time is not an absolute entity but is itself relative. Time beats at different rates depending on how fast one moves. This notion seems absurd to most of us since all of our experiences occur in an environment where objects move at speeds so far below the speed of light that we don't notice these effects. When traveling on a commercial jet, our speed is only about one millionth that of light. Even space shuttles never achieve speeds greater than forty-thousandths light speed. On the other hand, we all travel at speeds near c all of the time relative to far away galaxies from which we separate at near light speed. There is a distinction in this case however; we do not travel through space, but we travel with expanding space.

Figure 9 Light Clocks and Time Dilation

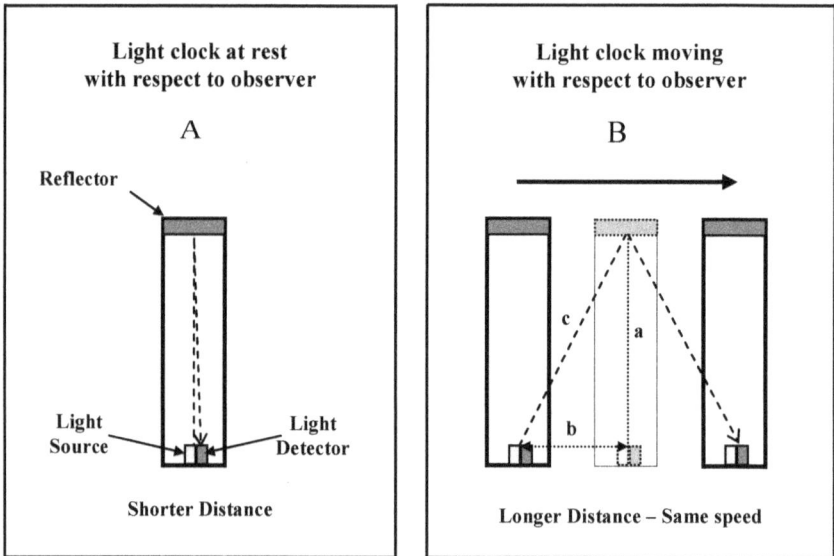

Figure 9 illustrates the concept of time relativity to motion via a "light clock," a hypothetical device whose usefulness is limited to explanatory purposes. The "light clock" consists of a box with a light source at one end and a mirror at the other. A flash of light leaves the source, travels the length of the box, bounces off the mirror, and returns to a light detector adjacent to the source. The process continues with each complete round trip of the light beam constituting the basic ticking of the clock. The purpose of the light clock experiment is to examine the length of this "ticking"—the time between the event of the light beam leaving the source and the event of its return—in two different reference frames. First, it is necessary to specify the meaning of an event, which is simply anything that occurs at a specific time and place.

In drawing A of figure 9, an observer is at rest with respect to the clock, so he sees the light make a round trip from the bottom to the top and back, covering a distance twice the length dimension of the box. It is important to note that the observer in A is at rest with respect to the clock, whether conducting the experiment while on Earth with the clock on a table in front of him, or if both he and the clock are traveling in a high-speed spaceship. The key factor is the observer and the timing device share the same motion. In B, where the box is moving with respect to the observer, the light takes a longer path following the two diagonals, each of which is longer than the length dimension of the box. Since the light travels a longer distance in B, and the speed of light is the same for both observers, the stationary observer B will record a longer time for the light beam to reach the bottom detector than the round trip in clock A; therefore there must be a different measure of time for the two observers. A stationary observer of moving clock B, will see time moving slower than an observer at rest with clock A. But here is where things get really strange. The observer traveling with clock B would see clock A moving slow; this is true since either observer has an equal right to claim to be the stationary one.

This phenomenon, where observers measure time differently based on the motion of clocks relative to the observer is called *time dilation*. The use of a "light clock" does not imply anything magic in the experiment; it simply makes the principle of time dilation more easily understood by incorporating the speed of light directly into the illustration. Use of ordinary clocks would bring the same result since time dilation is not about timepieces but is basic to time itself.

We can calculate the magnitude of the time dilation, the difference in the measurement of time, by the following simple equation:

$$t' = t \sqrt{1 - v^2}$$

Where: t' is the time measured by the observer relative to whom the clock is moving

t is the time measured by the observer relative to whom the clock is stationary

v is the fraction of the speed of light at which the clock is moving

The Pythagorean Theorem from high school mathematics easily determines the elements of the equation formed by:

1) The straight vertical line a in the central drawing of B extending from the point on the mirror   where the light beam reflects downward to the bottom of the clock, representing the length of the light clock.

2) The diagonal line c, formed by the path of light from the source to the mirror of clock B.

3) A line b, connecting points that represent the horizontal distance the clock moved while the light traveled from the source to the mirror of the moving clock B.

We can use the clock drawing information in the above equation to calculate the time dilation, t' between the moving and stationary clocks via the following equation from the Pythagorean Theorem.

$$a^2 + b^2 = c^2$$

The first element of the calculation t (the time measured by the observer relative to whom the clock is stationary), is calculated directly from the length of the time clock. If in the example, the observer measured the length of the light clock to be 1 meter, since light travels at 300,000,000 meters per second, the measured time for the light to travel from the source to the mirror would be 1 meter/300,000,000 meters/second = 0.0000000033 seconds, or 0.0033 microseconds.

For the second element v, let's assume the detector informed us the clock had moved 1.5 meters during the time the light beam traveled from the light source to the mirror, and then to the detector. Then our distance element, b is 0.75 meters, since at this distance the light that has traveled to the mirror is precisely one-half the distance it traveled from the light source to the detector.

Knowing two sides of a right triangle (a triangle with a 90-degree angle); we can calculate the third, the length of the path taken by the light beam on its way to the mirror, by solving for c in the equation:

$a^2 + b^2 = c^2$
Since a = 1.0 and b = 0.75

$$c = \sqrt{a^2 + b^2} = \sqrt{1^2 + .75^2} = \sqrt{1.562} = 1.25 \text{ meters}$$

Therefore, since the clock moved 0.75 meters in the same amount of time the beam of light moved 1.25 meters, the clock speed was:

Light: 1.25/300,000,000 = 0.0000000042 seconds or 0.0042 microseconds/per meter
Clock: 0.75/300,000,000 = 0.0000000025 seconds or 0.0025 microseconds per meter

Then, the velocity of the clock equals:
v = 0.0025/0.0042 = 0.60 or 60% of light speed

55

In addition, time measured by the moving observer is:

$t' = 0.0033 \quad 1 - 0.60^2 = 0.0026$ microseconds

The time dilation then is:

$(0.0033 - 0.0026)$ or 0.007 microseconds.

In summary, when Observer A glimpses Observer B's clock, he sees it running (0.0007 x 100/.0033) = 21 % slower than his own clock.

As relative motion becomes a greater fraction of c, time dilation increases logarithmically. For example, if the above clock traveled at 0.9 c, t' would be 0.0014 or a relative reduction of 56% in the speed of the clocks; at 0.95 c, the reduction is 0.0023 or 69%, and at 0.999 c, it is 95%. When the moving clock reaches 1.0 light speed, the reduction in clock speed would be the full 0.0033 microseconds, time dilation would be 100%, and relative to the motion of Observer A, time would stand still, and the light beam would never reach the detector.

Figure 10 Time Dilation - Relative Motion Clocks

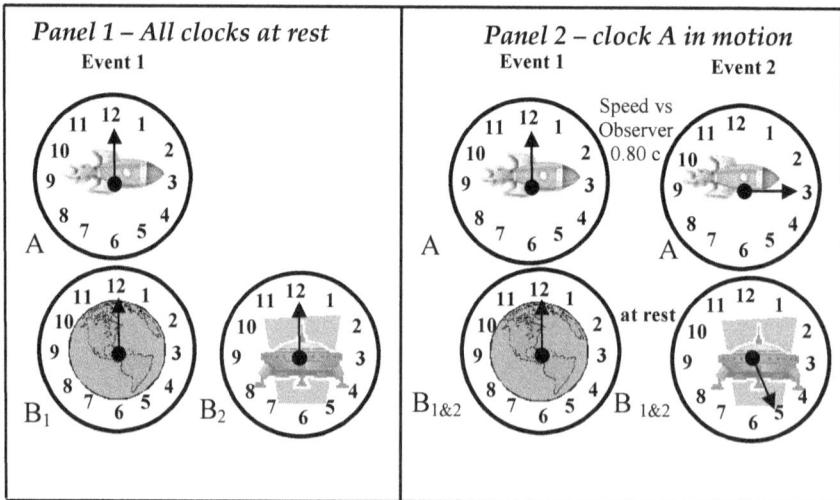

One of the effects of time dilation is that events—happenings at a particular place and time—will not occur at the same time for observers in motion relative to each other. Imagine two events that occur 2.93 billion miles apart. Panel 1 of figure 10 sets the stage at the time and site of event 1. Clock A will be aboard a spacecraft traveling from event 1 at a speed that will allow it to arrive at the distant space station just in time for event 2.

The travelling clock A is synchronized with two stationary clocks, B_1 and B_2, located at the sites of events 1 and 2 respectively. Panel 2 shows the change in time readings of the three clocks between the concurrence of clock A's departure from event 1 and the concurrence of its arrival at event 2. Due to time dilation, the elapsed time recorded by our traveling clock A must be less than the elapsed time measured by the other two clocks that remain at rest with respect to each other at their original locations. As shown in panel 2, clocks B_1 and B_2—remaining at rest with respect to each other—agreed, whereas A, being present at both events, measured the time t' between them as two hours, or 40% less than that of stationary clocks.

We can calculate the spacecraft speed that produced 40% time dilation by rearranging the time dilation equation and solving for v (velocity):

$$t' = t\sqrt{1 - v^2}$$

$$v = \sqrt{1 - t'^2/t^2}$$

$$v = \sqrt{1 - 3^2/5^2}$$

$$v = \sqrt{0.64}$$

$$v = 0.80\ c$$

This aspect of special relativity is not confined to effects on clock speeds. Since light governs interactions that occur at the atomic level, it applies to all phenomena, even human biological processes. This concept is often explained by a pair of space traveling twin astronauts, one of which is in a spaceship moving away from Earth at 0.8 c, to a star 5 light years from Earth, while the other twin remains at home. A single light year is the distance light travels in a year at 670 million miles per hour or about 5×10^{12} or 5.4 trillion miles. Since distance = speed x time, from the point of view of the earthbound twin, his brother's trip took:

t = 5 light years / 0.8 light years per year = 6.25 years

Due to time dilation however, the travel time aboard the spaceship was:

$t' = (6.25 \text{ years}) (\sqrt{1 - 0.8^2}) = (6.25 \text{ years}) (0.6) = 3.75 \text{ years}$

Now imagine that the spaceship reaches the star, then turns around and makes the return trip home at the same speed, 0.8 c. For observers on Earth, the return trip takes another 6.25 years for a total elapsed time of 12.5 years; while on board, only another 3.75 years passes giving a total of only 7.5 years. The twin who remained behind now finds himself 5 years older than his space traveling twin brother. This situation creates a paradox since relativity gives either twin an equal right to claim that he is stationary and the other is in motion. For example, the twin in the spaceship has the right to claim at rest status and the Earth took a 5 light year trip at 0.8 c, making him the older of the two. The answer to the paradox is that the situation is not symmetric. The earthbound twin remains in a single, uniformly moving reference frame for the entire duration of the round trip, while the astronaut twin reverses direction halfway through the trip, thereby occupying two different reference frames, each moving uniformly but in opposite directions. This example elucidates the "special" aspect of special relativity; it applies only to frames of reference in uniform motion. In relativity, it is meaningless to say, "I am moving." However, "My motion has changed" is meaningful.

Simultaneity

The statement that a clock in motion runs slower than one at rest with respect to an observer who is at rest, seems at first glance to raise a contradiction to the special relativity concept that either of two clocks in motion relative to each other, has an equal right to being the one at rest. This means that the observer of clock A in figure 10 should be able to say with equal accuracy that while sitting motionlessly in his spaceship, clock B1 zoomed past at 0.8 c and was "running slow." The answer is it can. This is due to another concept of special relativity that states: Two events that are simultaneous in one time frame are not simultaneous in another frame moving relative to the first.

The concept that *simultaneity* itself is relative allows the seemingly impossible result where two observers in motion relative to one another, see each other's clock running slow while their own clock is completely accurate. Returning to the star trip, it is obvious that from the traveling twin's point of reference, the distance to the star was considerably less than 5 light years as he made the journey in 3.75 years while traveling at 0.8c. In his frame of reference, the distance calculated at:

(0.8 light years/year) x 3.75 years = 3 light years.

Therefore, like time, distance is also relative; and this applies not only to space but to the length of objects occupying space as well.

Just as the distance of a journey in a spaceship shortens in proportion to its speed, so the length of the spaceship (or any object) shortens in the direction of its motion.   Furthermore, it does so according to $t = \sqrt{1 - v^2}$, the same factor that describes time dilation. This phenomenon is called the *Lorentz contraction*, named for the physicist Hendrik Lorentz, who hypothesized that the failure of the Morley/Michaelson experiment to show that "ether winds" affect the speed of light stemmed from the apparatus shrinking due to its motion through the ether.   Lorentz and his collaborator George Fitzgerald hypothesized correctly, but it took Einstein to explain why.

Figure 11 Simultaneity of Passing Airplanes

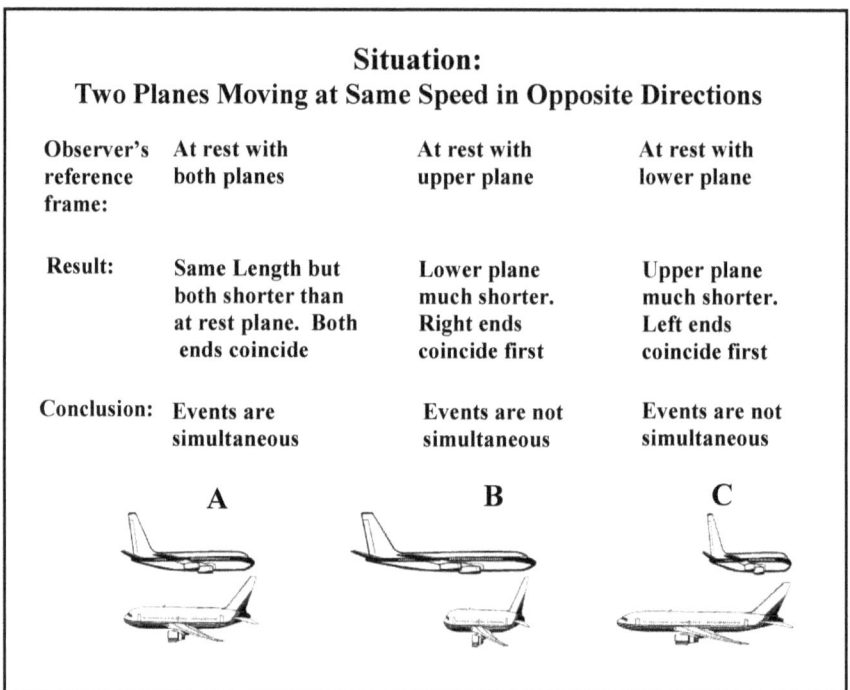

Situation:
Two Planes Moving at Same Speed in Opposite Directions

Observer's reference frame:	At rest with both planes	At rest with upper plane	At rest with lower plane
Result:	Same Length but both shorter than at rest plane. Both ends coincide	Lower plane much shorter. Right ends coincide first	Upper plane much shorter. Left ends coincide first
Conclusion:	Events are simultaneous	Events are not simultaneous	Events are not simultaneous
	A	B	C

We can use the Lorentz contraction to illustrate the concept of simultaneity by considering two very-high-speed airplanes flying past each other in opposite directions.   Figure 11 illustrates how three different observers, each occupying a different frame of motion, see the same events at different times and in different sequences. Situation A shows two equal length airplanes, viewed from the perspective of a hovering helicopter that is not at rest with respect to the motion of either plane. The two planes travel past each other at the same speed, and are therefore the same length, although shorter than when viewed from a perspective at rest with the planes (note the height and thickness of the planes remain the same). In this reference frame, the left ends of the two planes coincide at the same time as the right ends; therefore, the events of the two ends coinciding are simultaneous.

Situation B gives a perspective at rest with the upper plane, which would correspond to the view of a passenger on a plane traveling parallel to, and at the same speed as, the upper plane. The Lorenz contraction does not apply to the upper plane as it shares the same motion with the observer. However, the Lorentz contraction of the lower plane is greater than in A since the combined speed of the lower planes, and the observer's plane, is involved. From this perspective, the observer would see the right ends of the two planes coincide (nose of the upper plane and the tail of the lower plane) before the left ends, so the events are not simultaneous. Situation C shows the same view from a passenger on a plane traveling parallel to and at the same speed as the lower plane. This observer would also see the events as non-simultaneous although he would see the left ends of the planes coincide first. In effect, the time order of the two events is reversed.

Just a word of clarification about the Lorentz contraction; aboard the plane moving at 0.8 c, things don't become jammed together. Rather, they all become contracted just as the planes exterior. The seats, food trays, overhead bins, and serving carts, all shrink in the same direction of the plane's motion. In fact, a lady who happened to be five feet, six inches tall and weighed one hundred and ninety pounds, who wanted to show a potential suitor how she would look as a svelte one-fifteen, would need only to become a passenger and zoom past her admirer at 0.8 c to be so appreciated. However, she would not be able to share the admirer's view of her contracted self, since on board; everything would appear just as it did on the ground.

But she would be able to glimpse how he might look minus a similar number of pounds since he has an equal right to claim that she is stationary and he is moving at 0.8 c.

The Lorentz contraction is not just about space travel and warping objects but is fundamental to the very concepts of space and time, and must be accounted for in things we use and take for granted on a daily basis. The effect is important in items as basic as a 1960s TV picture tube in which electrons project to the screen at about 0.3 c. The resulting picture would not be in sharp focus if engineers had not accounted for length contraction when designing the picture tube.

The Elsewhere

This aspect of simultaneity—that the time order of events can change depending on your motion—seems to imply that cause and effect relationships are alterable; for example, a person's death could precede his birth. While the concept has provided inspiration for dozens of science fiction thrillers, the truth is it cannot happen. The only events for which the time order can be reversed are those far enough apart in space that there is not enough time for a light signal to travel from one event to the other. Since the speed of light limits the transmission of information that can affect such events, they cannot influence each other, so they cannot be causally related. All events that fall into this category belong to another realm of time called the *elsewhere*, giving us four time-related categories for identifying events: the past, the present, the future, and the elsewhere.

In relativity, the past consists of all events that can effect the present. This would include your birth since light would have had plenty of time to travel from the birth event to any other event in your life. Observers, in all other frames of reference, will agree that your birth is in your past, although they will disagree on your age. There can be other events that occurred simultaneous to your birth in some galaxy sufficiently far away (based on the Earth's frame of reference) that they are not in our past, since light from that galactic event hasn't had time to reach us. Similarly, in relativity the future consists of events that both the past and the present can influence. Events happening on Earth tomorrow belong to our future, but those happening tomorrow in the Andromeda galaxy—Earth's nearest galactic neighbor—are in our elsewhere since light cannot travel there from Earth in a day.

The elsewhere is not a place that is forever inaccessible to us; it's simply inaccessible to us from the present. It consists of events not in our future since we cannot influence them, and those not in our past since they cannot influence us. There are however, events now in our elsewhere, that were once in our future and others that later on will be in our past.

The Mars Rover project of the late 1990s presents an understandable example of the elsewhere. Mars was located about 107 million miles, or 11 light minutes away at the time; therefore, an event in the control room on Earth and a problem 5 minutes later in the Rover, could not be causally related since neither light nor any other signal can reach Mars in 5 minutes. Therefore, the problem event would be in the control room's elsewhere; however, 6 minutes later, the event would be in its past, and the event aboard the Rover will have effected operations in the control room.

Spacetime

Relativity does not imply that everything is relative. Obviously,

63

since the speed of light is the same for all observers regardless of their motion, light is not relative but is invariant. Another factor on which all observers will agree is *spacetime*, a kind of four-dimensional "distance" that incorporates both space (three dimensions) and time. In spacetime, observers in different frames of reference will not obtain the same values when measuring the distance between, and the times of, events. However, they will agree on their spacetime interval. We can explain this concept by considering two travelers using the two maps in figure 12 below, which represent two different reference frames to guide them from point A to point B. The two maps are analogous to the different states of motion of the travelers in the figure 11 example that involved planes moving at near light speed. The north direction of one map is based on a magnetic north reference, or the north we would detect from a simple compass, while the other map—the one used in most map coordinate systems—represents the Geographic North Pole, the point where the Earth's axis of rotation intersects the Earth's surface.

Figure 12 Invariance of Distance with Reference Frame

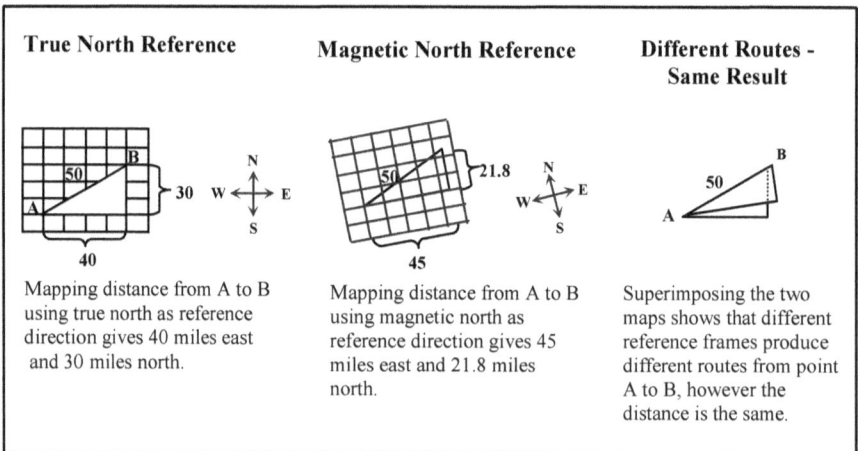

True North Reference

Mapping distance from A to B using true north as reference direction gives 40 miles east and 30 miles north.

Magnetic North Reference

Mapping distance from A to B using magnetic north as reference direction gives 45 miles east and 21.8 miles north.

Different Routes - Same Result

Superimposing the two maps shows that different reference frames produce different routes from point A to B, however the distance is the same.

The Magnetic North Pole is not a constant location but is forever wandering about the Arctic. In 2005, it was situated about 520 miles from the true north pole at 82.7° N longitude and 114.4° W latitude having moved nearly 700 miles in the last Century. Obviously, the difference between "true north" and magnetic north measurements will vary from different parts of the globe. A measurement taken in Salt Lake City Nevada, near the 114.4° W latitude will be in agreement with true north since it is located on a straight line drawn through the magnetic and true north poles, whereas one taken in London England near 0° longitude will disagree significantly.

The true north corrected map shows the route from point A to point B travels 40 miles to the east, then 30 miles to the north. The uncorrected magnetic north map however, would be tilted relative to the corrected one, skewing it so the same directions would not apply, telling the traveler to go 45 miles east and 21.8 miles north. The straight-line distance measured by both maps however is the same—50 miles.

This mixing of spatial directions from two different map-making reference frames is much like the mixing of time and space information by observers in different motion reference frames. When using different map-making criteria, we obtain different values for the east and north legs of the trip, but identical straight-line paths. In the space journey example, we get different measures of space and time for the at-rest and traveling twins, but the same measures of spacetime.

Spacetime diagrams perform a function similar to the map example of figure 12 by mixing space and time measurements. Figure 13 shows two versions of space and time diagrams. The diagram on the left, labeled Newtonian, represents the way scientist visualized space and time before relativity, while the one on the right accounts for relativity effects. As you might expect, spacetime diagrams blend the two factors implicit in their title, namely time and space. The vertical, or Y-axis, of the diagram represents time while the horizontal, or X-axis represents space. Together, the Y and X axes form a plus sign with the center identifying an event at the present time and location. All points on the horizontal space line denote events happening now—in event E_1's frame of reference—in three-dimensional space while points above and below the horizontal line represent events in the future, and in the past of the present time respectively. The diagrams are simplified by expressing both the time and spatial coordinates in time-related terms. Time is in years, and since the main application of spacetime diagrams lies in describing situations involving enormous distances, the space element is measured in light-years, the distance light travels in a year. Using this choice of terms, the lines separating the past and future from the elsewhere naturally form 45-degree angles to the time and space lines.

Figure 13 Spacetime Diagram

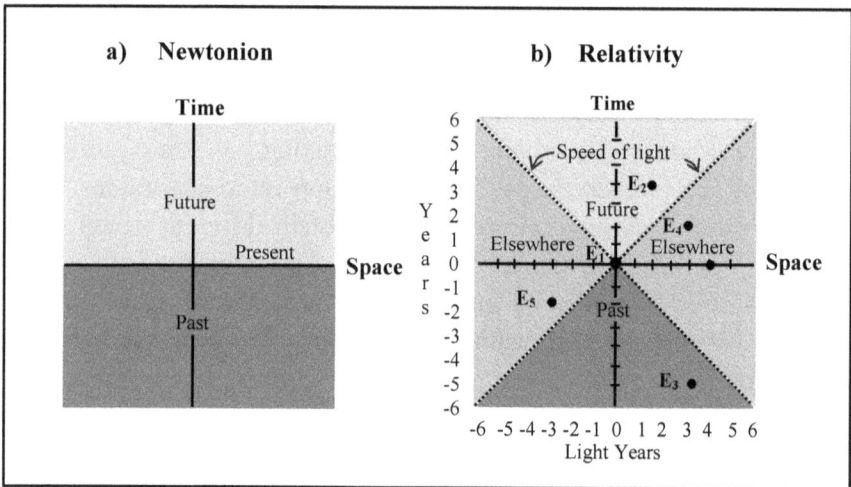

Another, and necessary simplification, is the manner in which space is portrayed. Normally, location of the place component requires three spatial coordinates. However, the act of representing space as one of the components of spacetime in its normal three dimensions is a formidable task to say the least, and therefore requires simplification. As a result, in the spacetime diagram, space is a single dimension yet it represents all three. This simply means ignoring the fact that space in our Universe permits us to move in three distinct directions, and accept that we can only move back and forth along a

66

single line.

A spacetime diagram is all about events, and a particular event always anchors it at its center, the intersection of the time and space coordinates. This central point identifies the here and the now, and is normally the time and location of the event to which another event or events is compared. Since three-dimensional space is "squeezed" into one dimension, the horizontal space line represents all of the events happening in the entire Universe at this particular moment. Conversely, the vertical time line is the entire history and the future of the location of the current event. The extent of the line below is the history of that location and above the future. A line drawn in any direction on the diagram is a *worldline* that could designate the history and/or future of any one of an infinite number of possible scenarios.

To identify the place of events we wish to compare, move across the horizontal axis the correct number of units (in this case light-years) to establish where the event will, or in the case of a past event, *has*, occurred. Similarly, we mark the time of the event by moving up or down the vertical time line. The event in spacetime is then the intersection of perpendicular lines drawn from the representative points on the space and time lines.

In Newtonian physics, simply going through the procedure discussed thus far provided all of the information one could expect from a spacetime diagram, and the drawing of figure 13a explains it perfectly well. A point in time for one person appeared the same as for anyone else, regardless of the states of motion. One could describe any event happening at this instant in New York City as occurring in the present and it would be the same present as any event happening right now in the Andromeda galaxy. The present, future, and the past would be the same for all events throughout the Universe. In the 13a diagram, the present is the horizontal line; all the area below is the past, and above is the future. Relativity however, changed all of this through one simple statement, that light travels at a constant speed and nothing can exceed that speed. In addition to the familiar present, past and future, the elsewhere is necessary to clarify events.

Figure 13b shows a post-relativity spacetime diagram describing five events within Earth's region of the Milky Way Galaxy. Tick marks on the X and Y axes identify increments of space in light-years, and time in years respectively. Every point on the diagram marks an event within the resulting spacetime, and identifies it with respect to the present time and location of an observer at event 1 (E_1), the intersection of the time and space lines. The dotted lines are paths of light beams traveling through E_1, the event to which we compare the other events. All events within the upper triangle are close enough that E_1 would have time to send a message at the speed of light or less, which would reach the event in time to influence it; therefore, they belong to E_1's future. For example, event E_2 in figure 13b is located 2 light-years in space and will happen 5 years into E_1's future. This means there is time for E_1 to send a signal, either through electromagnetic waves or a spaceship moving at one-third the speed of light or greater, that would reach E_2 in time to influence the event; therefore E_2 is in E_1' s future. Similarly, E_3 happened 5 years ago and 4 light-years away, so a light signal from E_3 could have arrived in time to affect the outcome of E_1 placing E_3 in E_1's past.

Events E_4 and E_5 on the other hand occur in different frames of reference with respect to E_1. E_4 will take place 2 years from now but is 4 light-years away. In other words, no form of information about the event could reach and influence E_1 since it would need to exceed the speed of light; therefore, the two events are in each other's elsewhere and cannot be causally related. The same applies to E_5, which took place 2 million years ago, 3 light-years in the opposite direction of E_3 and E_4. Had E_5 sent a signal to E_1, it would not have arrived in time to influence the event and therefore was not in E_1's past.

However, events are not eternally bound to occupy their present time relationships to event observers. Only the event relationships remain permanent. Two years from now E_2 will move into the elsewhere of an individual at the location of E_1, and in 4 more years will be in his past. Similarly, in 1 year E_5, and in 6 years E_4, will have moved from the elsewhere into his past where they shall forever remain.

Relativity changed the present, past, and the future. Instead of

68

the present being a line as long as the Universe, it became a point in the Universe that is different for each individual depending on his state of motion. The future for an individual shrank from including all events of the Universe happening later in time, to only those events accessible from the present time and location by a beam of light. In figure 13b, the future is the wedge shaped area bounded by the speed of light lines above the time and space line intersection, while the past is the corresponding wedge below. The two wedges to the left and right of center define the elsewhere—all the events of the Universe that cannot effect the present moment.

As indicated earlier, spacetime diagrams relate to events, but they also involve people since only events that affect us concern us. So how do we fit into the picture? As mentioned earlier, on a spacetime diagram, each of us is a path, called a *worldline*. Each moment in a person's life occurs at some time and some place, so each moment represents an event, or a point on the spacetime diagram. Since your worldline is confined to one dimension, it does not portray your left-to-right and up-and-down movement as you progress through life; instead, your life events form a continuous sequence of spacetime points. While you have freedom to wander about in three-dimensional space as you please, your worldline sequence moves steadily into the future on a single line.

Figure 14 Worldline

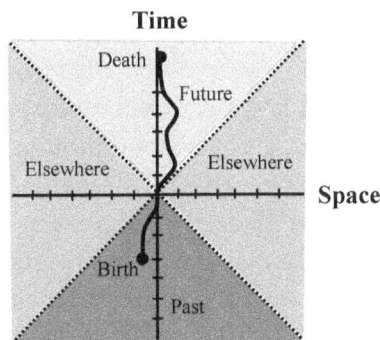

69

Figure 14 shows a spacetime diagram with the worldline of an individual human life. Note that no part of the worldline can be inclined at 45 degrees or more to the vertical since that would imply travel at the speed of light or faster. The worldline of a light beam is simple; it is the 45-degree line advancing 1 light-year in space each year. It is necessary to exaggerate the sharpness of the turns taken by the worldline of a person not fortunate enough to own a spaceship, since due to the magnitude of the light-year distance measurement, any movement in space by an earthbound person from his place of birth would be too small to detect. It would essentially appear to be a straight vertical line superimposed over the spacetime diagram time line. The exaggerated worldline of figure 14 probably belonged to a twenty-five year old individual who never left his hometown until he enrolled at the nearest university where he remained seven years to earn his PhD. He might not appreciate the future we have taken the liberty of plotting for him, but it has him spending the next forty years moving up the corporate ladder of a major corporation, then retiring to a place where he could be near his alma mater, of which he was very fond.

The enormous distance and speed factors make the concept of spacetime diagrams difficult to understand. Light travels 670 million miles in an hour, so a light-year is $670 \times 10^6 \times 24 \times 365 = 5.4 \times 10^{12}$, or 5.4 trillion miles. We could of course reduce the distance factor in our example to a level more consistent with our earthly experience by substituting light-years with 1,000-mile increments. But in the example above, the elsewhere would almost disappear since the time required for light to travel the six one-thousand mile increments is less than .04 seconds, reducing the angle of the speed of light line to the parallel space line from 45 degrees to 3×10^{-6}, or 3-one millionths of a degree. As a result, for the spacetime diagram to have meaning, the spatial distance must be of the same order of magnitude as the speed of light. To do so while comparing distances measured in thousands of miles, we would also need to reduce the vertical time scale proportionally. This would mean examining events separated in hundredth-of-a-second increments. Then, instead of dealing with distances outside our range of experience region, we would be concerned with an equally unfamiliar time frame. Therefore, like many aspects of relativity, the elsewhere does not normally affect our daily lives, since we do not travel at the speed of light or concern ourselves with time differences measured in hundredths of a second. Thus, we experience a worldview more like the Newtonian drawing. Only when we consider great distances or short time periods, does the speed of light line move toward 45 degrees, revealing the elsewhere separating the past from the future.

Again, this does not mean that events from the elsewhere remain forever inaccessible, only that they cannot effect our present. On August 27, 2003, Mars was 34.65 million miles, or 3.5 light minutes from Earth, the closest it has come in the previous 59,619 years. You could neither influence, nor be influenced by an event happening that day on Mars within 3 minutes since it would have been in your elsewhere. However, had you known of it 1 minute earlier, it would have been in your future, so you could have influenced the Martian event. Similarly, an event that happened 3 minutes ago on Mars cannot influence you at this moment so it is in the elsewhere of your here and now. But had had it happened 1 minute sooner, it would have been in your past, and you could have taken steps to alter it. In summary, an event once in your future is now in your elsewhere and seven minutes from now will be in your past. Its relation to you changes because you are not an event, you move through spacetime.

As illustrated in figure 15, the spacetime diagram is also useful for demonstrating the concept that two observers in motion relative to one another may disagree as to the order of events they both witness. Ship B is an early model spacecraft built in the latter half of the twenty first Century that travels at only one-third the speed of light at 195,330,000 miles per hour, or 0.333 c, while C is a late 2130 model that moves twice as fast at 0.667 c. Since in this example, we are not concerned with the past and only with a single direction, we have modified the diagram to include only the quadrant representing the future, and a single direction.

Figure 15 Spacetime Diagram, Reference Event A

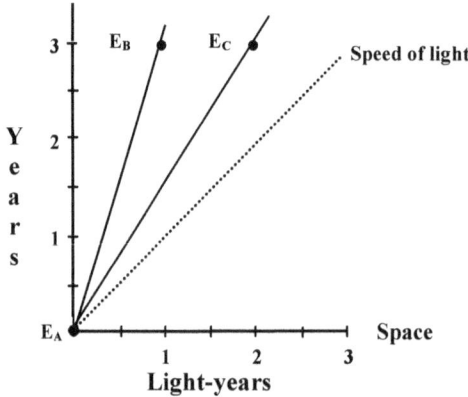

Both spaceships began their journey from the Kennedy Space center at the same time. Since all spacetime diagrams have a point of reference, we have designated the mission control room labeled event A, or E_A, and placed it at the intersection of the time and space axes. Events B and C, E_B and E_C, took place on board the two ships, and the captains immediately notified the Kennedy control room by sending an electromagnetic signal at the speed of light. Kennedy received the E_B message alerting them of the event exactly 4 years after the two spaceships departed Kennedy. Although spaceship B had launched 4 years ago, it was obvious that E_B had occurred sometime earlier since the light message had to travel a considerable distance. However, knowing the speed of light c, and that B traveled at 1/3rd c, determining the time of the event was simple algebra.

If x is the travel time for spaceship B, then the travel time for light, which is 3 times faster than spaceship B, must be x/3. Since the total trip is 4 years, then:

$$x + x/3 = 4 \text{ years}$$

Multiply both sides of the equation by 3:

$$3x + x = 12 \text{ years}$$
$$4x = 12 \text{ years}$$
$$x = 3 \text{ years}$$

Therefore, spaceship B spent 3 years traveling to event B, while light, being 3 times faster, required only 1 year returning, thus accounting for the 4 years between the launch at Cape Kennedy and the receipt of the message.

Exactly 1 year later, Kennedy received a message describing event E_C, also transmitted at the speed of light. Using the same algebra, as 5 years had passed since C launched toward Proxima, 3 years must have been spent traveling at $2/3$ c when E_C occurred. Therefore, the light message would take 2 years to reach Earth, accounting for the 5 years. Since, according to Kennedy's frame of reference, both spaceships had traveled for 3 years when events E_B and E_C occurred, spaceship B, moving at $1/3$ c would have been 1 light-year away while spaceship C, at $2/3$ c would be 2 light-years from Kennedy, or almost half way to Proxima.

$$x + 2\,x/3 = 5 \text{ years}$$
$$3x + 2x = 15 \text{ years}$$
$$x = 3 \text{ years}$$

Everything so far is consistent with "our common sense" notions of Newtonian time and space. Aside from spaceships traveling at hundreds of millions of miles per hour, all of our observations and calculations fit with earthly experience. From Kennedy's point of reference, E_B happened 3 years after liftoff and 1 light year away, with E_C 3 years and 2 light years away. But what about an observer aboard one of the moving spaceships? Since relativity allows each event to declare itself stationary and all other events in motion, we can question how the events would appear to an observer on the slower spaceship B. Figure 16 describes this situation. Since B is the frame of reference, we assign it the stationary status and place it at zero light-years on the space line, with the Earth falling behind at 0.333 c, while spaceship C is moving ahead, also at 0.333 c relative to B. We accomplish this in figure 16 by tilting the lines AB and AC (the lines connecting E_A with E_B and E_C respectively) to the point where AB is vertical while maintaining the angle between AB and AC. Note that in figure 15, the location of point E_B means line AB has both a space and time component. However, in figure 16, AB is vertical along the time axis, thus removing the space and movement component and establishing E_B as the at rest observer.

Figure 16 Spacetime Diagram, Reference Event B

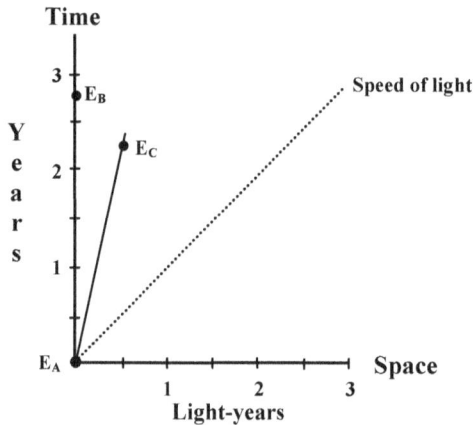

From the figure 15 scenario, both spaceships A and B reported events 3 years after liftoff according to the Kennedy Space center, or E_A. However, due to time dilation, the clocks belonging to observers on the two spaceships did not agree with the E_A clock. Since, according to the Lorentz contraction, time and space shorten for objects in motion, the accurate clock and calendar aboard spaceship B indicated they had traveled less than the 3 years measured by the space center:

$$t'_B = 3 \text{ x} = 3 \sqrt{1- (0.333)^2} = 3 \text{ x } 0.9428 = 2.83 \text{ years}$$

Likewise, the clock and calendar of spaceship C moving at 2/3 c showed:

$$t'_C = 3 \text{ x} = 3 \sqrt{1- (0.667)^2} = 3 \text{ x } 0.7454 = 2.24 \text{ years}$$

A comparison of figures 15 and 16 reveals the surprising observation that three different observers of the same two events can each correctly assign different times to their occurrence. The stationary observer at E_A received accurate information that events on spaceships B and C occurred simultaneously at 3 years following their departure, while the observer on the slower ship at E_B calculated an equally accurate value of 2.83 years, and the faster ship at $E_C = 2.24$ years, or more than nine months less. The fact that the perception of time is different for observers in different states of motion is obvious from a brief examination of figures 15 and 16. A line connecting E_B and E_C in figure 15 (line BC) would be parallel to the space axis, meaning there is no time element involved; the relationship between the two events is completely spatial, so the events occurred at the same time in E_A's reference frame. A line connecting the same points in figure 16 however, is not horizontal, meaning it does contain a time element, so the events could not have happened simultaneously.

Since time is not a fixed parameter that marches forward in lock step, identical for all observers and events throughout the Universe, the door is left open for individuals to attend events in the past according to their frame of reference. First, though, a definition of time travel: *experiencing the same event twice*. If the observer of E_B could return to the time a few moments prior to the event, he could witness the same event again (an instant replay without video, if you will). An appropriate analogy to the phenomenon would be to begin a westward journey on a specific longitudinal line and continue traveling around the globe until you returned to your point of embarkation. Your trip would constitute a "closed path". Similar journeys in spacetime, in which you return to the event from which you began is called a *closed time-like curve*. On such a journey, your world is not moving backwards; in fact, you and the hands on your watch continue moving forward. All of your body functions remain the same as now and, unfortunately, you continue to age. So if you happen to be revisiting your tenth high school reunion, you return to the event as the oldest member of your class.

Several schemes exist for time travel, most of them quite complex including one based on a relativity concept called *gravitational time dilation*, which we will discuss later. As you might have guessed, we have chosen the easiest to describe and will begin with figure 17, which is based on information borrowed from figure 15. E_A is not relevant to the description, so only E_B and E_C, which occurred simultaneously (the same vertical height) but at different spatial points (one light-year apart on the horizontal axis) in E_A observer's reference frame, repeat on the left side of figure 17 below.

Figure 17 Wormhole Shortcut between Simultaneous Events

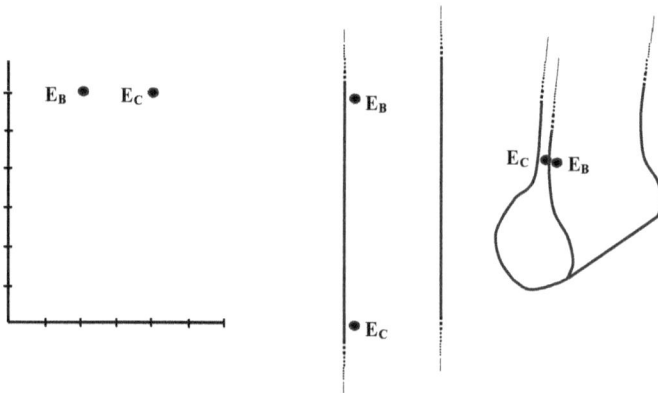

Now we need to begin using our imaginations to picture "spatial shortcuts" called *spatial wormholes*, or more technically closed timelike curves that will rapidly take us to different points in space without exceeding the speed of light. While we will discuss wormholes in more detail in chapter 8, let it suffice for now to describe one as an extremely short tunnel, or throat, with a mouth at both ends that connects distant events in time and three-dimensional space. To demonstrate how a wormhole might connect E_B and E_C, we have plotted them at the center of figure 17 on a two-dimensional sheet of paper representing a "slice" of infinitely long three-dimensional space. On the right side of the figure, the paper is folded in such a manner as to bring E_B and E_C together at a point representing a wormhole. While a sheet of two-dimensional paper is a poor representation of a wormhole, it does allow us to visualize its function, which is to "draw together" distant points in three-dimensional space.

Now imagine being aboard spaceship B when E_B occurs. Imagine next that we could induce a wormhole to appear at the 1 light-year location of E_B. and that we could enter through the mouth, then immediately be deposited on spaceship C just as E_C began. Thus deposited aboard spaceship C, the first thing we see is another spaceship. Call it spaceship D, passing by at 0.333 c, the same speed as spaceship B in the control room's reference frame. Finally—as luck would have it—D just happens to have a wormhole connecting it to spaceship C. We then immediately enter the wormhole entrance on C and exit aboard D at the same location as B, 0 light-years, only at an earlier time. Now let's see how E_B and E_C look from our perspective aboard the new spaceship. To do this, we need to redraw E_B and E_C of figure 16 with spaceship D as the frame of reference.

Figure 18 Spacetime for the Returned Time Traveler

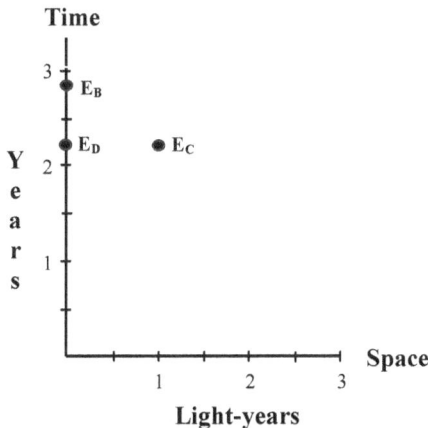

As shown from the points on the space and time axes of figure 18, we arrive at the same location, or point in space as E_B but at the same point in time as E_C. Therefore, we have time-traveled 1 light-year from E_B to E_C and returned before E_B happened. Now we only need to wait $2.83 - 2.24 = 0.59$ years, or 7 months, and we can experience E_B for the second time; we have returned to the point in space where we began, but have completed a closed time-like curve and returned at an earlier point in time.

While the fundamental scientific aspects of time travel are well established, this does not mean we have the capability to make such a journey—or for that matter that we ever will. Serious scientists agree unanimously with the concept of time dilation through the Lorentz contraction, the key to the concept that timing of events is different for observers in different states of motion. Therefore, the feasibility of placing ourselves in a position wherein we could return to an event that had taken place prior to our departure is fact. The problem lies in the return route. As already mentioned, simply turning the spaceship around and traveling back at the same high speed is not an option since special relativity requires that the state of motion be symmetrical. Changing our craft direction violated that requirement. This leaves open only return transportation modes that involve "shortcuts" such as a *wormhole*, and science is not in such a high state of agreement on this theoretical phenomena of space travel. It is not that scientists do not believe wormholes exist; only that travel through them may well be impossible. Some scientists believe that many, if not all, black holes are actually wormholes; however, those large enough to permit entry to humans may take billions of years to exit the other side. On the other hand, wormholes sized to permit rapid passage could be so small that only tiny streams of particles could enter.

The Large Hadron Collider (LHC) at CERN, near Geneva Switzerland, may answer many of the questions concerning wormholes and their application to time travel. Mathematical physicists, Professor Irina Aref'eva and Dr Igor Volovich, at the Steklov Mathematical Institute in Moscow believe that subatomic

78

collisions in the LHC will generate energies powerful enough to create tiny wormholes.

Kinship of Mass and Energy - E=mc²

The most well known aspect of Einstein's work, the equation E=mc², was not the primary goal of special relativity and not even part of the original paper on the subject. The realization of a fundamental equivalence between mass and energy emerged logically from his theory, almost as an afterthought. In fact, he did not present it to the *Annalen der Physik* until three months later on September 27, 1905, under the title *Does the Inertia of a Body Depend on Its Energy Content.* The equation states that the quantity of energy E, in any object, is a function of its mass m, multiplied by the square of the velocity of light c. Since the velocity of light squared, is such a tremendously large number, conversion of a small quantity of mass results in the liberation of a vast amount of energy. So much so, that if we could convert all of the mass of a single peanut to energy, there would be enough to power an entire city for a day. However, in practice, such complete conversion of mass to energy is not achievable; even nuclear fusion and fission reactions convert only a small fraction of the available mass.

Einstein's equation is normally associated with nuclear reactions employed in atomic weaponry or electrical energy production; however, it applies to all energy transformation, including the chemical reactions involved in burning gasoline and coal. Any activity that adds energy to, or liberates energy from a substance changes its mass; even a rubber band weighs ever so slightly more when stretched than when it is relaxed.

Therefore, this equivalence of mass and energy means that a body in motion has more mass than when at rest since the mass equivalent of the energy required to attain its motion is added to its at rest mass. This too only becomes significant at speeds representing a significant portion of the speed of light. The mass of an object traveling at 58,600,000 miles per hour, or 10% of light speed, would only increase 0.5%; however, at 90 % of the speed of light, its mass would double. The mass of an object nearing the speed of light increases more and more quickly since the additional energy needed for acceleration is added to the increasing mass of the object, causing it to require even more energy in order to speed it up further. No object, regardless of how small, can ever attain the speed of light since this accumulation of mass would become infinite and require an infinite amount of energy. This fact provides one of the best explanations as to why it is impossible for any massive object to go faster than the speed of light. Since energy converts to mass in this manner, Einstein reasoned that the opposite must be true as well. Mass must convert to energy. Because of *mass/energy equivalence*, energy, like mass, manifests itself as inertia making an object more and more difficult to accelerate as its velocity increases.

The knowledge—that the energy required to move a massive object converts to mass —allowed Einstein to establish the now famous relationship, although he formatted the equation differently in his September 27, 1905 paper. He began with the statement "...If a body emits the energy L in the form of radiation, its mass decreases by L/V^2..." Expressed as an equation, this became $L=mV^2$ and remained the construction until 1912. At that point, E, which had emerged as the representation for energy replaced L, and V, the shorthand for the velocity of light, changed to the more familiar c for constant; resulting in the world's most famous equation, $E=mc^2$.

General Relativity

By bringing motion into the picture, Einstein brought about radical changes in the way we view the Universe. Special relativity was revolutionary in its recognition that concepts of time and space as old as humankind itself were not the fixed parameters dictated by common sense but in fact were variables that could take on any value depending on the observer's state of motion. However, Einstein was still not satisfied since his theory was only a "special" case of what he was certain was a much larger whole. His goal was to eliminate the "uniform motion" qualifier from his special relativity theory and simply state: The laws of physics are the same for all observers regardless of their motion. He recognized that the Achilles heel of special relativity was an inconsistency between two facts: (1) that nothing in nature could travel faster than the speed of light and (2) the instantaneous nature of what Newton had called "the force of gravity". In Newtonian gravity, the gravitational force between two objects is an instantaneous phenomenon. This means for example that if the Sun, 11 light minutes away from Earth, should suddenly and totally disappear, we would feel it instantly as well as apocalyptically. On the other hand, relativity proves that nothing can travel faster than light, so any information about the Sun's disappearance could not reach Earth in fewer than 11 minutes; therefore, Newtonian gravity and special relativity must be out of sync.

Einstein's answer: Newton was wrong. Gravity is not a force at all but a consequence of the fact that spacetime is not flat but warped or curved by the distribution of mass and energy within it; and gravity is simply that curvature of spacetime. Objects, including the Earth travel through space on the straightest possible line in 4-dimensional spacetime. In the case of the Earth, the warping of spacetime created by the Sun's mass causes this straightest possible line to be what we observe as a near circular orbit. An often-used analogy to this description of gravity depicts four-dimensional spacetime as a two-dimensional elastic membrane on which small spheres are rolled in a straight line from one side to the other. If we place a large spherical object near the center that is sufficiently heavy (e.g. a bowling ball) to depress the membrane surface, then roll a smaller sphere (a marble) near the vicinity of the larger one, the resulting depression will shift the marble's direction toward the large sphere. In fact, if we design the conditions just so, we can cause the small sphere to "orbit" the larger one for a few passes before spiraling inward to a halt.

The Principle of Equivalence

In general relativity, Einstein once again considered the motion of objects to account for inconsistencies between Newtonian and special relativity aspects of gravity. This time however, he did not limit his analysis to uniform motion but included accelerated motion as well (motion in which the speed or direction of an object is changing). Newton's laws had established that the mass components of his gravitation and motion equations, was the common factor in the close relationship between gravity and acceleration. Newton's second law of motion says the force exerted by an object is equal to its *inertial mass* times its acceleration, while his law of gravitation states that the gravitational force between two objects is proportional to the product of their *gravitational mass* and inversely proportional to the square of the distance between them. Inertial mass is the measure of a body's resistance to acceleration, while the measured mass of an object results from the presence of a *gravitational field*. Furthermore, Galileo had already established a one to one relationship between gravitational mass and inertial mass in experiments where he rolled different size balls down an inclined plane (not by dropping them from the tower of Pisa as is commonly believed). All of the balls rolled down the plane at the same speed regardless of their size showing that acceleration due to gravity was indifferent to the mass of objects. This outcome could only happen if gravitational mass and inertial mass are equal.

Newton's 2nd Law of Motion:

$$F_{(inertial)} = M_i \, a$$
$$M_i = \text{inertial mass}$$
$$a = \text{acceleration}$$

Newton's Law of Gravitation:

$$F_{(gravitational)} = GM_A M_B/[d_{AB}]^2$$
$$M_A, M_B = \text{gravitational mass of objects A and B}$$
$$G = \text{gravitational constant}$$
$$d = \text{distance between objects A and B}$$

Einstein's brilliant breakthrough came from the observation that acceleration and gravitation are not only related via the mass component; in a closed environment, their effects are indistinguishable. Since the M components of both equations are based on the same definition of mass, the other components— acceleration a in the motion equation, and the gravitational constant G along with the distance component $[d_{AB}]^2$ in the gravitation equation—must be related.

To Einstein, this meant that gravity and accelerated motion are more than just similar; they are the same. We see a graphic example of this equality when a space astronaut undergoes acceleration, either through rotation of his spacecraft or increasing its speed at the precisely correct rate. He cannot tell the difference between the "artificial" gravity thus created, and the "natural" gravity he feels while on the launch pad at Cape Kennedy. (In fact, g-forces that measure the acceleration experienced by astronauts during launch are simply multiples of the Earth's gravitation.)

In addition, we can make the effects of gravity disappear by properly directed offsetting acceleration. Imagine a man in an elevator with a broken cable that is falling freely, accelerated by gravity from the top floor of a very tall building. In fact, why not make the building as tall as the mountain from which Newton launched his orbiting satellite in order to provide time for him to work a few experiments on the way down. If he dropped his ballpoint pen during the ever rapidly increasing descent, instead of falling to the elevator floor or rising to the ceiling, it would continue to float along beside him at the same level it left his hand, as if the law of gravity had been repealed. His speed would increase continuously as he fell but, in the absence of sound or other outside influences, he would feel no sense of motion whatever. There would be no difference between his sensation (other than perhaps pure panic) and one felt by an astronaut in space with no simulated gravity. Unfortunately, his voyage would be more brief and tragic than the astronauts.

Not only can we use acceleration to offset gravity, we can use it to create the identical properties of gravity. Picture the same man in an elevator located in gravity-free outer space, and some force is pulling him upward at a rate identical to Earth's gravitational acceleration. The man feels a downward force against the elevator floor, which can only be due to his inertial mass being pulled upward in an accelerating manner; however, if the elevator was located at rest in an equivalent gravitational field (at rest on Earth), he would experience exactly the same feeling except it would be due to gravitational mass.

It was through pondering these indistinguishable characteristics of accelerated motion and gravity that Einstein made one of the great mental leaps in scientific history, calling it his "happiest thought", when he realized that gravity was indistinguishable from accelerated motion; and from this seemingly simple insight, proposed the *principle of equivalence*. The principle states simply: "*The laws of nature in an accelerating frame are equivalent to the laws in a gravitational field.*"

While Einstein was convinced that the equivalence of accelerated motion and gravity was the key to eliminating the "special" aspect of special relativity, proving it was another matter. In special relativity, Einstein had progressed from the theory's conception in early 1905 to publication and acceptance of results by the scientific community all within the same year. However, general relativity was a different animal. It began innocently enough with his happy thought in 1907 but the scientific community had not fully accepted it until twelve years later in 1919. He first published the new idea, as a fifth article addendum to his original relativity papers in November of 1907 in the *Yearbook of Radioactivity and Electronics*, in which he stated almost as an afterthought "…is it conceivable that the principle of relativity applies to systems that are accelerated with respect to each other?"

However, going was slow over the next two years. The far more difficult mathematics brought about by accelerated frames of reference coupled with Einstein's somewhat deficient math skills (probably stemming from the lack of respect he had given the subject in his university days) added up to slow progress. Furthermore, in spite of the success of his special relativity theory, Einstein remained a lowly patent clerk at the end of 1907 and spent much of the next two years trying to turn his newfound fame into employment within the academic community. By 1909 however, he had received a professorship and was able to return his full attention to general relativity. Realizing his mathematical shortcomings, he enlisted the aid of Marcel Grossman, a former classmate and mathematics professor at the Zurich Polytechnic, to aid in the complex calculus and geometry needed to describe the new theory. The search for a general relativity theory completely consumed Einstein for the following six years, with numerous stops and starts while looking for the right *metric tensors* (mathematical tools that calculate the distance between points in space) to fit the theory; but by 1915 everything had fallen in place and the theory was complete. All that remained to gain acceptance for the new general relativity theory was successful testing of its predictions. Right away, testing of several of these proved them correct; however, it was not until four years later in 1919 that the prediction that would convince science and change the world was proven.

Planetary Orbital Precession

Almost immediately, general relativity unleashed a slew of strange and exotic predictions about the content and workings of the Universe. Among the first proved was a problem that had puzzled scientists since the middle of the 19th Century regarding the instability of the planet Mercury's orbit. Orbits of all planets are subject to *precession*, a phenomenon predicted by Newton's theory that arises from effects planets have on each another. Precession is a change in the elliptical orbit of a planet caused by the gravitational effects of other planets that gradually shift the points when it will be closest to or furthest from the Sun. When orbital precessions are calculated using Newton's equations, the values for all planets except Mercury match observations. However, the measured value for the precession of Mercury's orbit is 5,600 *arcseconds* per Century (an arcsecond is 1/1,296,000 fraction of a circle) while Newton's equations predict a precession of 5,557 arcseconds per Century, a discrepancy of 43 arcseconds.

While 43 arcseconds per Century seems a rather small deviation, it was significant and something yet unknown had to be causing it. The most obvious answer was the presence of a yet to be discovered small planet located between Mercury and the Sun. Belief in this answer to the problem was strong enough that astronomers had given the hypothetical planet a name, Vulcan. Einstein's general relativity equations however, provided the correct answer. They predicted there should be a very small reduction in the *transverse velocity* of Mercury (velocity along its path) due to the increased gravitational field as it came nearer the Sun during the perihelion phase (the point in the orbit when the planet is closest to the Sun) of its orbit. The reduction in velocity was of the precise order of magnitude needed to add 43 arcseconds matching Mercury's observed 5,600 arcsecond orbital precession.

The reason Mercury was singled out for this orbital inconsistency comes from the fact that compared to the other planets its orbit is highly elliptical, a consequence of Mercury being a more recent addition to solar family of planets. All of the planets in the solar system, whether they originated from matter ejected from the surface of the Sun or grew from material captured from galactic space, initially had highly elliptical orbits. However, in time gravitational effects eventually made them more circular. When a planet with a highly elliptical orbit passes close to the Sun during its perihelion, velocity and gravitational forces are at their greatest level causing the planet to experience an increased gravity drag effect, slightly slowing the planet as it passes near the Sun. As the planet continues its orbit, the slightly lower velocity will shorten its distance from the Sun when it reaches its *aphelion* (the furthest point from the Sun) making the elliptical orbit slightly more circular. The continuation of this process for billions of years gradually produces the nearly circular orbits of older planets. Younger planets like Mercury still possess more elliptical orbits and therefore are more subject to small but significant changes in the perihelion phase.

Gravitational Bending of Light

The prediction that had the greatest impact, not only on the scientific community but also on the world public, was that the presence of a massive object should bend the path of a light beam causing it to travel a curved path. The first mention of the prediction came in1907, but a formal announcement had to wait until *Annalen der Physik* published it in a June 1911 edition titled *On the Influence of Gravity on the Propagation of Light*. Gravitational Bending of Light does not mean that gravity is pulling on photons of light since light has no mass and cannot be subject to Newton's law of gravity. Bending of light comes from the fact that the presence of massive objects curves spacetime and light simply follows the shortest path between two points. It was through pondering this aspect of relativity that led Einstein to one of his most startling discoveries; namely, that space itself must be curved, and a ray of light merely follows that curvature of space. Einstein, being a theoretician who never attempted to prove his own predictions, proposed that physicists could prove relativity by observing light traveling to the Earth from a distant star on the other side of the Sun. As light from the star passes by the Sun, the warping of spacetime in its path should bend the stream of photons making the star appear to be from a different spot in the sky. The path of light is not curving as it travels through space; it is following the straightest possible line through locally curved spacetime.

To illustrate this relationship of light to gravity, Einstein used a thought experiment based on the equivalence principle. It again involves a person in an elevator being pulled upward in an accelerated fashion when a beam of light enters through a small hole in one wall. By the time the light reaches the other side, the elevator will have advanced upward causing the light to strike at a lower point on the opposite wall. If a person plotted the trajectory of the light as it passed through the elevator, she would see that it took a curved path; and, since according to the equivalence principle, accelerated motion and gravity are indistinguishable, gravity must bend light as well. If the elevator had been moving at constant speed, the light would still have struck the opposite wall at a point lower than its entry; however, plotting the trajectory would have yielded a straight line.

Experimenters were unable to prove the prediction for several years since the Sun's brightness made it impossible to detect the comparatively dim stars that might align with the Earth and Sun due to their apparent close proximity. However, a solar eclipse was due on May 29, 1919 when several bright stars would be aligned with the earth and behind the Sun, making conditions ideal for such an observation. Praying for clear weather, Sir Arthur Stanley Eddington, an English Quaker and one of the most influential astrophysicists of the early 20th Century, led an expedition to a site in Africa with the goal of measuring the apparent change in position of the stars caused by the impact of gravity on the stars' light as it passed by the Sun. If Einstein's prediction was correct, the stars, even though located behind the Sun, would be visible since their light would be bent by gravity as it passed the Sun deflecting it on a path toward Earth. Eddington's party picked up the stars with their telescopes in the precise locations predicted by Einstein's calculations providing the most prominent confirmation of all of relativity's predictions. When asked what he would have said if Lord Eddington's eclipse observation hadn't confirmed his theory, Einstein, showing more faith in the elegance and beauty of his creation than in experiment, replied "…I would have had to pity the dear Lord. The theory is correct."

Figure 19 Bending of Light by Gravity

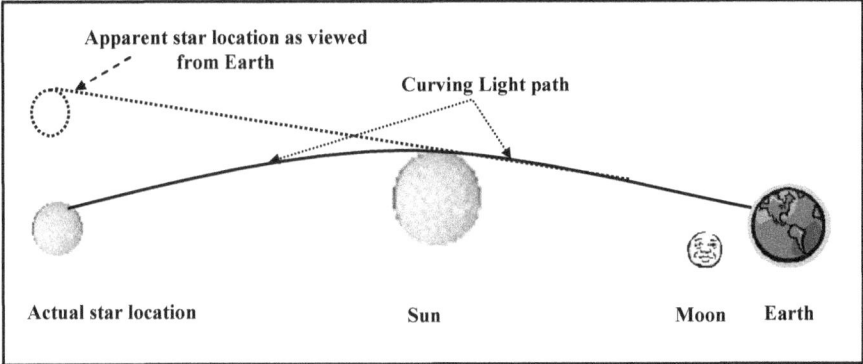

Even before Eddington's proof of gravitational bending of light, Einstein's fame had grown to worldwide proportions. This was different however; this major confirmation of general relativity brought him rock star status. The *New York Times* ran the story on November 10, 1919 under the headline "Lights All Askew in the Heavens"; with a second deck line announcing, "Einstein Theory Triumphs"; and another assuring its readers they needn't bother trying to comprehend it since in the world, there are only "twelve wise men" known to have the ability. Einstein continued to make the newspapers on an almost daily basis for the remainder of 1919 alerting the public to this outrageous new theory.

The mania surrounding the strange new world of relativity, with bending light rays and warped space coupled with the madhouse of quantum mechanics wherein everything exists only as probabilities, spilled over into art, literature, and theology. Philosophers erroneously subjected accepted ideas of right and wrong to relativistic interpretation wherein, like space and time, they were dependent on frames of reference. The quantum world seemed to dismiss the idea of certainties, which translated to license for questioning faith in all absolutes. Relativity became a springboard for a new *relativism* that Paul Johnson wrote about in his novel on a history of the Twentieth Century, titled *Modern Times* "…It formed a knife to help cut society adrift from its traditional moorings…"

Gravitational Time Dilation

While the 1919 proof of the gravitational bending of light gained nearly universal acclaim and acceptance for general relativity, testing many of its claims had to await the development of new experimental techniques. One of these was a prediction that required further alterations to our concepts of time. In special relativity, time dilation makes clocks in uniform motion run slower than when at rest. General relativity predicts another region of variable time called *gravitational time dilation*, in which clocks will run slower when undergoing accelerated motion; or, by virtue of the equivalence principle, through being near a source of gravity. It stems from the fact that overcoming gravitational force requires spending of energy; this applies not only to objects with mass but electromagnetic particles (light) as well. Gravitational time dilation results from the energy lost by light while escaping the gravity of a massive object. Since the pull of gravity cannot affect the speed of the escaping light, the energy loss is manifested by a lower wavelength frequency, which causes a clock seen by an observer in a different reference frame (e.g. far away from the gravitation source) to be running slower.

One of the first verifications of this prediction of general relativity came from an experiment performed by R. V. Pound and G. A. Rebka Jr at Harvard's Jefferson laboratory in 1959. Since the Earth is a relatively massive object, a sufficiently accurate clock near its surface will run measurably slower than one above it. Using nuclear radiation experiments, physicists were able verify gravitational time dilation predictions by measuring differences in the frequencies of photons emitted from radioactive decay experiments conducted on two iron

(^{57}Fe) samples. One sample was located in the laboratory basement and the other on the roof, a vertical distance of 47 feet. Since the gravitational pull of the Earth increases as you approach its center (the basement being nearer than the roof), the photons emitted from the basement experiment should have lower wavelength frequencies than the roof photons. The original experiments confirmed the gravitational time dilation prediction at the 90% confidence level, which was improved to better than 99% in later experiments.

While the affect on clock speeds encountered in our earthly environment is extremely small, accounting for gravitational time dilation is still critical in some applications. The Global Positioning System, depended on so heavily by our military and airlines—and now even to guide our automobiles around the country and find restaurants—would be in error by hundreds of yards if gravitational time dilation were not taken into account.

Singularities

Einstein was not the first to predict the existence of black holes, but warpage of time and space around massive celestial objects is inherent in his general relativity equations. According to Newton's theory of gravity, a singularity occurs when gravitational pressure acts on an object with negligible opposing pressure, causing it to collapse inward, compressing to a zero radius, and therefore to an infinite density. In general relativity, a singularity corresponds to an infinite curvature of spacetime (stemming from the fact that spacetime is curved in the presence of massive bodies; therefore, it becomes infinitely curved in the presence of infinitely dense ones). In 1930, Subrahmanyan Chandra, a nineteen-year-old Cambridge graduate student from India, predicted that very massive stars that no longer contain sufficient quantities of elements needed to support the nuclear core reactions would eventually undergo gravitational collapse (to be discussed in the following chapter). With no nuclear reactions in their core providing outward pressure to counter the ever-increasing gravity, the huge star would continue collapsing to the point of infinite density and zero radius; it would become a singularity. Since the density of these objects is so great that nothing including light can escape their gravity, they were given the name black holes.

The inability of light to escape from an object can be explained by the concept of *escape speed*. To overcome Earth's gravity and place an object on the Moon, one needs to accelerate the object to a speed of 7 miles per second, its escape velocity. The Moon, having considerably less gravity than Earth, has an escape velocity of only about 1.5 miles per second, while the more massive sun needs 400 miles per second. Gravity is so strong in a black hole that the escape velocity is greater than the speed of light, or 186,000 miles per second; therefore since nothing can exceed light speed, nothing can escape from a black hole. They also represent the nearest thing in our current Universe to the conditions existing at the time of creation, the Big Bang.

3   The First Dawn

Georges Lemaître, a Belgian born ordained priest and cosmologist who lived in Springfield Missouri, first proposed the two major tenets of what would become the Big Bang theory in a 1927 paper published in the Belgian journal *Annales de la Societe de Bruxellos*. In the paper, titled *Hypothese de l'atome Primitive*, Lemaître held that the Universe must have begun in a tiny hot dense phase that he called the *primeval atom*, and that it must be expanding; the galaxies should be racing away from each other at speeds proportional to their distances from each other. The scientific establishment initially ridiculed the hypothesis suggesting that Lemaître's theology motivated him to "...sneak a Creator into science". Einstein on the other hand, who by that time had become the unofficial but final arbiter of scientific debate, refrained from religious implications, while at the same time banishing the theory to the backburner of science, calling Lemaître's physics "abominable."

Scientists prior to 1917 had long agreed with the philosophical and religious view of the Universe that described it as a perfect unchanging place where all the stars, planets, and other celestial objects had eternally been just as they viewed them; it had forever been in a steady state. The steady state model claimed that new atoms constantly sprang-up from non-matter sources. The new particles then agglomerated into new stars and galaxies, replacing old ones as they burned themselves out, thus maintaining a Universe that always appeared constant. They believed that someone marveling at the majesty of the heavens ten billion years ago or ten billion years in the future would enjoy essentially the same sky as they.

This view so pervaded scientific and philosophic consciousness that it compelled Einstein to disbelieve what equations from his general theory told him; namely, that the Universe is not static but should be either expanding or contracting. In an expanding universe, stars and galaxies continue drifting further and further apart, eventually producing a dark desolate place. Should the Universe be contracting, a "Big Crunch" is its fate. Gravity pulls all of the galaxies closer and closer, continually increasing their mutual gravitational attraction until they are eventually on top of one another and become a most monstrous black hole. Since neither scenario left room for a static Universe, Einstein reluctantly concluded that in the absence of any known outward force that could cause expansion, the mutual gravitational attraction of the galaxies means the Universe really wants to contract.

However, Einstein was so certain of the static perfection of "God's creation" that he felt this theory must somehow be wrong, or at least it must be incomplete in this respect. He reasoned that the very fact contraction had not already occurred means there must be a compensating energy creating an outward force that opposes the inward gravitational pull. This firm belief in a static Universe compelled him to commit what he later termed his greatest error; he modified general relativity, adding a new term λ (lambda), which he called the *cosmological constant*. The constant essentially assigned an inherent pressure to empty space itself that tended to push the Universe apart. The magnitude of the lambda force precisely offset the gravitational attraction exerted by galaxies on each other, thereby maintaining a static universe.

Lemaître's view on the other hand, held that two forces have been at work since the beginning of time, each trying to move the Universe in opposite directions. One is the outward velocity of all the galaxies, initiated by the exploding primeval atom energy that wants the Universe to continue expanding forever. In the second, the accumulated gravity from the combined mass and energy of the Universe, counters expansion energy in an attempt to bring about contraction. Lemaître also invoked the logic that since the Universe exists; energy from the primeval atom must have over-powered gravity, giving an expanding universe.

However, Einstein's cosmological constant remained part of the Universe description for only a few years. In the mid 1920s, Edwin Hubble, working with improved telescopic techniques at Mount Wilson Observatory in southern California, demonstrated that our Milky Way was not the total Universe; that other galaxies existed beyond our own. He mounted a 100-inch telescope in a room built on a platform that rotated at a rate precisely calibrated to a clock. This arrangement compensated for the Earth's spin on its axis and orbital rotation around the sun, permitting him to lock in on a particular region of the sky for extended periods in order to take time-exposed photographs. The results astounded the scientific world revealing objects so distant their light had been too dim for conventional telescopes to capture. To Hubble, and the rest of the world's amazement, he discovered there was far more to our Universe than previously believed; he photographed what turned out to be the first distant galaxy.

Hubble had been photographing what earlier astronomers had called the *Andromeda Nebula*. Prior to development of telescopes, their naked eyes had seen only a few Nebulae, and these appeared as curious smudges of light explained away as oddities within the Milky Way galaxy. By the latter part of the Eighteenth Century however, debate had already begun as to whether nebulae were individual stars within the Milky Way or in fact distant galaxies themselves. During October of 1923, Hubble noticed a previously unreported speck on one of his negatives near the Andromeda Nebula. After examining other photographs from the same area of the sky, he found the speck could be seen quite well in some negatives but barely visible in others, a characteristic of *Cepheid variable stars*. The discovery ranked as the greatest achievement of Hubble's career at that point since Cepheid variables have a unique characteristic that enables Astronomers to measure their distance from Earth. Since the Cepheid belonged to the Andromeda Nebula, knowing its distance would settle a long-standing debate as to whether nebulae were curiosities of the Milky Way or distant galaxies.

Cepheid variables, first discovered by John Goodricke and Edward Pigott in 1784, are stars that do not always appear the same but have periodic variations in their brightness levels. This peculiar aspect of Cepheids results from the fact that, unlike other stars, they are not in stable equilibrium. Stars, being extremely massive objects, generate correspondingly massive inward gravitational pressure wanting to crush themselves. In turn, this inward pressure raises the star's core temperature, thereby increasing the rate of nuclear fusion reactions, causing a countering outward expansion pressure. In most stars, these two forces (internal gravitation and external expansion heat) balance one another, allowing them to shine at consistent levels of brightness until they begin to deplete their nuclear fuel source millions to billions of years later.

This equilibrium is out of sync in a Cepheid variable. It experiences relatively cool periods, which allow gravity to take over, causing the star to contract and making it visibly dimmer. Eventually however, the greater inward crushing force increases core temperature, and therefore the rate of nuclear reactions, generating more heat, causing it to again expand and brighten. Depending on the size of the Cepheid star, its period (the length of time between its dimmest and brightest state) can vary from several hours for smaller Cepheid's to four months for larger ones.

An astronomer named Henrietta Leavitt had discovered in 1912 a relationship between the luminosity (brightness) of a Cepheid variable and it's cycling period. The problem with establishing this relationship earlier had stemmed from a paucity of information, and that which did exist came from Cepheids located in widely different regions of the sky. Therefore, one could not tell whether a faint star appeared dimmer due to intrinsic dimness or if it was equally bright but simply further away, making any relationship between brightness and period meaningless.

Using self-developed techniques and dogged determination, Leavitt set about accumulating a more coherent database that would include a statistically meaningful number of Cepheids, all located approximately equidistant from Earth. While studying photographic negatives of the Small Magellanic Cloud—a stellar formation only visible from the southern hemisphere, and first discovered by explorer Ferdinand Magellan during his 16th Century circumnavigation of the globe—she was able to identify twenty-five Cepheid variables with cycle periods ranging from less than one day to 120 days. Leavitt reasoned that since all twenty-five Cepheids belong to the same grouping, variations in the distance from Earth to individual stars within the group would not be significant when compared to the much greater distance from Earth to the group itself; therefore, one could consider all their distances equal. Based on this assumption, any differences Leavitt would find among luminosity measurements must be due to real, rather than perceived luminosity differences.

Astronomers base stellar brightness measurements on methods called *apparent magnitude* and *absolute magnitude*. Apparent magnitude tells how bright a star appears when viewed from Earth without considering its distance. Obviously, shiny objects viewed from close range will appear brighter than equally shiny but more distant ones. Absolute magnitude on the other hand measures a stars intrinsic brightness—the amount of light it actually emits. It puts all stars and other celestial objects on an equal footing by determining how bright they would appear if viewed from the same distance. Specifically, absolute magnitude is the apparent magnitude a star or galaxy would have located 10 parsecs (32.6 light years) away from Earth.

The stellar magnitude brightness rating is an unusual measurement method since brighter objects have lower, or even negative, numbers—the lower an object's absolute magnitude value, the greater its intrinsic brightness. For example, the Sun has an absolute magnitude of +4.83, while the brightest star in our sky Sirius has a much higher intrinsic brightness rating since its absolute magnitude is +1.43. A star with an absolute magnitude brightness one unit lower than another (e.g., +3 versus +4) is intrinsically 2.5 (more precisely 2.512) times brighter; 5 units of absolute magnitude brightness translates to 100 times intrinsic brightness, while 7.5 units means 1,000 times and 10 units, 10,000 times more intrinsic brightness. This peculiar characteristic of the magnitude scale comes from a convention first established by Ptolemy in the 2nd Century AD. Ptolemy classified stars in his catalog according to their apparent magnitudes by establishing the brightest star he could see as magnitude 1 and the dimmest as magnitude 6. Stars of apparent magnitude higher than +6 are not visible from Earth without the aid of telescopes; a convenience not available to Ptolemy. Over time, as Astronomers from different regions of the globe discovered more and more stars, they found many brighter than magnitude 1, which required negative values (e.g. Sirius has an apparent magnitude of − 1.4).

Perhaps Ptolemy would have come up with a different scale, had he known the Sun was just another star. Obviously, on an apparent magnitude basis, our Sun has the brightest rating of all celestial bodies at −26.7 while the Moon is second at −11 and Sirius, although intrinsically 40 times brighter than the Sun, places a distant third at − 1.4. The apparent magnitude difference between the Sun and Sirius is about 25 (26.7 − 1.4 ≈ 25). Since, as stated above, the ratio for every five magnitudes of difference in brightness is 100, Sirius is 100 x 100 x 100 x 100 x 100 = 10,000,000,000 or ten billion times dimmer than the Sun in apparent magnitude.

	Sun	Sirius	Brightness vs. Sun
Distance, light years	0.000015	8.6	
Apparent Magnitude	−26.7	−1.4	10^{10} x dimmer
Absolute Magnitude	+4.83	+1.43	40 x brighter

Note that Sirius has a more positive absolute magnitude than apparent magnitude rating. This results from the nearness of Sirius, at

8.6 light years, relative to the absolute magnitude standard distance of 32.6 light years. All stars closer than 32.6 light years from Earth will have more positive (dimmer) absolute values while those further away will have more negative (brighter) ones. Since the Milky Way measures 100,000 light years across, the overwhelming majority of its stars are more distant than the 32.6 light year absolute magnitude basis, so their absolute magnitude brightness values will be more negative than their apparent values.

When Leavitt graphed her data, she found a very definite relationship between brightness and periods of her 25 Cepheids. In the left graph of figure 20, Cepheid brightness is plotted on the y-axis, versus their period in days on the x-axis; the upper line gives measurements at their brightest peak, and the lower curve the dimmest point of their period cycles. Leavitt's brightness measurements were of the apparent magnitude type since no one had yet been able to measure the absolute distance to a Cepheid in order to calibrate the absolute magnitude scale. However, the graphed data did clearly show that brighter Cepheid stars from the Small Magellanic Cloud had longer cycles of fluctuation between their dimmest and brightest states than less luminous ones; and since all 25 Cepheids were approximately the same distance from Earth, brightness variation within the grouping was intrinsic.

Figure 20 Cepheid Stars: Brightness vs. Period

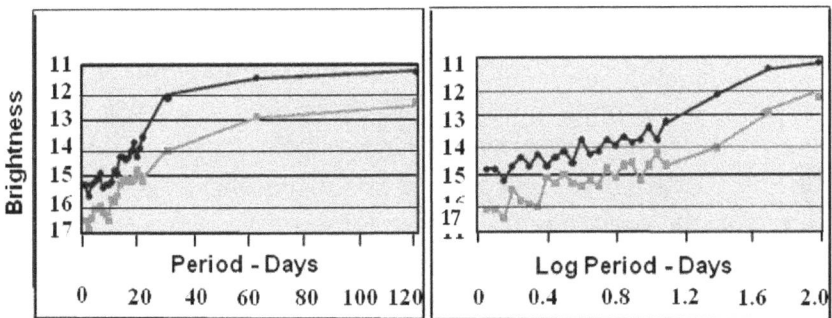

In figure 20, brightness measurements for both the maximum and minimum cycles rise rapidly for Cepheids with shorter periods before gradually tapering off as periods lengthen. We are not surprised by the curved brightness/period relationship since the brightness scale is a logarithmic function. The right side of figure 20 plots the brightness data against the logarithm (the base 10 logarithm (log) of a number is essentially its exponent, the value to which 10 is raised, to produce the number) of the period duration, which places both variables on the same basis and brings the data into a more straight-line configuration.

Although discovery of the Cepheid brightness/period relationship did not enable astronomers of Leavitt's time to calculate actual distances from Earth to stars, it did give them a tool for determining their relative distance from Earth. An astronomer could now measure the period of a Cepheid variable from any section of the sky and determine its distance from Earth relative to a Cepheid with a similar period in the Small Magellanic Cloud. Since its true brightness must be the same as that of the SMC Cepheid with the similar period, any difference in its measured brightness must be due to the difference in its distance from Earth. Furthermore, since the brightness of a star diminishes with the square of its distance, a Cepheid 9 times dimmer must be 3 times further away; and one 144 times dimmer, 12 times further away.

Astronomers now only needed to establish the absolute distance to just one Cepheid; then distances to all of the others could be determined as simple ratios of their distance to the known one. Techniques involving *parallax* turned out to be the solution to this phase of the problem.

Parallax is the perceived motion of a fixed object we get when we view it from a different position. You can see this effect quite readily by focusing on a nearby stationary object (say 2 feet away) and alternately opening and closing your eyes one at a time, and watching the apparent left and right movement of the object relative to other objects in the background. Each eye essentially sees portions of the background behind the object hidden from the other eye. Since you see greater portions of background when viewing closer objects than further ones—causing them to appear to "jump" more relative to the background—you can use a simple trigonometric technique to calculate their distances.

Scientists use parallax to measure stellar distances; however, to observe the effect from such distant objects, they need a wider separation of stereoscopic sight than that provided by our two eyes. To obtain this separation, they take photographs of the celestial object six months—and some 186 million miles—apart when the Earth is on opposite sides of its orbital path around the Sun (see figure 21). The distance the object "jumped" relative to stars and galaxies in the background of the two photographs estimates the parallax effect. Just as near objects appear to move more than distant ones when you close one eye and then the other, the relationship between the angle of parallax and stellar distance is an inverse one, the greater the movement, the greater the angle, and the shorter the distance to the object.

Figure 21 Parallax and Distance to Stars

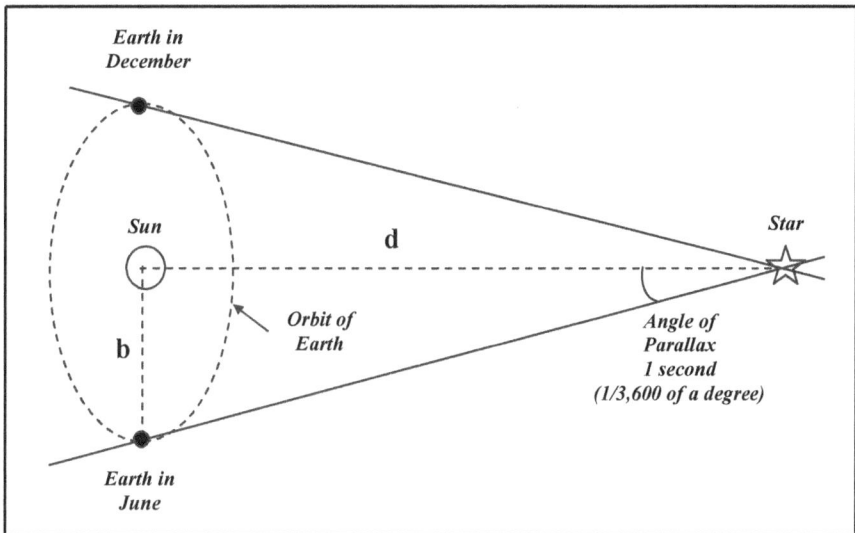

Astronomers find it more convenient to use a quantity called a *parsec* to measure stellar and galactic distances. It is the distance at which the **par**allax of a hypothetical star equals one **sec**ond of arc—hence the shortened term parsec—when measured from a triangle formed by a base line with a distance equal to the radius of the Earth's orbit around the Sun and a point at the center of the object. This distance that produces one second of arc from a base line equivalent to the distance from Earth to Sun is 19,170,000,000,000 miles (19.17 trillion, or 1.917×10^{13} miles). There are no stars exactly 1 parsec from Earth; all are either fractions or multiples of parsecs. More distant objects have less parallax than nearer ones, so a star with 0.5 arcseconds of parallax will be 2 parsecs, or 38.34×10^{12} miles away while one with 2 arcseconds of parallax will be one half a parsec, or 9.585×10^{12} miles away.

The tiny value of this parallax angle—1/1,296,000th part of a circle—makes an extremely long triangle. If you draw such a triangle with a base line equal to the width of your easy chair—say 3 feet—and the other point far enough away that the intersecting lines form an angle of 0 degrees, 0 minutes, and 1 second, the object would be 117 miles away.

Referring to figure 21, the following simple trigonometry function calculates the distance that produces 1 second of arc in the triangle.

Tangent of angle = opposite side (b)/ adjacent side (d)

Where the opposite side is the radius of the Earth's   orbit and the adjacent side is the distance from the Sun to the star.

opposite (b)   = 92,600,000 miles
tangent of 1/3,600 = 4.848×10^{-6}

Therefore the adjacent side (the distance to the star) is:

adjacent (d)   = opposite (b)/tangent

$$= 92.6 \times 10^{6} / 4.848 \times 10^{-6}$$
$$= 1.917 \times 10^{13} \text{ miles}$$

You may better appreciate the magnitude of this number if we compare it to astronomical units (AU). An AU is the average distance from the Earth to the Sun, or 93 million miles. Since a single parsec has 206,265 AU's, one would need 103,000 round trips to the Sun to earn a single frequent flier parsec. However, the parsec is still small enough to be cumbersome for galaxy distance calculations so megaparsec, Mpc (or 1 million parsecs) is the unit of choice.

Using the parallax method, Astronomers accurately determined the distances to numerous nearby Cepheid variable stars thus providing the means for measuring other more distant ones. The Small Magellanic Cloud from which Henrietta Leavitt gleaned so much of the information that gave scientists the Cepheid yardstick turned out to be a mini-galaxy about 230,000 light years away.

Hubble's identification of a Cepheid variable star near the Andromeda nebula meant a monumental step forward for Cosmology and Astronomy. Since scientists could measure the distance to the neighboring Cepheid, they could now calculate the distance to Andromeda. Hubble determined the Cepheid period to be 31.415 days, and using Levitt's brightness-period relationship calculated its absolute brightness to be 7,000 times greater than our Sun. Then by comparing its apparent brightness with its absolute brightness, they placed it 900,000 light years from Earth. Since the Milky Way is only 100,000 light years across, Andromeda could not be part of it; Andromeda must be another galaxy, and based on its brightness, must contain many billions of stars. Hubble had finally settled the long lasting debate: nebulae were not strange little curiosities of the Milky Way but full-grown distant galaxies.

Subsequent findings regarding Cepheid behavior and brightness greatly increased the Andromeda distance estimates to 2.3 million light years; however, these only served to strengthen the conclusion that it did not belong to the Milky Way. Astronomers further established that since the Andromeda Nebula was actually the Andromeda Galaxy, all of the thousands of other nebula known at the time must be galaxies as well. The size of the Universe had just exploded.

As Hubble zoomed his telescope further and further out, he

began to grasp the awesome scope of the Universe, galaxies as numerous in the greater Universe as individual stars within our own galaxy. Further, he noticed that as more and more galaxies came into view, a sense of universal order, not obvious when viewing individual stars, began to emerge. Just as stars aggregate into galaxies, galaxies in turn pack into clusters. Zooming further revealed an even more amazing phenomenon. The combination of all the galaxies and clusters of galaxies formed a somewhat boring pattern of sameness that appeared very similar throughout all areas of the sky. This combined uniform agglomeration was dubbed the *cosmological fluid*.

The discoveries pouring forth from Hubble's new invention changed the way scientists viewed the heavens and revolutionized the science of Cosmology. The plethora of stars that Astronomers encountered when training telescopes skyward on clear moonless nights had always been an intimidating factor, suggesting a complexity in the Universe too great to understand. Hubble's work rejected the belief that the Universe consists of countless stars spread willy-nilly throughout; order existed within that complexity after all. Our own star was simply one of billions forming the Milky Way galaxy, a grouping that would appear similar to any of the specks in Hubble's photographs if viewed from a vantage point within one of these distant galaxies. From an earthly perspective however, it appears a pattern-less, complex entity that reluctantly lends itself to examination and comprehension.

As world-shaking as these discoveries were, Hubble's greatest contribution to cosmology came from observations proving that the Universe was not static as Einstein had wanted it, but had to be expanding as his general relativity equations had originally told him. Following Hubble's initial discoveries, astronomers began examining light emitted from stars in other galaxies and found the spectra to be the same as in light coming from our Sun, except in nearly every case the wavelength was longer, or shifted to the red end of the spectra.

This change in apparent wavelength of light, or shift, is aptly called the *redshift*, also known as the *Doppler Effect*. It is the same phenomenon we notice with sound waves: e.g., when a train approaches, the sound is high pitched, and low pitched when it moves away; or when the crests of the waves pushed in front of a small object as they move across a still pond, are closer together than the trailing waves.

Figure 22 Doppler Effect

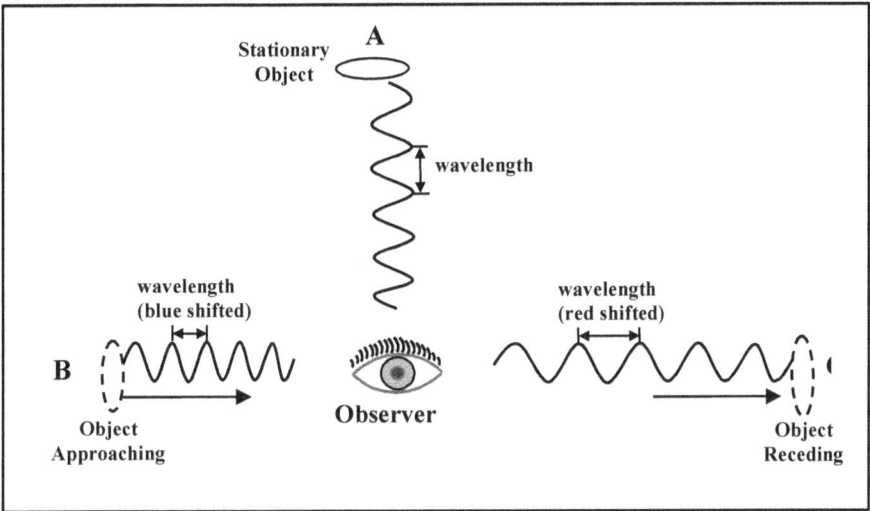

Figure 22 illustrates this effect. An observer sees a particular wavelength of light emitting from three objects in different states of motion: a stationary object A, an approaching object B, and a receding object C. Since object A is stationary relative to the observer, the wavelength of its light will be the same as that measured in the laboratory for the source material. The observer measures a shorter wavelength for the approaching object B (blue Doppler shifted) and a longer wavelength for the receding one (red Doppler shifted). Of course, our eyesight cannot detect these effects in objects moving at earthbound speeds since such objects travel at tiny fractions of the speed of light. Even with spectrometric equipment, objects need to travel at ten million miles per hour or more to detect a wavelength difference.

The longer wavelength of the light emitted by the galaxies conclusively proved the expanding universe theory. Had all wavelengths been identical, it would mean a static unchanging Universe; or if shorter—a shift toward the blue end of the spectrum—galaxies would be approaching one another, signaling a shrinking universe, and a Big Crunch would be our fate. The only exceptions to the red-shift observations came from nearby galaxies such as Andromeda, which are slightly blue-shifted, meaning they are approaching. This effect is consistent with other information that shows local gravitational attraction causing the Milky Way and Andromeda to drift closer together and eventually cross paths some eight billion years from now.

Even more intriguing, Hubble's redshift was not uniform but greater for more distant galaxies than for nearer ones. This meant that other galaxies not only moved away from ours and from each other but also, the greater their distance, the faster they receded. In fact, galaxies receded at a rate directly proportional to that distance; e.g., a galaxy twice as distant would retreat twice as fast. This is the description of an explosion; particles from a grenade exploding in space would follow the same pattern. Applying simple math to these observations produced the inescapable conclusion that all galaxies must have at one time been at the same place, or essentially on top of one another. These simple observations formed the major evidence for the Big Bang theory, an expanding universe, and as a bonus, provided a way to estimate the Universe's age.

In 1930, Sir Arthur Eddington, recalling the paper proposing the expanding universe model Georges Lemaître had written three years earlier—that Eddington and the rest of the scientific establishment had ignored—published his own paper in the prestigious journal Nature, in which he referred to Lemaître's model as a brilliant solution and a complete answer to the problem. The approval of Lemaître's model by a respected authority like Eddington convinced the rest of the science community to revisit his theory in light of the confirming evidence emerging from Hubble's discoveries. A year later the model received nearly complete acceptance when Albert Einstein, during a personal visit with Hubble at the Mount Wilson Observatory, recanted his cosmological constant calling it his "biggest blunder" and dubbed the Big Bang theory "...the most beautiful and satisfactory explanation of creation to which I have ever listened."

When first confronted with the expanding universe concept, one might think of galaxies separating from one another by their movements through space. The theory states however, that space itself expands and galaxies just go along for the ride. Eddington explained this concept by comparing three-dimensional space to the two-dimensional surface of an expanding balloon with dots on its surface representing galaxies. As the balloon inflates, all of the dots (galaxies) move apart from one another in direct proportion to the increase in size of the balloon. If one adds enough air to double the balloon circumference, distances between all of the surface dots will likewise double; therefore, dots separated by greater distance will move apart faster than closer ones. The dots do not move across the balloon surface, the surface itself expands.

Age of the Universe

Hubble's discoveries not only answered many of the questions as to how the Universe came to be. They also held the key to when it happened. Armed with the information that all of the galaxies in the Universe are accelerating away from each other, the only thing needed to determine when they all occupied the same place—or the moment of the Big Bang—is the distance and recession speed of any single galaxy relative to the Milky Way. The time needed for that galaxy to have traveled the separation distance at the recession speed is the age of the Universe. Calculations based on accurate data from more distant, or from nearer galaxies, all give a similar answer due to the proportionality of recession speed and distance.

A recently resurrected cold case file provides an easily understood example of the calculation. It involves an infamous 1960s crime where a band of renegade litterers wantonly defaced and disrupted the aesthetic beauty of a small Massachusetts town's city dump. As it turns out, police officer Obie of the Stockbridge PD had been resting in his cruiser at the crime scene, located at mile marker 70 of the throughway, following a grueling session producing 8 x 10 glossy photographs of the desecrated dump, when he received information regarding two suspicious vehicles traveling on Rt. 183. One, a flower painted Volkswagen microbus, was passing the 160-mile marker at a constant speed of 40 mph; the other, a red Corvette was doing 80 mph at the 250-milepost. While Obie settled for eighth place honors in the high school math medal competition, he nevertheless had no trouble determining the VW went by 2 hours and 15 minutes earlier using the equation:

$$t = d/v$$

t = time
v = velocity
d = distance

Microbus: $t = (160m - 70m) / 40m/h = 2.25$ hours

When he made the same calculation for the Corvette, he came to the startling conclusion that the two vehicles had passed him at exactly the same time, placing both at the scene and proving once-and-for-all the dastardly crime was a two-vehicle job.

Corvette: $t = (250m - 70m) / 80m/h = 2.25$ hours

Determining the age of the Universe, or the length of time since the Big Bang when all of the galaxies occupied the same location, is only slightly more complex than officer Obie's calculation.

First, however we need to define the terms used in the measurement. The distance factor (d) of the equation tells the distance to the galaxy measured in parsecs, or megaparsecs (Mpc = 10^6 parsecs), while (v), the speed at which a galaxy moves away from Earth, is based on the magnitude of the redshifts of light emitted from hydrogen burned by stars within the galaxy. Hydrogen emits light with a wavelength of 0.6563 Å (Angstrom = 10^{-8} centimeter). With the exception of Andromeda and other nearby galaxies, wavelengths of their hydrogen spectra as measured from the Earth, all vary and always slightly longer, or shifted more toward the red end of the spectrum. Even though our Sun is 93 million miles away, the wavelength of its hydrogen spectra measures precisely 0.6563 Å since the distance between the Sun and the Earth remains nearly constant.

While hydrogen is not the only element burned as fuel in stars, it is by far the main source for the Sun as well as all other stars in the Milky Way and other galaxies. In fact, hydrogen accounts for 90.84% of all the atoms in the Universe with helium picking up most of the remainder at 9.08%. That leaves 0.08% to account for all other elements, of which oxygen takes 0.05% to make H_2O to fill the oceans and 0.02% for carbon to construct all living things. The remaining 0.01% accounts for all other elements combined—including silica, calcium, and iron to make the rocks, sand and magma of Earth and other planets throughout the Universe.

The magnitude of the difference between the wavelength of light from a celestial object moving away from us, and light from the Sun, measures its red shift, and is used to calculate the speed at which the object recedes, and hence, its distance. Hubble developed an equation describing this relationship in the 1920s using distance measurements based on redshifts from twenty galaxies, varying in distance from about one hundred light years to 7 million light years away. The equation, known as Hubble's Law not only provides a technique for accurate measurement of galaxy distances, it also serves as the basis for determining the age of the Universe.

Hubble's Law: $d = v/H$

v = radial (recession) velocity

d = distance in parsecs

H = Hubble's constant expressed in km/second x parsecs

Transposed, Hubble's equation is the same as the one used by officer Obie with the t term replaced by $1/H$.

$$t = d / v$$
$$1/H = d/v$$
$$d = v/H$$

Scientists have revised the value of H (and therefore the age of the Universe) numerous times since its introduction due to advances in spectroscopic measurement techniques. It was recently estimated at 70 km/sec/Mpc, based on data comparing thousands of galaxy redshift measurements with their known distances, although numbers ranging from 50 to 100 belong to the data collection. Uncertainty in the number stems more from difficulties in measuring redshift values for more distant galaxies than from galaxy distance measurements. The following is an example of the calculation for estimating Hubble's constant using redshift and distance data for the Corona Borealis galaxy:

Hubble's Law: $d = v/H$

(1) $H = v/d$

Where v is the radial velocity (the recession speed) and d is the galaxy distance

Distance to the Corona Borealis   = 290 Mpc

Corina Boealis H spectra wavelength, λ = 6963 Å

Sun hydrogen spectra wavelength, λo = 6563 Å

Radial velocity, v is equal to the redshift, z multiplied by the speed of light, c

     (1) $v = cz$

c is the speed of light (3×10^5 km/sec) and redshift z is defined by:

     (3) $z = (\lambda - \lambda_0) / \lambda_0$

          Where: λ = receding galaxy light wavelength
                      λ_0 = sunlight wavelength

Calculating the redshift from (3) based on spectral data for the Corona Borealis galaxy:

  $z = (6963 - 6563)/6563 = 0.06095$

Calculating the radial velocity from (2) based on Corona Borealis galaxy distance:

  $v = cz = 3 \times 10^5(0.06095) = 18{,}284$ km/sec

Calculating Hubble's constant from (1) where d = 290 Mpc:

  $H = v/d = 18{,}284$ km/sec/290Mpc = 63 km/sec/Mpc

Having determined Hubble's constant, the age of the Universe (based on data from one galaxy only) is only a step away, beginning with the same equation that Obie remembered from math class.

     $t = d/v$

Substituting Hubble' Law, d = v/H in terms of v into (1), we get:

     $t = (v / H)/v$

     or

Eq (4) $t = 1/H$

Where t is in seconds, H is 63 m/sec/Mpc, and 1 Mpc = 10^6 parsecs:

$t = 1/H$
  $= 1/63$ km/sec/Mpc
  $= 1$sec x Mpc x 10^6 parsecs x 3.086 x 10^{13} km/63 x Mpc x parsec

$$t = \frac{1 \text{ sec x Mpc x } 10^6 \text{ parsecs x } 3.086 \text{ x } 10^{13} \text{ km}}{63 \text{ km x Mpc x parsec}}$$

$t = 4.9$ x 10^{17} seconds

$t = 4.9$ x 10^{17} seconds/60/60/24/365 = 1.5 x 10^{10} years
        or 15 billion years

Applying the same calculation to other galaxies yields similar results, usually ranging from 13 to 15 billion years. This demonstrates that all the galaxies, regardless of their current separation, were at one time, at the same location.

Measuring the redshifts of light emitted from receding galaxies provide a reasonably accurate estimation of the age of the Universe; however, variations in the results from galaxy to galaxy still left room for disagreement among scientists as to the accuracy of individual calculations. Results from NASA's Wilkinson Microwave Anisotropy Probe (WMAP), released in 2007 changed all of that, providing the most direct and accurate measurement of the Universe's age to date. The WMAP satellite, launched in 2001, orbits approximately 900,000 miles from Earth where it is constantly in Earth's shadow, thereby avoiding electromagnetic interference from the Sun. There, the WMAP constantly scans the cosmos with an ultra sensitive microwave receiver, mapping microwave background radiation that originated during recombination 380,000 years after the Big Bang. The temperature of the Universe at that time had cooled to 3,000 C, which allowed electrons to unite with protons to form neutral hydrogen atoms, permitting photons trapped in the particle haze to escape and travel throughout the Universe.

WMAP's age of the Universe comes from information representing its condition at intervals beginning a mere 380,000 years after the Big bang, compared with conditions of the current Universe. These include the density, composition, and expansion rates of the Universe during each period and calculate with an accuracy of better than 3%, which translates to 1% in the currently accepted age of the Universe, 13.73 ± 0.12 billon years. You may be curious about how conditions from different time periods can be known. The answer is based on the speed of the messenger service supplying the information; namely, light. None of the light reaching us from stars and galaxies represents their current condition since its travel time depends on the object's distance. Therefore, a galaxy one billion light years away tells us conditions from one billion years ago, while one ten times as far away gives information from ten billion years ago.

The Unknown

Physicists have described a scientifically consistent succession of events beginning 10^{-43} of a second (10 millionth of a billionth of a billionth of a billionth of a billionth of a second) after the Big Bang, and continuing through the 13.7 billion year history of our current Universe. The brief preceding instant is the Planck era, or more aptly, the unknown period, since the current state of physics is inadequate for developing estimates of factors such as its size, temperature, or the nature of its content. Physicists believe however, that the four forces of nature were of equal strength and united into a single *"superforce"* brought together by an equally mysterious phenomenon called "*supersymmetry*". They believe the size of the Universe at the beginning of the unknown period was equal to zero and therefore infinitely hot. At precisely 10^{-43} of a second, the temperature of the Universe was 10^{32} degrees, and its size equal to the Planck length, or 10^{-33} centimeters across. This small length (100 billion billion times smaller than the nucleus of the smallest atom, hydrogen) divided by the speed of light defines the smallest measurable time interval, the Planck time, or 10^{-43} seconds.

In the study of matter and energy, the present state of scientific knowledge necessarily divides the laws of physics into two major categories. One is quantum theory, or quantum mechanics, which describes matter and energy phenomena at the very small, or micro scale; and the other relativity, which governs the very large, or macro level of the Universe. During the unknown period or prior to the Planck time, these physical laws break down, precluding the ability to obtain information about the very beginning of time and the Universe. This happens due to the merging of the two branches of physical law, quantum mechanics and relativity. Since at the Planck time, the macro-universe was submicroscopic, attempts to understand the physics of that moment, forces these two disciplines reluctantly together and exposes their incompatibilities. They describe different domains that do not intersect. Quantum mechanics invokes Heisenberg's uncertainty principle and insists it is impossible to know both the location and velocity of a particle. It presents the quantum micro-world as a hectic, highly irregular place governed by probability equations that allow, and in fact insist, that particles will not be localized and states that the more accurately you want to locate a particle, the more energetic the packets of probing energy (quanta) must be. Locating a particle with a precision equal to the Planck length therefore would require an infinite amount of energy.

Relativity on the other hand governs the domain of the gigantic, where probability issues, while they still exist, are ameliorated by the huge number of particles involved, resulting in a smooth predictable Universe. A gambling casino operates in much the same manner, assuring themselves a profit even though there is considerable uncertainty of winning individual bets. When many gamblers make bets, the gains and losses cancel out, and the favorable house odds prevail. This does not say that separate laws rule different aspects of the Universe; they must have been unified at Planck time. It does however highlight a yet to be explained incongruity between quantum mechanics and relativity that is currently the premier challenge of research physics, the development of a theory that is consistent within both physical realms.

First Second

The period for which the two realms of physics do function quite well, namely all time following the first 10^{-43} seconds, began with a tiny Universe measuring slightly more than 10^{-30} centimeters with a temperature of 10^{32} degrees (one hundred thousand times a billion, times a billion, times a billion). The gravitational force had separated. Still the other three fundamental forces of nature, the electromagnetic, the strong nuclear, and weak nuclear forces, remained unified. The first indication of this emerged from a discovery that earned Princeton physicists David Gross, Frank Wilczek, and David Politzerthe, the 2004 Nobel Prize in physics for their 1973 discovery of a principle they called *asymptotic freedom*. They showed that as quarks move closer together, the strong nuclear force that binds them into protons and neutrons, actually becomes weaker. Shortly after, a group of Harvard physicists found that the reverse applied to the electromagnetic and weak forces; interactions become stronger at close distances. Since the strong nuclear force becomes weaker, and the weaker forces become stronger at close distances, this suggests, and mathematics supports that at the 10^{-39} seconds point, with a Universe temperature at 10^{32} degrees, and distances between particles billions of times smaller than a hydrogen nucleus, these forces may have been the same. They were simply different manifestations of a single underlying force in much the same way electricity and magnetism are two separate sides of the same electromagnetism coin. These developments generated much excitement that perhaps physicists could unite the quantum and relativity theories and the modern holy grail of physics, the *Grand Unified Theory* (GUT) seemed within reach. Not only would the sciences of particles and the natural forces be united, gravity would be included as well—all into a single comprehensive definition of the Universe.

The primary component of the Universe at this time was not matter and energy but a substance called *dark energy*, which exhibited a repulsive, negative gravity-like property much like the lambda of Einstein's cosmological constant. Dark energy effectively broke the supersymmetry of the forces and overwhelmed the weaker gravity force, causing the baby Universe to expand at a rate many times faster than the speed of light (this does not violate Einstein's dictum that nothing can exceed the speed of light since it is empty space that expands). This caused an exponential runaway inflation of the space that housed the Universe, causing it to double and redouble its size at such a rate that within 10^{-34} seconds, it had grown by a factor of 10^{50}, making it larger than the entire region of the Universe that we can now see. The rapid expansion also meant rapid cooling, with the Universe's temperature dropping by a factor of 100,000 to 10^{27} degrees.

Credit for the discovery of the inflation concept, lauded as one of the most important contributions to the Big Bang theory, belongs to MIT physicist Alan Guth based on his work in the late 1970s. Big Bang theory had been losing momentum prior to Guth's discoveries due to inconsistencies with the particle physics of quantum theory, considered the more credible and mainstream of the two disciplines at the time. Particle physics relies heavily on information from high-energy collisions in particle accelerators that break matter apart into sub-atomic particles. The early Universe environment was not that

116

different from the large high-energy accelerators used by particle physicists and should therefore have produced some of the same particles observed in their experiments; however, physicists had not observed many of the particles predicted by cosmology's Big Bang theory, casting doubt on the validity of cosmology and its theories.

One such particle predicted by cosmology was the magnetic monopole, an extremely massive particle best described as "a magnet with only one pole". Since particle physicists have not observed magnetic monopoles directly, many refer to them as hypothetical particles. However, relativity predicts their existence, and monopole-based ideas stemming from string theory (a branch of particle theory to be discussed in chapter 8) have proven accurate in particle accelerator experiments. Furthermore, the extreme conditions of the early Universe should have produced them. This led scientists to seek answers to several questions, including: what quantities of magnetic monopoles emerged from the Big Bang? Were they stable; or did they decay immediately after being formed? If stable, are some still around for scientists to detect?

By the late 1970s, all attempts to answer these questions met with total futility. Based on particle theory calculations using temperature, pressure, and other conditions existing during the first microseconds of the Big Bang, the entire Universe should be composed of nothing but magnetic monopoles. This ridiculous answer meant that either particle physics or Big Bang cosmology was badly in error; and since between the two disciplines, particle physics of the time had a far superior track record, cosmology lost the argument. The loss of credibility stemming from the fact that our Earth is composed of the kind of matter we see around us instead of magnetic monopoles meant a big setback for cosmology, sending many physicists in search of projects in different areas. Guth however, opted to stake his career on the Big Bang and turned his attention to the study of particle physics, only from a different point of view than that of traditional particle physicists. The standard approach used particle physics to learn about cosmology; but Guth reversed the process opting to use cosmology to study particle physics, a choice that marked the beginning of a new branch of physics appropriately named *particle cosmology*—and earning for himself the distinction of being the first particle cosmologist.

Guth's entry into particle physics led him into a sub-field of particle theory research called *phase transition*. We observe phase transition every time we convert liquid water to ice or to steam. We freeze water by changing it from a liquid to a solid in the former case and boil it by changing it to a gas in the latter. These three states of matter—liquid, solid, and gas—are called *phases* and can be changed from one to the other via a phase transition by changing conditions; in this case temperature, but the same effects may be obtained by changing pressure.

Particle physics says that in the early Universe, "stuff" that would later form the primary particles of matter and energy underwent phase transitions to produce magnetic monopoles. The problem arose from estimated properties of magnetic monopoles, which dictated they should "freeze" into a massive (extremely dense) solid at the sizzling temperature of 10^{27} degrees (1 followed by 27 zeros). The temperature of the Universe exceeded 10^{27} degrees centigrade during the period from 0 to 10^{-19} seconds after the Big Bang so monopoles would have been in a non-solid state—some say like a "liquid lava". However, as the Universe expanded, the temperature would very soon drop below the magnetic monopole melting point, freezing the liquid lava destined to be the material that would later form particles, and filling the Universe with supermassive monopoles.

The question now became; why don't magnetic supermassive monopoles permeate the Universe? Guth reasoned that he could avoid the problem if magnetic monopoles somehow did not solidify at such an extremely hot temperature but remained briefly in a liquid or vaporous state while other materials froze. Guth's research of the subject led him to discover that in some particle physics models, the Universe would *supercool* for brief periods. Supercooling is not an unusual phenomenon but simply the process of cooling a liquid below its freezing point without it becoming solid. By carefully dropping its temperature under tightly controlled conditions, very pure water can be cooled to as low as minus 42 degrees centigrade without freezing; however, supercooling produces an extremely unstable state; in fact, the entire liquid will immediately turn to ice crystals if subjected to the tiniest disturbance.

Guth next considered the gravitational properties of the supercooled matter, a search that would lead to one of the most important discoveries of Big Bang theory, and the solution to nearly all of its outstanding mysteries. He found that the supercooled magnetic monopoles would be gravitationally repulsive. Much like lambda of Einstein's cosmological constant, it would "fight" normal gravity— repelling all of the matter now condensing from the primordial "stuff", and make the infant Universe expand at a phenomenal rate. The key factor to Guth's "antigravity" however, stemmed from his version of lambda, which was not a permanent factor, but a temporary one. It would activate, causing rapid inflation only for the tiny fraction of a second when the Universe supercooled, then end as soon as it transitioned from the liquid to solid phase and the particle "stuff" had frozen out.

One of the effects of this brief inflationary period was a highly segregated Universe; in fact, we can see or contact only a tiny fraction of it, regardless of the power of the telescope or sensitivity of the detection device. The rapid inflation divided regions so greatly that the separate sections became unreachable from one another. Neither light nor anything else can ever reach from one region to another; each region is in the elsewhere (see chapter 2) of all the others. The part of the Universe to which we belong stretches about 10 billion light years. The outer limit for the total Universe however, including all the regions separated during inflation, could be 10 billion billion light years or more. To visualize the impact of the inflationary period as it relates to our tiny share of the Universe—i.e. the visible Universe—the inflated balloon analogy is again useful. Imagine rapidly inflating the balloon with the galaxies painted on its surface. When fully inflated, draw a microscopic circle on the balloon. This tiny circle represents the entirety of our visible Universe—that accessible by our most powerful telescopes. It is fascinating to consider the vastness of such a Universe, and the number of stars identical to ours that might have planets with earthlike biospheres and all the other prerequisites of life. Regardless of how improbable the accident of life's creation might be, it would seem even more unlikely that other beings, many lesser and others greater than us, would not exist among the decillions of these unreachable stars.

As quickly as it began, the repulsive force halted, with the strong force separating from the remaining two forces, and dark energy converting into ordinary energy. This newly liberated energy supplied the heat to power continued expansion, but at a greatly reduced rate. However, only a relatively small quantity of the dark energy converted. Dark energy still accounts for about two-thirds of the mass-energy (the energy component of mass-energy comes from the equivalence of mass and energy via $E=mc^2$) of the Universe and is responsible for its continuing expansion. Dark matter accounts for most of the remaining third of mass-energy (dark stemming from the fact it does not emit light and therefore cannot be seen), with only 2% normal matter contained in all of the atoms that make up the planets, stars, and galaxies, and another 2% in diffuse intergalactic gas that has never condensed into stars.

Scientists first proposed the existence of dark matter in the 1930s to explain the motion of galaxies, which seemed to behave as if the Universe contained more matter than they could account for by their constituent stars. The phenomenon of *gravitational lensing*, first reported by Orest Chwolson in 1924 and brought to public attention by Einstein in 1936, offered further evidence of dark matter by providing an explanation of light from stars and galaxies in sections of the sky where no other ordinary matter exists. Gravitational lensing normally results when light from a quasar (distant regions of compact gasses surrounding galactic-size black holes) or some other distant source approaches a massive object situated between the Earth and the source. The gravity of the massive object warps spacetime in its vicinity and bends the light around the entire circumference of the object. This alters the time required for the light to reach Earth and magnifies the apparent image of the source. The Hubble Space Telescope confirmed the effect through an image that showed a ring of dark matter measuring 2.6 million light-years across surrounding a galaxy cluster located 5 billion light-years away. The galaxy cluster labeled ZwC10024+1652, had formed from a collision between two separate galaxy clusters. Computer simulations revealed that following the collision, the dark matter in the two clusters fell to the center of the merged galaxies then rebounded to the outer edge where gravity prevented its further escape, forming a ring of dark matter around the newly formed cluster. The presence of such accumulations of dark matter prevent clusters of galaxies from drifting apart since the quantity of visible, ordinary matter is not great enough to keep them together. Of course, astronomers could not see the ring of dark matter but they could detect its presence through gravitational lensing, which revealed light from galaxies on the other side of the combined galaxies bending to form arc-shaped streaks of light around the perimeter. The photograph revealed the galaxy pair at the center of a dark ring, surrounded by the halo of light formed through gravitational lensing.

Definitive proof of the existence of dark energy had to wait until the 1990s when scientists found it responsible for the increase in the expansion rate of the Universe. While today, scientists still know relatively little about dark energy and dark matter, they know it must exist due to such effects on the motion of stars and galaxies, and have developed a number of ways to test for their presence. Dark matter and energy have become the most avidly studied subjects in particle physics and astronomy, and have even been explained in terms of extra dimensions and the existence of other universes (more on this in chapter 8). Professor Sean Carroll of the California Institute of Technology placed the existence of such vast quantities of dark matter and dark energy, combined with such meager knowledge of their properties, in proper perspective as follows: "...Not only are we not, like Aristotle would have it, sitting at the center of the cosmos, we are not even made of the same stuff as the cosmos. The kinds of things we are made of amount to only 5 percent of the energy density of the Universe. This is a big deal!"

As the Universe grew, its temperature halved each time it doubled in size. When about 10^{-12} (a millionth of a millionth) seconds old, its size had doubled another 26 times, and its temperature had plummeted to 10^{20} degrees, allowing particles of matter to condense from the high-energy plasma soup that described the early Universe. However, nearly half of those early particles were not the kind of "stuff" with which we are familiar. For every type of particle (proton, neutron, electron, meson, quark etc.) in existence now, or at the beginning, there is a corresponding antiparticle. These are identical to particles but opposite in charge, e.g. antiprotons have negative charges and antielectrons (or positrons) have positive charges. The Big Bang produced nearly equal quantities of particles and antiparticles, which initially took the form of quarks and antiquarks. Fortunately, however, quarks enjoyed a slight advantage of one additional quark per billion quark/antiquark pairs. Since all particles and antiparticles combine immediately and violently to annihilate each other, producing a photon of radiation energy for each of the reacting particle pairs, only the one-billionth excess of particles over antiparticles remained to form the present Universe. The resulting radiation from this annihilation that still permeates the Universe today turned out to be a key clue to solving the riddle of how the Universe came to be, and the single most important confirmation of the Big Bang theory.

Two workers from the Bell Labs in Holmdel, New Jersey named Arno Penzias and Robert Wilson, both untrained in cosmology, provided the first hard evidence supporting the Big Bang in 1965 through a serendipitous discovery that validated the matter/antimatter annihilation aspect of the theory. While using a microwave detector designed to receive transmissions from artificial satellites, they noticed the antenna picking up annoying static that interfered with their observations. After making numerous equipment checks, the radiation proved real and even more interesting, it appeared equally intense regardless of the direction their telescopes aimed. Even though they could not explain the source of the radiation, they published their results in the Astrophysical Journal in an article titled "Excess Antenna Temperature at 4080 Mc/s." Following the publication, other scientists picked up on, and realized its significance. Penzias and Wilson had discovered the remnants of the radiation caused by the extinction of matter and antimatter in the tiny Universe at the moment of creation.

In 1990, measurements from the Cosmic Background Explorer

(COBE) confirmed the type and order of magnitude radiation predicted by the particle/antiparticle annihilation. Since there were a billion times as many particles and antiparticles annihilated as surviving particles, producing two photons per annihilation—one for each type of annihilated particle—the Universe should contain two billion photons for every extant matter particle. Measurements showed 412 million photons of this radiation in every cubic meter of the Universe compared to a matter density equivalent to 0.2 particles of matter (calculated as protons) per cubic meter. This provided a very precise match with the predicted 2 billion to 1 excess from the matter/antimatter extinction phase of the Big Bang (412,000,000 / 0.2 = 2,060,000,000), confirming the total Universe was once a compressed gas many times hotter than the core of the Sun.

By the time the Universe had aged to 10^{-10} seconds, and had expanded to nearly 50 billion miles across with a temperature of 10^{15} degrees, protons and neutrons began forming from quarks. The extreme temperature and density of matter at this point caused continuous high-energy collisions between nuclei resulting in constant inter-conversions between protons and neutrons. The fact that neutrons weigh slightly more than protons insured that more collisions of quarks produced protons than neutrons, resulting in a ratio of about 10 protons for every 2 neutrons, a ratio that remains the same today throughout the combined atomic mass of the Universe.

When the Universe, now composed of elementary particles, each surrounded by two billion photons of radiation, had aged to one second, it had expanded to five trillion miles, or one light year across and its temperature had cooled to 10^{10} degrees. This stage has been referred to as the Goldilocks era—not too hot, not too cold. At higher temperatures, particles would be too energetic—moving too fast—to combine, while lower temperatures would have prevented protons and neutrons from fusing into atomic nuclei. Furthermore its average density had dropped to about 1 gram per cubic centimeter, or about that of water. Less frequent and energetic particle collisions halted proton/neutron inter-conversion and stabilized conditions to the point where adjacent proton/neutron pairs did not have enough energy to escape the attraction of the strong nuclear force. This allowed some of them to bind into deuterium nuclei that further collided with protons and neutrons to form helium nuclei.

The fact that all of the neutrons in the Universe reacted in this way makes calculating the proportion of helium to hydrogen in the early Universe an easy chore. First, reduce the entire Universe to 12 nucleons—10 protons and 2 neutrons—consistent with the composition of the early Universe. The two neutrons combined with 2 of the 10 protons to produce a helium nucleus leaving 8 protons to serve as hydrogen nuclei. Therefore, the atomic weight of helium (4) divided by the sum of the weights of helium and hydrogen nuclei (4+8) equals 33% by weight (4/(4+8) = 0.33). The mass fraction of helium to hydrogen actually observed in the stars and interstellar gas of the Universe today ranges from about 24 to 30% indicating that some of the neutrons decayed into protons and electrons. This left 70 - 76% of the protons available to form hydrogen, the main fuel of the stars.

Except for extremely small quantities of lithium and barium nuclei— AN (atomic number) 7 and 8—there would be no further formation of atomic nuclei in the early Universe for another billion years or so. This is due to: (1) all of the neutrons had fused with protons to form helium nuclei, and (2), an inherent instability of nuclei containing 5 nucleons (protons or neutrons) that prevents the single nucleon hydrogen from fusing with four-nucleon helium (2 protons and 2 neutrons). Since the obvious path for transforming hydrogen and helium into heavier nuclei must go through a 5-nucleon intermediate step, the route to elements such as carbon and oxygen (normally 12 and 16 nucleons respectively) is blocked. The intermediate step for the fusion of hydrogen and deuterium to form helium goes through a 3-nucleon intermediate step as shown in figure 23. In the first stage, deuterium and hydrogen nuclei fuse to form light helium. Two light helium nuclei fuse in stage 2 to form the nucleus of the more stable helium isotope plus two protons, or hydrogen nuclei. Two deuterium nuclei could also fuse directly to form stable helium; however, since deuterium is quite rare, the probability of two of deuterium nuclei coming together is extremely slight. As a result, most of the fusion takes place through combination with abundant hydrogen nuclei.

Figure 23 Hydrogen Conversion to Helium

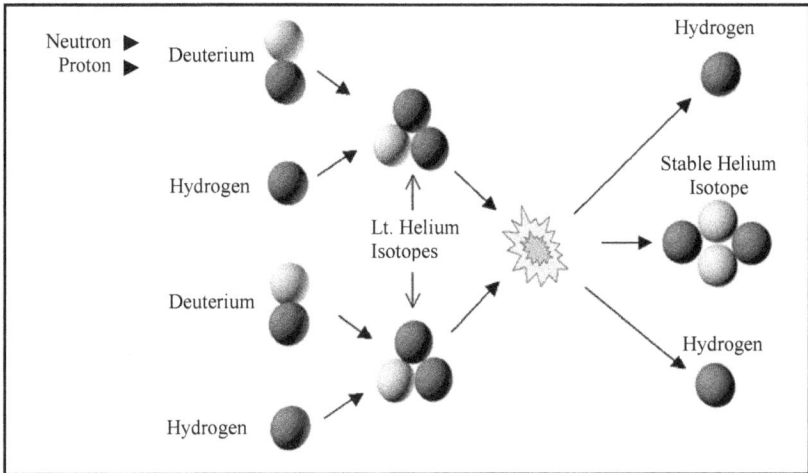

Another route for creation of heavier elements in the early Universe would seem to be the combination of two 4-nucleon helium nuclei to yield an 8-nucleon structure that could further combine with hydrogen, deuterium, and helium nuclei to make a variety of heavier nuclei. Eight-nucleon structures however suffer from the same instability problem as those with five, effectively blocking all avenues for heavy element production during the first moments of the Big Bang. Fortunately, further opportunities to produce the other elements needed to build planets and all forms of life would present themselves in another billion years or so, and the factories that manufacture them would be the stars.

Light to Darkness

For the next 380,000 years, the Universe remained a mass of atomic nuclei and free-floating electrons with temperatures too high to form atoms. The mass of electrically charged free electrons presented an impenetrable barrier to photons created from particle/antiparticle annihilation. Photon interactions with electrons prevented their passage through the nucleon/electron plasma in much the same way fog disperses light from an approaching car at night, resulting in an opaque Universe. However, compared to the chaotic first seconds of its life, a boring period of expansion and cooling ensued until the Universe was about 2×10^{19} miles (or 4 million light years across). At this point, the temperature of the Universe had cooled to 3,000° C, reducing the energy of electrons to the point where they could bond with nuclei to form atoms.

Coupling of electrons with atomic nuclei marked a major milestone (or more appropriately, light-year-stone) since the Universe became transparent and dark for the first time. Electrons, ceasing their chaotic random movements and taking their places in orbit about the atomic nuclei, allowed photons to pass unhindered through the inter-orbital space of the atoms and proceed on through space.

As mentioned earlier, the event of electrons joining nuclei to form atoms generally referred to as *recombination*, provided cosmologists with another very convincing proof of the Big Bang theory. In 1948, George Gamow, Ralph Alpher, and Robert Herman predicted that if the light radiation from the particle/antiparticle annihilation could ever be detected, it would have a wavelength of about 0.001 millimeters as it emerged from the opaque plasma. They postulated that since space itself has stretched over the ensuing billions of years, so too would the emerging light; it would have lengthened by a factor of about 1,000 and should now be about 1 millimeter. This placed the radiation wavelength squarely in the radio frequency region of the electromagnetic spectrum matching the frequency of the radiation subsequently discovered by Pinzias and Wilson.

Stars and Galaxies

The visible Universe contains over 100 billion galaxies, each housing about 200 billion stars. Essentially all of the matter that now comprises the Universe formed in the first 380,000 years following its birth; however, with the exception of small quantities of lithium and barium, all of it was in the form of hydrogen and helium, the two simplest elements. Fortunately, not all of the matter within the Universe distributed uniformly during expansion; some areas had slightly more matter than others. The gravitational attraction of matter within these denser regions caused their expansion rate to lag slightly behind the more dilute regions, allowing the density difference between them to increase slowly over time. Had this not been the case, gravity could not have acted on particles of matter to bring them together to build stars and galaxies. An entirely homogenous distribution would mean that all atoms would be attracted equally in all directions. With no higher density areas to initiate gravitational attraction of individual hydrogen and helium atoms into larger accumulations, the Universe would have remained forever dark, the continued expansion serving only to dilute the un-coalesced matter even further.

About 1 billion years after the Big Bang, galaxies began to form from the localized areas of higher matter concentrations, each with hydrogen and helium masses equivalent to billions of our Suns. The combined mutual gravitational attraction of all the matter within the galactic regions prevented them from expanding as rapidly as the "empty" space. Individual galaxies contained even greater localized hydrogen and helium concentration regions stretching several hundred light years across, serving as precursors to star formation. Gravity acting on the clouds caused them to begin collapsing and spiraling inward while friction from more frequent particle collisions increased their temperature. The collapse went very slowly at first, taking about 10 million years to shrink by a factor of 10 until they measured only about 40 million light-years across. However, after another hundred years or so, the collapse became a free fall creating a protostar. Internal pressure continued to increase the protostar temperature, causing molecular hydrogen (H_2) to dissociate to its atomic state (H·) releasing infrared radiation that further heated matter from the outer regions of the star, accelerating the dissociation of H_2.

Finally, when the outward pressure at the core had risen to the point where it counteracted the inward gravitational attraction, the collapse halted, and the protostar became a full-fledged star.

Temperature at the star's center had risen to 10 million °C, and all the hydrogen and helium ionized returning them to their nuclei state. Conditions at this point met the requirements for nuclear fusion reactions, which provide the fuel for the long-term life of the star; namely by combining hydrogen nuclei (protons) with deuterium (a proton and a neutron) to form an isotope of helium. Helium nuclei have slightly less mass (about 0.3%) than the hydrogen and deuterium nuclei that formed them. This difference in mass, known as the *mass defect*, measures the bonding energy of the atom, and is released as energy according to Einstein's $E=mc^2$. This equation, the same that powers the hydrogen bomb, will provide fuel for the star for billions of years.

Star-Death, and Elements

Despite the fact that early scientists identified most of the elements that make up our planet Centuries ago, where and how they came into being remained a mystery until relatively recently. Near the beginning of the 20th Century, Sir Arthur Eddington suggested that stars were the "crucibles" for creating all but the smallest, most basic elements, namely hydrogen, helium and lithium. Eddington felt certain that the heavy elements formed through violent collisions of lighter ones that welded them together. He also knew that extreme pressure and temperature conditions would be necessary to provide a sufficient mass of particles moving at great enough speeds to cause such collisions. Primarily by process of elimination, he concluded that the centers of stars provided the only possible source of such severe conditions; however, it took another two decades for science to provide the tools needed to prove his hypothesis.

The problem involved bringing two extremely tiny, positive charged protons (hydrogen nuclei) close enough together to overcome their electromagnetic repulsion and allow the strong nuclear force to dominate and bond them together through nuclear fusion. Eddington's initial calculations showed stellar conditions in stars too mild to synthesize elements; nuclear fusion needed a temperature of 1 billion degrees to make protons collide, and the actual stellar core temperatures reached only 10 million degrees; one one-hundredth of the necessary temperature.

As it turned out, Eddington proved correct with his stellar core temperature calculation but grossly in error regarding the required fusion temperature. Fusion can take place at a lukewarm 10 million degrees. Again, a plea of ignorance would justify Sir Arthur's forgiveness since physics had not yet provided him with all the facts needed for his calculation; namely the existence of quantum tunneling (see chapter 1). In the absence of this phenomenon, 1 billion degrees would be the correct answer; however, the probabilistic nature of quantum tunneling assures that a small fraction of protons will ignore energy barriers and locate themselves where they should not be—in this case adjacent to particles of like charge.

Furthermore, not all protons at the center of a 10 million degree star have the same energy. A phenomenon called *thermal distribution* assures that about 1 in every 10 million protons will have 10 times more energy than the average proton around it, pushing it further up the probability scale. Higher temperatures and more densely packed particle environments increase the probability of fusion; however, a threshold temperature is necessary for sustainable reactions to proceed. But once started, tremendous energy releases; enough that even with the low probabilities of individual nuclei undergoing fusion in any given period of time, the quantity of hydrogen nuclei that do "tunnel through" can keep the star burning at a rate that permits it to shine for billions of years. To get a picture of the magnitude of this energy release, consider that the process of converting 4 protons to helium nuclei gives 15 million times more energy than the combustion of an equal weight of gasoline.

Stars come in all different sizes, and size greatly affects their life span as well as their eventual fates. Our Sun, an average size star (one solar mass by definition), will have a life span of about 10 billion years and is now entering middle age at 5 billion years. A somewhat surprising aspect of stars; when it comes to stellar longevity, smaller is better. Low mass stars have life expectancies of hundreds of billions of years; many times longer than giant ones. This proves true in spite of the fact they have a much smaller supply of hydrogen available for nuclear fuel. The lower gravity of stars smaller than the sun translates to lower pressure and temperature in their inner core, which causes nuclear burning to proceed at a slower rate, making them much dimmer but longer living stars. At the end of their lives, they simply become dimmer and dimmer, eventually fading into *Black Dwarfs*. Black Dwarfs are by necessity theoretical objects since their lifespans exceed the age of the Universe (13.7 billion years), therefore no black dwarfs yet exist.

Nuclear burning (conversion of hydrogen to helium) continues in a medium size star like the Sun until it exhausts its core hydrogen fuel and becomes a *White Dwarf* with an absolute brightness about 1/1,000 that of its original value. When a medium sized star exhausts its core hydrogen, the internal nuclear fires that had served to counter the gravity trying to crush it, die and it begins to collapse inward. The resulting increased pressure causes temperature in the star's core to rise, generating enormous quantities of heat, which travel away from the core toward the stars outer regions. There, even though pressure is considerably lower than core levels, temperatures become so high that the star begins burning outer region hydrogen, "puffing up" the star. In the case of our Sun, in another five billion years the expansion will bring its surface out beyond the orbit of Earth and cool the solar surface to about 4,000 degrees C giving it a reddish glow. As viewed from other solar systems, it will appear more than 1,000 times brighter than at present and will be categorized a *Red Giant*. Eventually, the Red Giant, whose core consists almost entirely of helium, will spend its outer region hydrogen fuel as well then begin to cool and once again contract inward. Since medium sized stars do not have sufficient gravity to generate pressures and temperatures that support nuclear fusion of helium, they continue contracting inward to a radius of only a few thousand miles and a density measured in hundreds of tons per cubic inch. They will remain in this state tens to hundreds of billions of years; a tiny ball of degenerative matter called a *White Dwarf*, emitting heat and light until they eventually cool to become black dwarfs like their smaller cousins.

Gravities of high mass stars on the other hand cause very high core temperatures that produce brighter but short-lived stars. Stars 10 times heavier than the Sun burn their core hydrogen 100 times faster giving life spans of only a few hundred million years. Still heavier stars (100 solar mass) can burn themselves out in a mere 1 million years. While lives of heavy stars are too brief to permit complex life systems like those on Earth to evolve on their planets, earthly life could not have evolved without them since they created most of the necessary atomic elements.

During the frenetic early seconds of the Universe, when nearly all of the hydrogen and most of the helium formed, temperature and pressure conditions had been changing too rapidly for significant quantities of heavier elements to form. High mass stars, however, provided a second opportunity to create heavy elements to fill out the periodic table and make possible the formation of planets and life to inhabit them.

Figure 24 Helium to Carbon - Nuclear Route

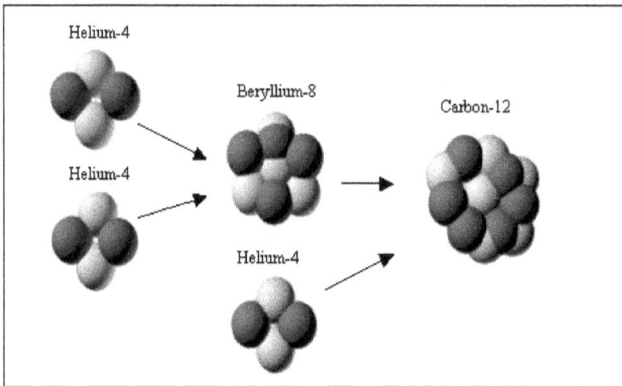

The first proof that carbon and all heavier elements originated in the stars came from Cosmologist Fred Hoyle. Hoyle concluded the only possible route to carbon formation went through the fusion of an 8-nucleon beryllium nucleus (beryllium-8) from two 4-nucleon helium nuclei (helium-4) as shown in figure 24. Hoyle faced two major hurdles in gaining acceptance for his hypothesis. One, the instability of the beryllium-8 nucleus, known to have an average life of only 10^{-15} seconds, and the other, mass/energy improbability; the combined weight of beryllium and helium is significantly greater than the carbon-12 nucleus.

Hoyle believed that the existence of an excited state of carbon-12, with a mass precisely equal to the combined mass of beryllium-8 and helium-4, would allow helium-4 to react more quickly and within the brief lifetime of the unstable beryllium-8. An atomic nucleus achieves an excited state by taking on energy, which alters the geometric arrangement of its nucleons. The added energy converts to mass via $E=mc^2$ thereby increasing the mass of the nucleus. Hoyle calculated that in order for an excited state of carbon-12 with a mass equal to the combined beryllium-8 and helium-4 to exist, it would need to absorb energy amounting to 7.65 MeV (megaelectron volts) above the ground state energy (lowest energy state) of carbon-12.

While creation of higher mass excited states of nuclei is feasible by adding energy, nuclear physicists at the time, having performed exhaustive research on excited states of carbon nuclei, concluded that no excited state carbon-12 nucleus with the required mass existed. Hoyle invoked what later would be named the *anthropic principle* to convince experimental physicists to renew their search for the excited carbon atom. In 1953, the researchers found Hoyle's excited state carbon-12 and it had an excitation energy of 7.65 MeV, exactly as he had predicted. Hoyle had reasoned that since the only path to formation of carbon nuclei had to pass through a 12 nucleon high energy excited state, coupled with the fact that carbon exists, and we are here to argue it, means the excited state must exist. There are numerous versions of the anthropic principle, but the one applicable here would be:

We are here to study the Universe, so the laws of the Universe must be compatible with our own existence.

Having solved the 8-nucleon stability problem, physicists had little trouble identifying pathways for forming the remaining elements of the periodic table. The limitless availability of protons, neutrons, and helium nuclei in trillions of stars that varied widely in mass, temperature and pressure, provided ample opportunity to create all of the periodic table elements from boron to iron.

Dmitri Mendeleev originally listed the known elements of his time in order of increasing atomic weight, which is roughly equal to the sum of the protons and neutrons in the nucleus, then arranged them according to similar chemical properties. He organized them according to their atomic number, which is simply the number of protons in an element's nucleus. However, as scientists discovered new elements, some of them did not fit well into this scheme. Due to the variation in the number of neutrons among elements, the atomic weights and chemical properties of some—like iodine and tellurium—suggested a reverse order. However, even though variation in the number of neutrons in an atom prevented the atomic number from being a direct indicator of atomic weight, it was closely related; more importantly, it held a more consistent relationship to physical and chemical properties.

Table 4 Composition of Selected Atomic Elements

Element	Symbol	Atomic Number	Protons	Neutrons	Electrons	Atomic Weight*
Hydrogen	H	1	1	0	1	1.007
Helium	He	2	2	2	2	3.999
Lithium	Li	3	3	4	3	6.934
Beryllium	Be	4	4	5	4	9.005
Boron	B	5	5	6	5	10.801
Carbon	C	6	6	6	6	12.000
Nitrogen	N	7	7	7	7	13.994
Oxygen	O	8	8	8	8	15.985
Sodium	Na	11	11	12	11	22.969
Silicon	Si	14	14	14	14	28.060
Chlorine	Cl	17	17	18	17	35.427
Iron	Fe	26	26	30	26	55.799
Uranium	U	92	92	146	92	237.862
Lawrencium	Lr	103	103	159	103	262

Atomic Weight of an element calculated by adding the combined mass of the protons, neutrons, electrons, and force particles will not always agree with the values listed in the Periodic Table since these values are weighted averages of all of the known isotopes of each element. For example, there are three isotopes of hydrogen, each containing one proton and one electron; however, each contains different numbers of neutrons in their nuclei. Protium, which accounts for 99.9851 % of hydrogen, has no neutrons, while deuterium, which accounts for 0.0149 % has 1 neutron, and tritium, which does not occur in nature, but does exist near the surface of the Sun, has 2.

Table 4 below shows the atomic weights and numbers of several elements along with their nucleon and electron compositions.

The story of element formation in the stars goes as follows: After a heavy star has spent its supply of convertible hydrogen to helium, it is temporarily out of fuel (actually some hydrogen remains, but not enough to sustain the reaction). It then begins to collapse inward since there is no outward pressure from nuclear reactions inside its core to counter the inward pressure of the massive star's gravity. This causes temperatures deep within the core to rise above 100 million degrees where conditions become favorable for helium nuclei, atomic number 2 (AN 2), to combine with other particles (other helium nuclei, protons, and neutrons) to form the higher atomic number elements, lithium, AN 3 and beryllium, AN4.

Heat released from conversion of helium to lithium and beryllium continues to counter the inward gravity induced pressure until it spends the helium near the core center. Again the star begins to collapse, causing a further increase in core density and pressure, which again raises temperature until conditions become right for nuclear conversion of lithium and beryllium to boron, carbon and nitrogen nuclei (AN 5, 6, and 7). Then in a subsequent phase, helium and carbon merge to form oxygen, AN 8.

This process continues in heavy stars forming onion skin-like layers of elements of increasing atomic number until nuclear conversion reaches the element iron, AN 26. Instead of liberating energy, formation of elements with atomic numbers higher than iron require input of energy. Therefore, the sequence ends when a star reaches the iron core stage. Without the core nuclear reactions to balance the star's gravity, it begins to implode, becoming a *neutron star*, or in the case of even heavier stars, a black hole.

Neutron stars are a billion times denser than an ordinary solid and result from the immense gravity ensuing the dying star's implosion, which causes the electron/nucleus structure of the atoms to collapse. In its normal atomic state, the atomic nucleus contains nearly all of an atom's mass while the electrons occupy most of the space. The enormous pressure brought on by the stars gravitational collapse forces the negatively charged electrons to be "absorbed" into the protons, canceling their positive charge, and converting them to neutrons. This produces a star composed entirely of neutrons, hence the name, neutron star. In the case of still heavier stars, gravity from the collapse is so huge the entire mass of the star compresses to energy of infinite density; producing an object with gravity so great, that not even light can escape its attractive force.

The collapse of heavy stars, however, releases tremendous quantities of energy, enough in fact to cause an explosion called a supernova that blows off the outer layers of the star containing the matter formed during its lifetime of generating elements via nuclear fusion reactions. Light generated by Supernovae is so intense that they will outshine the combined light of the remaining 200 billion stars of their residence galaxy for many years. A Chinese astronomer named Yang Wei-De recorded the light from one such supernova now known as the Crab Nebula that reached the Earth in the year 1054 and still shines today. At the time of its occurrence, it would have been visible even in the daytime sky for several weeks.

Without supporting nuclear fuel, gravitational attraction causes the former giants to collapse to a neutron star with a radius of only about 10 miles and density measuring hundreds of millions of tons per cubic inch. The result is an object with gravitational force a million million times greater than that of Earth. Neutron stars have such fierce gravity that dropping a golf ball from pocket height would generate an explosive force equal to 10 tons of TNT. The amount of energy needed to place a projectile into orbit from Earth would only raise the same object 1 millimeter above the surface of a neutron star. The golf ball would need to achieve a speed in excess of 90,000 miles per second (about one-half the speed of light) to escape neutron star gravity.

The material from the explosion, including the nuclei of the heavy elements formed during its life, ejects into the galactic space where it combines with other gasses and stellar dust. The heat of the supernova, which reaches trillions of degrees also supplies energy to form the heavier than iron elements such as uranium. These elements are found on Earth and throughout the Universe but in much smaller quantities than iron and lighter elements.

After the gasses and stellar dust cool, and the nuclei recombine with electrons to form their respective elements, chemical reactions begin to occur among the elements, producing water and a host of organic chemical compounds. The unique structure of carbon allows it to form multiple bonds, either with other carbon atoms or with other elements such as hydrogen, oxygen, nitrogen, sulfur, etc., making it the base of a seemingly limitless variety of compounds. Many of these such as carbon monoxide, carbon dioxide, methanol, and ethanol, form through reactions that take place in the stellar dust.

The typical galaxy contains about 200 billion stars with many constantly dying, and their remnants combining with primordial gasses from the Universe to make new stars. This continuing birth and death cycle provided the elements that made up our Earth and the other planets of the Universe. Since heavy stars have shorter lifetimes, the Universe has already witnessed trillions of such supernova events, over 200 million of which resulted from deaths of heavy stars in the Milky Way galaxy alone. Even before the formation of the Sun, there were 8 billion years of time devoted to birth and death of stars. Stars of size similar to the Sun that formed at the beginning of this period still burn, and smaller ones will continue for 100s of billions of years more. Those massive enough to be neutron stars, however, have had sufficient time to produce 90 generations of supernova while the still heavier black hole size stars have had time for 9,000 generations. The churning of matter within the galaxy over these many eons provided the elements that became the building blocks of life on Earth.

Black Holes

The next and ultimate rung of the stellar mass ladder hosts stars with solar masses greater than 20, those destined to become *black holes*. The great mass of such a star causes even higher core temperature, making it burn its nuclear fuel much faster, producing a far brighter star. Otherwise, it behaves essentially like a neutron star during its lifetime, and its death follows the same pattern until it begins the final phase of collapse. Like neutron stars, absent the outward pressure from nuclear fusion reactions, the immense gravity squeezes orbiting electrons into their atomic nuclei, momentarily creating a neutron star. However, with black hole size stars, the mass of the remaining core becomes so great that not even the forces holding atomic nuclei together can stop the inward implosion. The inward gravitational pressure is so great that the mass constituents of the star, namely neutrons at this point, shift to the energy side of the $E=mc^2$ equation. Equivalent quantities of mass and energy have identical gravitational effects. However, since gravitational force depends on both the mass of an object and the distance from its center, concentrating the resulting energy to an infinitely small point greatly increases gravity, producing the black hole *singularity*.

In order to visualize a black hole, it is helpful to consider the two-dimensional elastic membrane analogy used earlier. But this time, in addition to being a gravitationally massive object, it is infinitely small resulting in extreme warping of spacetime. Rather than simply causing a curvature of space, the huge gravity "tears" the fabric of space, totally altering space and time in its region, producing the effect illustrated in figure 25.

Figure 25 Black Hole Geometry

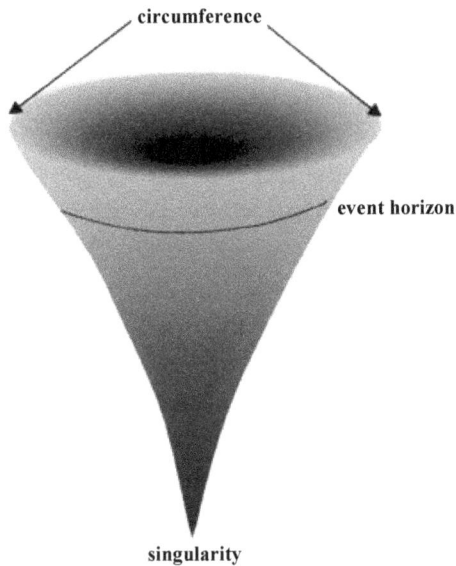

If you could safely observe an explorer entering a super-massive black hole (due to the hazards of his mission, we could call him a kamikaze cosmonaut or KC for short) from a distance, he would initially appear quite normal (not that anyone exploring a black hole could be normal). However, you would notice that his actions and his clock move slower due to gravitational time dilation brought on by the increased gravity. KC however would communicate back to you that his clock is OK and everything copasetic as he continues into the upper section of the black hole. As he descends deeper, his rate of descent accelerates; however, being in free fall, he does not notice this. You don't notice it either since from your vantage point, KC and his clock continue to move even slower the deeper he sinks. Early into his descent, he encounters the black hole's event horizon, the boundary where gravity becomes so great that escape velocity exceeds the speed of light. This is the point of no return for KC's journey since the gravity pulling him to the singularity becomes so great that no quantity of energy can overcome its pull.

Further into the hole, KC begins to notice a feeling of being stretched due to the presence of extreme tidal forces. These tidal forces are identical to the ones that create tides at the seashore caused by oceans on the side of the Earth facing the Moon, feeling its gravity more than oceans on the away side. The forces manifest as both horizontal and vertical differentials. *Vertical differentials* in this case arise from KC's feet being nearer the center of the black hole than his head, and therefore feel the gravity forces stronger. *Horizontal differentials* come from the fact that gravity draws all parts of KC's body toward a single point at the center of the massive body. To picture horizontal forces acting on KC, draw lines from each side of his shoulders to the point where they intersect at the black hole singularity. The sides of the resulting isosceles triangle would define the horizontal shrinkage he would undergo along his path toward the center. The closer to the singularity, the more the horizontal differential squeezes and the vertical differential pulls KC, the overall effect being to stretch him into a taller, leaner cosmonaut.

Let us assume KC is falling feet first into the hole. As he nears the center of the black hole, the distance from KC's head to his feet becomes a significant part of the total distance from his head to the singularity. Since the pull of gravity is a function of the distance between the masses of the objects, KC's feet, being the object nearer the singularity, will be experiencing a much greater pull than his head. This means the vertical differential between the gravity force at his feet and his head is great enough to begin pulling him apart, while at the same time, the horizontal differential shrinks his width dimensions. As he gets nearer and nearer the singularity, tidal forces become greater than the chemical bonds that hold human tissue together and tear his body in half. As these two pieces descend, ever-increasing tidal forces continue to make smaller, thinner portions of our friend via a process called—believe it or not—*spaghettification.* Eventually, gravity becomes so severe that all structure of body tissue is lost and further still, electrons surrounding the atoms making up his body compress into the protons converting them to neutrons. Finally, he reaches the black hole singularity. The immense gravity has converted all of the matter that once belonged to KC's body into energy; time has stopped and KC has forever separated from the Universe.

The amount of time you would have to observe your friend would depend on the size of the black hole. A stellar mass black hole has a surface far smaller than KCs' body so tidal forces would destroy him before entering, and he would advance to the singularity within a few thousandths of a second. A supermassive black hole lurks at the center of our Milky galaxy with mass equivalent to 3 million Suns that continues to grow by feeding on galactic gasses and stars within its region at the rate of about one star per year. Since tidal forces are less severe at the entrance of such wide surface, large black holes, you would have several hours to observe KC exploring this monster.

Tidal forces are not limited to black holes but apply to all massive objects including Earth. The tidal forces of Earth's gravity pull more at our feet than our head. They also generate greater horizontal shrinking pressures at sea level than when on a mountaintop; however, at 8,000 miles or so from Earth's center, the differences are too small to notice. If some tremendous pressure force squeezed the Earth to the point where it became a neutron star, it would retain its present mass but would be about the size of a marble, and tidal forces would become destructive near the new surface. If KC descended toward the new neutron Earth surface from a point equal to its distance from the current Earth surface, he would not notice tidal forces for a considerable distance. However, at eight miles inward, horizontal compression would shrink his 20-inch shoulder width by one one-thousandth, or to 19.98 inches. At 800 miles however, he would begin looking more like a fashion model, having been compressed a full 10 percent to a width of 18 inches and would definitely be feeling it. By 7,200 miles, he would be only 10 percent of his original self; he would resemble a piece of string when 8 miles from the new planet and a tiny mass of neutrons at ground zero.

Even though the mass of the planet had not changed, the overall force of gravity would be many times greater than on the original surface. This is because gravity is not only a function of mass, but increases by the inverse square of the distance from its center; e.g. 100 times greater 1 foot from the center than at 10 feet. At this point, with his feet twelve inches from the center, the head of his elongated body is hundreds of feet above the marble sized Earth; therefore, gravity at his feet would be thousands of times stronger than at his head, stretching him even further into a tiny thread.

Supermassive black holes with solar masses in the billions often occupy the centers of galaxies. In addition to intergalactic gases, their growth diets sometimes include complete star systems drawn to the galaxy center by the black hole's enormous gravity. In fact, stars from our own Milky Way could suffer such a fate several billion years from now when it crosses paths with the Andromeda galaxy, which boasts a 30 million solar mass monster as its centerpiece. While quite obviously we do not see black holes, astronomers have identified six hundred of them. Extrapolating this information suggests that more than 300 million must exist in our neighborhood of the Universe. Since they emit no light, we can only detect them indirectly by observing the bending of light rays passing near them, or by the gravitational influence they exert on other stellar objects. Black holes turned out to be the answer to the puzzle of rotating stars that appear to be circling about in empty space; they are in fact orbiting around a black hole.

Solar System

Most of the material from supernovae permeates the star's home galaxy having mixed with existing matter from earlier events. Occasionally, the shock wave of a new supernova, traveling at 500 miles per second, blasts into this gas/matter accumulation, causing it to lose energy and cool below 1300 degrees Celsius. This causes heavier elements in the gaseous mixture such as silicon, iron, and carbon to condense into microscopic dust grains. Such an event happened 5 billion years ago and led to the creation of our solar system. After about 100,000 years following the supernova shockwave, it had spent its energy and the supernova material combined with background interstellar matter that originated in the infant Universe.

As the dust and gas precursors of our solar system spiraled inward for the next 100,000 years, it formed a central plane of rotating material called a *nebula*. The Sun began to evolve near the center of the nebula, through a *T-Tauri contraction*, a stage in the development of stars where convection flows, moving from the inner core to the outer regions, cause rapid and erratic brightness fluctuations. Dust particles composed of elements from supernovae explosions began to collide in the outer regions, causing them to coagulate into grains, with larger ones cannibalizing smaller grains, eventually forming rocks. These highly energetic impacts generated enough heat to melt the smaller rocks welding them to the larger ones upon cooling. Following a hundred thousand years of such give-and-take, much of the mass within the solar system resided in kilometer-size objects. As the size of the objects grew, so did their gravity. Since the gravitational force of all objects concentrates at their center, the resulting incessant inward pull slowly shaped them into spheres. Collisions over the next 100,000 years placed many objects the size of our Moon, in orbits around the central star creating our solar system. As more of the available mass gravitated to, and became part of, larger bodies, the rate of collisions slowed. The arrested growth rate meant another 10 million years to advance from primarily moon-sized planetoids to Earth, Mars, and Venus-sized planets.

About one hundred million years from the beginning of the Earth's formation, the largest collision in its history occurred. A Mars-sized object from the asteroid belt grazed Earth's surface, ejecting billions of tons of molten material into orbit. This material coalesced into our Moon within a few years. It may be difficult to imagine any earthly benefit from such a collision, but without it, life as we know it, would never have evolved on Earth. The orbit of a single planet around a star follows a very orderly path since the only factors involved are the gravity exerted by the masses of the two objects and their velocity and orbital radii. However, when near other planets, in addition to the mutual attraction each planet has with the star, they also feel the gravity of each other, exerting influences that affect not only their orbits but the tilt of the axis about which they spin as well. The spin axes of all of the inner planets of the solar system change periodically due to resonances, brought about by complex gravitational interactions with one another. When these interactions happen at regular intervals, they can fall in step, essentially magnifying their impact. Pushing a child on a swing provides a good example of this kind of resonance. Timing the push at just the right point in the cycle causes the swing to travel through an ever-increasing arc, whereas the same effort exerted at a different point in the swinging cycle will either produce less benefit or cancel that of previous efforts. In the case of planets, gravitational tugs caused by the proximity of two or more planets provides the push and these can be magnified by the periodic conjunctions of their orbital positions.

The tilt of Mars' axis happens to be very similar to that of Earth; however, computer simulations show that it is subject to periodic variations of plus or minus 20 degrees. Dried-up riverbeds seen on the surface of Mars provide evidence that sometime in the past, a *resonance event* tilted the polar region toward the Sun melting its carbon dioxide and water ice cap. The spin axis of the Earth during its orbit around the Sun is about 23 degrees, a factor responsible for the seasonal cycle of our weather. However, among the inner planets, only the Earth has a massive moon stabilizing the angle of its spin axis, protecting it from chaotic resonance effects (Mars has two moons but they are too small to have significant impact). While our giant Moon has served us well providing inspiration for a thousand love songs, it has performed an even more endearing function— saving us from wild tilting gyrations. In the absence of moon-derived stability, resonance effects could alter Earth's spin axis by as much as 90 degrees, creating a six-month equatorial summer at the North Pole, while the South Pole endures complete darkness and unbearable cold, only reversing in the following cycle.

Jupiter, located far from the Sun with temperatures cool enough for water ice to form, grew much faster than the inner planets. Jupiter's gravity generated by its tremendous mass, allowed it to attract any material in its path, emptying this region of the solar nebula of extraneous material. Within only 10 million years, before *solar winds* (streams of charged particles—mostly electrons and protons that become sufficiently energized by the Sun's high temperature to escape its gravity) could eject all the hydrogen gas from the inner solar system, Jupiter had accumulated quantities of rock and ice amounting to more than 100 times the mass of the Earth. However, some of the coagulated material approached Jupiter at such an angle that rather than being drawn to its surface, it careened out into the solar system as comets. Other objects already in the embryonic stage of planet formation experienced collisions with large rocks with such enormous energy that they broke apart into smaller objects up to tens of kilometers across, and remain today as a large belt of asteroids called the *Oort cloud.* Without the massive Jupiter—and to a lesser extent, Saturn—running interference for Earth by accepting the brunt of the meteorite attack, complex life may never have had an opportunity to evolve on Earth.

Solar winds, over the next 10 million years, blew most of the gas and fine grains out of the solar system. The temperature of the disc-like material in solar orbit was about 1,700 C near the inner edge and zero °C outside. The space occupying the Earth was about 500 C at the time the dust began to coalesce, first into liquid, and then into solid rocks, eventually forming planets. All of this took only a few thousand years from the time the disc had initially formed. Since water, along with carbon dioxide, and nitrogen, are in gaseous states at this temperature, none could have been included in the original rocks that coalesced to form the Earth. Earth's gravity however, may have captured some water from vapor present before solar winds swept it away.

Still, most of these compounds came to Earth from meteors born of materials that had condensed further out in the lower temperature areas of the disc. The massive gravity of Jupiter attracted icy objects appearing much like giant snowballs containing condensed water, nitrogen and carbon dioxide. Many such objects approached Jupiter at an angle that avoided direct contact; then veered back into the solar system where they collided with other planets, including Earth.

For the next few hundred million years, all sizes of meteors containing the components of our oceans and atmosphere continued to bombard Earth's surface, the rate finally diminishing as the asteroid belts gradually depleted. Objects up to several hundred kilometers hit the Earth as often as every few million years. You get a clearer perspective of such an object's effect by considering the impact of a single 300-kilometer object. Such a massive meteor would generate enough energy to heat the Earth's entire surface to over 1,000 degrees C and evaporate all of its present oceans. Vaporized rock that had been the Earth's crust would displace much of the existing atmosphere. A millennium would pass before the Earth would cool enough for the liquid surface to solidify and liquid water begin to condense from the vapor.

While the rate of contact with comets and asteroids lessened with time, impacts did not completely stop. Even today, about 1,000 objects larger than 1 kilometer remain in or near earth-crossing orbits. A one-kilometer comet produces a crater about 10 kilometers across, but after 600 million years of erosion from wind, rain, and ice, their imprints become unrecognizable. A 10 K meteor on the other hand creates a crater 200 K across that can remain visible for 2 billion years. If a 10 K meteor hit Earth now, in addition to creating worldwide tidal waves, heat generated from debris would fill the sky with a glowing incandescence for hundreds of miles. Such an object did hit the Earth 65 million years ago, thus bringing about massive extinctions, including nearly every species of dinosaur. The Moon has experienced similar impacts, although correspondingly fewer due to its size and lower gravitational attraction. Most of these impacts remain visible due to the absence of surface plate activity and a weathering atmosphere. They also give the Moon its pock-marked appearance.

Continents and Air

Until 4.7 Billion years ago, the surface of the Earth was a magma-ocean—a pole-to-pole continuum of liquefied metals and rock. Its liquid state was sustained by heat generated from incessant bombardment by planetoid-sized meteors and radioactive decay of elements within its core. Periodically, areas of the liquid surface solidified then re-melted, releasing trapped volatile components that built up levels of carbon dioxide and water in the atmosphere. Over the next 600 million years, the frequency of meteor impacts fell precipitously to about 1/100 of its original rate, allowing the planet to cool and form a thin surface crust. Within the molten interior, the various components began stratifying as if in a giant fractionating column, each material seeking its own level according to its density. Iron became the predominate ingredient of the Earth's central core due to the combination of its relatively large quantity and high density. Liquid rocks, being approximately half the density of metals, remained at or near the surface, which if viewed on a solar scale would have resembled a translucent skin more than a crust. It was far from cool however, replete with open springs of boiling rock that inspired the name, *Hadean period.*

By 3.8 Billion years ago, the last of the 300K meteors had visited Earth and things had begun to calm, although smaller rocks capable of gasifying the upper few hundred feet of ocean continued for another few hundred million years. The cooling Earth surface caused atmospheric water vapor that had accumulated over the previous billion years of meteorite attack to condense to hot rain that fell continuously for more than 100,000 years—creating the first oceans. These early oceans would have presented a Spielberg dream sequence:

147

boiling springs created by the hot mantle just beneath the ocean floor buffeted by violent tides 30 times larger than tides of today, the result of a lunar orbit one-quarter its present distance from Earth, with a period of only five days.

Even after the ocean-filling flood, the atmosphere still contained some water vapor and nitrogen. However, carbon dioxide (CO_2), released from compounds such as silicon and calcium carbonates that had accumulated in the molten Earth from millions of years of comet bombardment became the main ingredient, accounting for 98% of atmospheric gases. This too proved fortunate for us, and all other life on Earth, since the Sun's luminosity was about 30% less than today. With the primarily nitrogen/oxygen atmosphere of today's Earth, temperatures would have been so low that all the oceans would have frozen. However, carbon dioxide, being a greenhouse gas absorbs infrared radiation that would otherwise reflect back into space. This compensated for the Sun's lower luminosity, thereby maintaining the warm planetary environment necessary for chemical reactions. This amount of carbon dioxide—about 10,000 times greater than exists on Earth today—would give our planet an atmosphere similar to present day Venus where the surface temperature is 470 Celsius (880 Fahrenheit). Even at these temperatures, water remained in the liquid state due to an atmospheric pressure 20 times greater than now caused by the extremely high carbon dioxide content.

As the Sun increased in luminosity, the presence of liquid water and its greater distance from the Sun protected Earth from runaway greenhouse effects that destroyed Venus's atmosphere. Runaway greenhouse gas events can happen because water also acts as a greenhouse gas. The higher the temperature, the more water evaporates into the atmosphere, which makes it hotter still, causing more water to evaporate, and so on. This feedback mechanism raised the temperature of Venus to the point where all of its liquid water vaporized, and radiation from the Sun gradually broke down the atmospheric water vapor into hydrogen and oxygen components. Then the lighter, energetic hydrogen escaped the planet's gravity evaporating into space.

However, Earth avoided this fate through a series of chemical processes that gradually reduced the concentration of CO_2 in the atmosphere and maintained Earth temperatures as the Sun expanded and its luminosity increased. The combination of carbon dioxide, water, earth-metals and minerals cooperate to maintain temperature balance via an atmospheric management cycle as follows: Atmospheric carbon dioxide dissolves in rainwater, producing carbonic acid ($CO_2 + H_2O > H_2CO_3$). This acid water attacks silica, as well as iron and other metals in rocks, forming metal carbonates like limestone and dolomite, which sediment out on the ocean floor ($H_2CO_3 + Si > SiCO_3 + 2H^+$). Removal of carbon dioxide through this process balanced the increased luminosity of the Sun, thereby maintaining the atmosphere below the runaway greenhouse temperatures.

Once in the carbonate form, CO_2 cannot return to the atmosphere unless the rocks first become subducted into the magma then subsequently returned to the surface by volcanic activity. Once the hot carbonates meet the comparatively low pressure of the surface, they decompose, liberating CO_2 back to the atmosphere. Over billions of years, this cycle maintained the planet at moderate temperatures by gradually reducing carbon dioxide levels by a factor of 10,000 as the Sun gradually increased in luminosity. This kept water in a liquid state while at the same time avoiding a Venus-like environment. This enormous cleansing of carbon dioxide caused the atmospheric pressure to drop to near current levels and, while CO_2 concentrations still greatly exceeded those of today, nitrogen became the atmosphere's main component.

As the Earth cooled, areas of skin began forming on the surface. At first, the thin crust islands proved too brittle for sub-surface dynamics to push them around en masse. However, as they thickened and became more rigid, they broke into large plates that floated on the mantle, and moved with the convective flows. These became the first *tectonic plates* and signaled the beginning of continents. Upward convection flows of magma broke through the mantle in some regions, partially melting the crust and pushing it downward. The volatile carbonate materials, that earlier had sedimented into the crust, were heated once again. This released carbon dioxide and water back to the atmosphere, thus completing the cycle. This cyclic process (beginning with carbon dioxide dissolving in rain water, reacting with calcium, silicates, iron oxides, and other Earth components, followed by melting, volatilization, and release back to the atmosphere) began in this era and continues until today. A complete cycle repeats on the average every 150 million years, providing a heat-sink mechanism driven by CO_2 control that minimizes gross atmospheric temperature variations and the accompanying freeze-fry cycles.

The continuation of this cycle over the next billion years gradually modified the composition of the atmosphere and the continental crusts. However, not all of the convection flow of magma reached the surface on continental landmasses. When material originating from deeper in the mantle, at depths below the carbonated sediments, cycles upward in mid-ocean regions, areas of ocean floor crusts melt, which brings on another important phenomena. Magma, absent any volatile gasses, pushes up less violently from mid-ocean vents than magma containing gaseous carbonate. This allows water to be cycled back into the vents and superheated to 400 degrees Celsius. The superheated water then recycles back into the ocean, rich in minerals providing warm mineral solutions capable of producing interesting chemical reactions. Such vents exist today; however, early oceans having much thinner crusts, undoubtedly harbored many more, providing billions of chemical laboratories to support reactions.

4   It's All About Chemistry

Just as a general understanding of nuclear physics, quantum theory, and relativity enhances the appreciation of how the Universe began, and a system of galaxies, stars, and planets developed, we cannot fully appreciate the genesis and evolution of life without some knowledge of chemical reactions and their endless complexity. While not an absolute necessity, a review of the basic concepts of chemistry will be helpful in appreciating the role carbon plays in terrestrial life. Readers with a background in elementary chemistry may want to proceed directly to chapter 5; others may prefer simply to scan the material to get a general feel for the subject, and a few of you may actually want to "study up". For those considering the latter, the following caveat is in order: I have not attempted to present a comprehensive chemistry text but to include only the minimum detail necessary to provide the reader with a feel for how complex chemical compounds can form from individual atoms. In the preparation of this chapter, this has meant overlooking many factors that are admittedly part and parcel of a chemistry education, but do not further the limited purpose of preparing the reader for this understanding. In other words, the least you need to know about chemistry in order to understand its contribution to life on Earth.

Whereas the physics responsible for the development of the Universe deals primarily with atomic nuclei and their constituent particles while electrons play a supporting role, the emphasis reverses when considering the science of living things. Electrons give elements the chemical properties that made possible the evolution of life itself. Chemistry through electrons touches every aspect of our daily existence—from the food we eat to the energy that runs our minds and our civilization.

As discussed in chapter 1, electrons exist in orbits around the nuclei of atoms in the form of wave functions with wavelengths corresponding to the specific energy levels of the electrons. The number of negative charged electrons carried by each atom is identical to the number of positive charged protons in its nucleus. In chemistry, the wave functions are called *orbitals* and these combine in *shells*, which further divide into *subshells* that house electrons of common wavelength and energy. Consistent with the uncertainty principle of quantum mechanics discussed earlier, the orbital wave function allows a level of probability that the electron can be located anywhere in space; however, the value of the probability function becomes extremely small at distances beyond a few angstroms (10^{-8} centimeters) from the nucleus. Orbitals subdivide into four energy levels denoted in order of increasing energy by the letters *s*, *p*, *d*, and *f*, and preceded by numbers from 1 to 7 that indicate the shell the electrons occupy. As elements increase in atomic number, and therefore in numbers of electrons, *s* subshells fill with electrons first followed in order by *p*, *d*, and *f*. There are seven shells available with different capacities for accommodating electrons. The first, or innermost shell can hold only two electrons; the second and third can each hold eight; the fourth and fifth eighteen apiece, while both the sixth and seventh shells can hold thirty-two.

Table 5 Periodic Table of the Elements (abbreviated)

	Group 1a									Group 0
	1 e' in outer shell	2 e's in outer shell								Noble gasses – Outer e shell filled
Period 1	1 **H** Hydrogen 1.008	Group 2a		Group 3a	Group 4a	Group 5a	Group 6a	Group 7a		2 **He** Helium 4.003
Period 2	3 **Li** Lithium 6.904	4 **Be** Beryllium 9.012		5 **B** Boron 10.81	6 **C** Carbon 12.011	7 **N** Nitrogen 14.007	8 **O** Oxygen 15.999	9 **F** Fluorine 18.998		10 **Ne** Neon 20.183
Period 3	11 **Na** Sodium 22.990	12 **Mg** Magnesium 24.312		13 **Al** Aluminnum 26.9815	14 **Si** Silicon 28.086	15 **P** Phoosphorous 30.974	16 **S** Sulfur 32.064	17 **Cl** Chlorine 35.453		18 **Ar** Argon 39.948
Period 4	19 **K** Potassium 39.102	20 **Ca** Calcium 40.08	Groups 3b, 4b, 5b, 6b, 7b, 8, 1b, 2a 2b (below)	31 **Ga** Gallium 69.72	32 **Ge** Germanium 72.59	33 **As** Arsenic 74.922	34 **Se** Selenium 78.960	35 **Br** Bromine 79.904		36 **Kr** Krypton 83.800
Periods 5, 6, 7 Not Shown			Period 6, Lanthanides	92 **U** Uranium 238.03						

	Group 3b	Group 4b	Group 5b	Group 6b	Group 7b	Group 8	Group 8	Group 8	Group 1b	Group 2b
Period 4	21 **Sc** Scandium 44.956	22 **Ti** Titanium 37.90	23 **V** Vanadium 50.942	24 **Cr** Chromium 51.996	25 **Mn** Manganese 54.938	26 **Fe** Iron 55.847	27 **Co** Cobalt 58.932	28 **Ni** Nickel 58.71	29 **Cu** Copper 63.546	30 **Zn** Zinc 65.37

More than half of the 112 elements of the periodic table are in some way necessary for the production and maintenance of the complex life forms on our planet. However, only seven of them: carbon, hydrogen, oxygen, nitrogen, calcium, phosphorous and sulfur—identified by the atomic symbols, C, H, O, N, Ca, P, and S respectively—account for the great bulk of its biomass, with lesser contributions from Na (sodium), Cl (chlorine), and about fifty other elements. In humans, oxygen represents the largest single contributor to body weight at about 60%, primarily due to the large volume of water in our bodies. However, carbon, which accounts for only 20% of body mass, forms the core of our structure, as well as that of every other living thing on this planet. Next to hydrogen, nature provides more known compounds of carbon than any other element and the majority of hydrogen compounds are carbon based. In fact, the prolificacy and importance of carbon compounds inspired assigning it an entire branch of chemistry, called organic chemistry.

Only six elements have completely filled electron shells, namely helium, neon, argon, krypton, xenon, and radon. These elements, called the *noble*, or *inert gasses* have exceptional stability making them reluctant to react with other elements or chemical compounds (actually, helium and neon are the only true elemental inert gasses since they are the only ones that form no known true chemical compounds). This configuration of fully occupied shells is the main driving force in chemistry. All of the other elements not blessed with filled outer shells, strive to find other means of achieving this noble configuration; either through charitable giving, larcenous taking, or mutually beneficial sharing of electrons with other elements or with their own kind. This tendency of atoms to donate, accept, or share their outer shell electrons, called *valence electrons*, is responsible for the great variety in chemical properties of elements. In their quest for nobility, they form alliances, the end result of which is the millions of chemical compounds found in nature and synthesized in laboratories.

The lowest energy of these orbitals is the s orbital, up to two of which can be located in each shell. s orbitals always fill with electrons first. Hydrogen, the smallest element needs just one electron to balance the single proton in its nucleus so it occupies an s orbital in the first shell. Helium with two protons has two s orbital electrons that completely fill the first and only shell thereby earning the royal designation of noble gas.

Table 5 shows an abbreviated version of the periodic table of the elements developed by Dmitri Mendeleef in 1869. It arranges the elements according to their atomic number into rows called *periods*, and columns called *groups* that combine those with similar chemical properties. Atomic number (AN) of an element simply identifies the number of protons in its nucleus as well as its number of orbiting electrons when in the elemental, or non-ionized state. Nuclei of all elements except hydrogen also contain neutrons, the number of which is equal to or greater than the number of protons. Furthermore, there are sub-species of some elements with the same number of protons and electrons as their parent element but different numbers of neutrons, and therefore have different atomic weights. These sub-species, called *isotopes*, have chemical behavior nearly identical to their corresponding element but are not as stable. This instability of isotopes means a tendency to gain or lose neutrons in order to become identical to the parent. Rather than assigning new names to isotopic elements, they are normally distinguished from the parent element by their *mass number*—the sum of the protons and neutrons in their respective nuclei. For example, the parent isotope of uranium, which accounts for 99.27 % of naturally occurring uranium, is ^{238}U; the next most prevalent isotope is ^{235}U at 0.72%.

While the nucleus of the parent hydrogen atom contains only a single proton, it has neutron-containing isotopes. Hydrogen isotopes are so important chemically that they have earned their own names; *deuterium*, also called *heavy hydrogen* has one neutron and *tritium* has two. Deuterium accounts for 0.0156% of naturally occurring hydrogen and the added mass provided by the extra neutron alters its physical properties enough that it will separate from water as D_2O with conventional fractional distillation. While the additional mass of the neutron changes the physical and chemical properties of hydrogen only slightly, the extra weight nevertheless has a much greater effect than a neutron would have on larger elements since the masses of heavy elements change only slightly whereas hydrogen's mass nearly doubles. Nuclear power plants produce deuterium oxide, HOD (better known as heavy water), commercially by the ton for use in nuclear fission reactors.

Over time, isotopes lose their excess neutrons and return to their more stable analog of the periodic table. Heisenberg's Principle predicts one cannot possibly know exactly when the nucleus of an unstable element will relinquish a particle (or decay) to achieve a more stable state; however it does allow that every isotope have its own statistical probability of decaying at a particular time. No one can estimate how long it will take an individual isotopic atom to decay but we can predict the rates of decay for large quantities of them. Scientists measure this rate in half-lives—the time needed for one-half of the isotope atoms to decay.

Half-lives of different isotopes vary from microseconds to tens of thousands of years making them useful for determining the ages of different substances. For example, carbon, which normally has 6 protons and 6 neutrons in its nucleus, has three isotopes; the parent atom carbon-12 (referred to as ^{12}C when emphasizing isotopic properties) accounts for 98.89% of all carbon while carbon-13 (^{13}C, 6 protons and 7 neutrons) accounts for nearly all of the rest at 1.11%. However, the upper atmosphere produces trace quantities of carbon-14 (^{14}C, 6 protons and 8 neutrons), an isotope with a half-life of 5,730 years. As we will discuss in chapter 3, geologists use carbon-14 to determine the ages of fossilized carbonaceous materials by the technique called *carbon dating*. Carbon-14, in the form of $^{14}CO_2$, mixes thoroughly with atmospheric CO_2 and as a result becomes part of all living things that depend on photosynthesis to convert CO_2 to their constituent carbon compounds. The moment a live organism dies however, it ceases to assimilate further CO_2 and its carbon-14 content begins to fall due to β (beta) emission, wherein one of the surplus neutrons emits an electron changing it to a proton. The added proton changes unstable carbon-14 to the stable nitrogen-14 (^{14}N, the parent atom of nitrogen with 7 protons and 7 neutrons). Therefore, by comparing carbon-14 concentration of a dead object with that of a living one, using its half-life, one can calculate when it died with an accuracy of 5%. If half of the carbon-14 is gone, the owner died between 5,400 and 6,000 years ago; if 75% is missing, death occurred between 10,900 and 12,000 years ago.

Hydrogen with AN 1 anchors the periodic table in the upper left corner in period 1 and group I, with other elements continuing in ascending order of atomic number from left to right. Period 1 concludes with the first element owning a completely filled outer shell, which in this case happens to be helium at AN 2 since it needs only two 1s electrons to fill its outer shell. All of the elements in the next row, period 2 can have two 1s electrons in their inner shell 1 but can accommodate eight electrons in their outer shell, which subdivides into subshells 2s and 2p. In all elements, the s subshells fill with electrons before the p, d, or f subshells receive any. Period 2 begins with the 3rd element lithium at AN 3, located beneath hydrogen in group I. Like helium, two of lithium's electrons fill the 1s shell;

however, the third occupies the unfilled 2s subshell. Like hydrogen, it has one electron in its outermost *valence shell* while element number 4 beryllium has 2 electrons, filling its 2s subshell. Element number 5, boron is the first to receive a 2p electron giving it 3 valence electrons. Electrons continue to add to the 2p orbitals for elements AN 6 - carbon, AN 7 - nitrogen, AN 8 - oxygen, AN 9 - fluorine, and AN 10 - neon giving them 3, 4, 5, 6, 7, and 8 valence electrons respectively. In the case of neon, its 8 valence electrons put it in the same inert or noble gas category as helium with all the orbitals in its outer shell filled.

Electron Behavior

The ability of atoms to transfer and share electrons provides the driving force behind all chemical reactions. G. N. Lewis developed the theory of valence during the second decade of the 20th Century based on the concept that elements transfer and share electrons in order to achieve stable electron shells. It is possible to remove one or more electrons from an atom of any element if it can be made to absorb sufficient energy. The energy required to do this, measured in electron volts, is the *ionization potential* of the element. An electron volt (e.v.) is the kinetic energy acquired by an electron when it passes through a potential difference of 1 volt; or stated differently, it is equal to one volt times the charge of a single electron. The lower an element's ionization potential, the more readily it reacts to form chemical compounds with other elements that need an electron to fill their outer shell. High ionization potential elements on the other hand, part with electrons more reluctantly but may be willing acceptors of electrons from other elements.

Figure 26 shows ionization potentials for removing the first electron from elements 1 through 36 (hydrogen–krypton). Figure 26 exhibits two discernable trends. First, as we move across the figure from left to right and compare elements from the same group, ionization potentials steadily decrease; for example, helium (He), neon (Ne), argon (Ar), and krypton (Kr) of the noble gas group 0. We can explained this by the fact that as we move down the Periodic Table, the negative charged valence electrons belong to shells further away from the positive nuclei. This added distance, relative to smaller elements, diminishes the attractive force nuclei have on the electrons, requiring less energy to extract them. The same applies to elements in groups II through VII; for example, beryllium (Be), magnesium (Mg), and calcium (Ca) of group II, and fluorine (F), chlorine (Cl), and bromine (Br) of group VII.

Figure 26 Ionization Potential versus Atomic Number

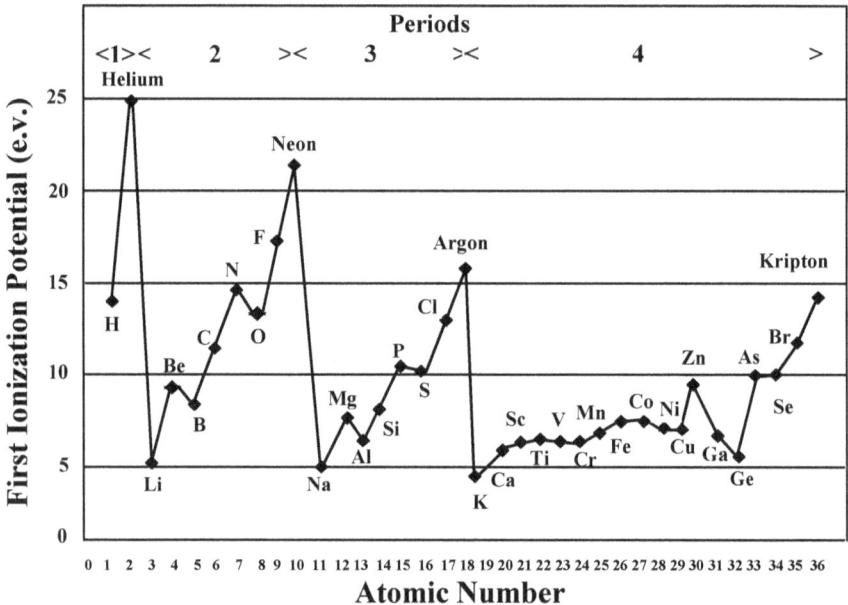

The second and most obvious trend; an almost steady increase in ionization potential within periods, which reach their highest values with the noble gasses of group 0. For example, consider the upward trend from Li to Ne in period 2, and sodium (Na) to argon (Ar) in period 3. The lone electron in the outer shell of group I elements, lithium (Li), sodium (Na), and potassium (K), make them vulnerable and easily picked off by a relatively low energy probe, hence their low ionization potentials. This same property makes them readily react with high ionization potential atoms such as fluorine (Fl), chlorine (Cl), and bromine (Br) to form chemical compounds.

Group I hydrogen is a special case due to the very close proximity of its lone electron to the nucleus. Electrons in the filled outer shells of group 0 on the other hand, protect one another through a shielding effect making electron removal more difficult and their ionization potentials higher.

The alter ego of ionization potential is a property called *electron affinity*, a measure of an elements willingness to accept an available electron. Technically, it is the measure of the energy released when an electron from another source attaches to an atom. In practice, high electron affinity simply means an element will more readily accept an electron. Intuitively one would expect values for electron affinity and ionization potential to be reciprocally related (e.g., an element with a high ionization potential would carry a low electron affinity) since a propensity to donate electrons should correspond to a reluctance to accept them. The problem however lies in historic convention; if the original researchers had opted to measure electron affinities in terms of energy absorbed—as they did with ionization potentials—rather than energy released, the inverse relationship would be more obvious. For example, the elements of group VII (F, Cl, and Br) have especially high electron affinities, just as they have high ionization potentials—about 3.3 to 3.8 electron volts—a result of the acquired electron providing the atom with an inert gas configuration. Hydrogen and other group I elements on the other hand have very low electron affinities ranging from near zero to 0.74 e.v.

Figure 27 shows two representations depicting the electron transfer chemistry of group I lithium and group VII fluorine; the upper figure details the presence of all the electrons while the simpler and more frequently used lower one, shows only those in the outer shell. The nucleus of lithium has 3 protons giving it a +3 charge. The +3 charge is neutralized by two 1s electrons in the filled first shell and one in the 2s orbital of a second shell that has the capacity to hold eight electrons. The 2s electron is shown in a different shade of gray to emphasize its role in the reaction. Lithium's 2s electron transfers to fluorine creating a +1 charge lithium ion which now has a filled outer shell that provides the more stable electronic configuration of helium. Fluorine accepts the electron to become a –1 charged fluoride ion and achieve a stable neon electronic configuration. Both +1 lithium and –1 fluorine are called *ions*; however positively charged ions assume the name *cations* and negatively charged ones *anions*.

Figure 27 Chemical Bonding by Electron Transfer

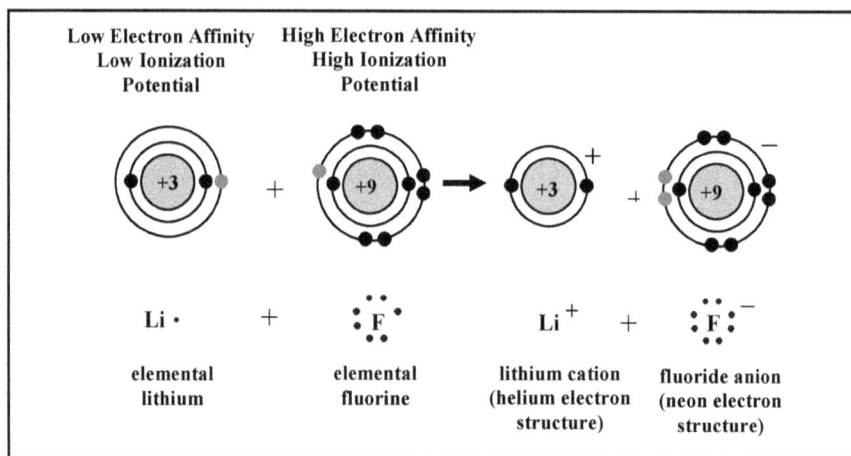

Low Electron Affinity Low Ionization Potential	High Electron Affinity High Ionization Potential		
Li ·	: F ·	Li⁺	: F :
elemental lithium	elemental fluorine	lithium cation (helium electron structure)	fluoride anion (neon electron structure)

Of course, this transfer does not produce two new separate products; you cannot make a bucket of lithium cations and a bucket of fluoride anions. In fact, the transfer is not even complete. Fluorine's affinity for the bonding electrons simply exceeds that of lithium, so on the average they stay much nearer the fluorine nucleus (or to be more quantum mechanically correct, the wave equation for the electron predicts a higher probability that it will be located closer to the fluorine nucleus at any point in time). The location of the bonding electrons nearer the fluorine nucleus makes the fluorine end of the molecule more negatively charged and the lithium end more positive. This electrostatic asymmetry, called polarity, keeps the positive lithium and negative fluorine ions bound together as lithium fluoride when in its purified gaseous state. However, the individual stability of the two ions permits them to readily dissociate and react with other polar compounds.

Once we understand the electron distribution among elements within a compound, it becomes unnecessarily tedious to draw the configuration each time we reference a compound, so we eliminate the electrons and identify lithium fluoride according to the symbols of the two elements simply as LiF. Lithium can form analogous binuclear compounds by combining with other elements from group VII. Lithium reacts with element AN 17(atomic number) chlorine and AN 35 bromine to form lithium chloride (LiCl), and lithium bromide (LiBr), as well with heavier group VII elements not shown in Table 5, AN 53 iodine and AN 85 astanine. Furthermore, as one might expect, other elements from group I such as sodium and potassium can react with elements from group VII, to form compounds like sodium chloride (NaCl), potassium bromide (KBr), etcetera. Group I elements below hydrogen in the periodic table are called *alkali metals* and group VII elements *halogens*.

Compounds formed by combining elements from these two groups, belong to a family of crystalline, salt-like compounds called *alkyl halides*. No two neighboring anions and cations within a crystal belong to a particular molecule. Each cation associates indiscriminately with six or more neighboring anions; and, anions are equally promiscuous. Polarity is responsible for much of their chemical behavior including solubility in water and other polar compounds. The salts dissociate into free ions that bond with the polar solvent due to the affinity of the cation for the negative end, and the anion for the positive end of the solvent molecule. This mutual electronic attraction takes credit for the solubility of salts in a wide range of solvents.

Group II elements have one more proton in their nucleus and

161

one more valence electron in their outer shell than do period 1 neighbors to their immediate left. First and second ionization potentials for removing both electrons from the period 4 and heavier group II elements, namely calcium AN 20, strontium AN 38, barium AN 56, and radium AN 88 are quite low making them willing donors of their two valence electrons for the sake of stability. They can form ionic compounds with right side periodic table elements from either group VI or VII that seek electrons for the same purpose. They can form ionic bonds in two ways; calcium for example can donate one electron to each of two group VII elements such as fluorine to make CaF_2. Or, it can donate both to a single group VI element, whose members need 2 electrons to fill their outer shells, like oxygen to make calcium oxide (CaO), or sulfur to make calcium sulfide (CaS). Beryllium, the lightest Group II element at AN 4, has a relatively high ionization potential of 9.3 giving it rather unique chemical properties among its Group II peers. This resistance to shedding its outer shell electrons results in bonding characteristics with other elements that resemble electron sharing more than ionic behavior, through *covalent bonding*, to be discussed next. The first ionization potential of 7.6 for magnesium AN 12 places it in an intermediate category compared to the other group II elements (calcium – 6.1, strontium – 5.7, barium – 5.2, and radium – 5.3 electron volts) giving it nearly equal amounts of ionic and covalent character.

Electron Sharing

Hydrogen also bonds with fluorine and other group VII halides but does so through a different mechanism called covalent bonding that gives a product, hydrogen fluoride HF—also called hydrofluoric acid. As you might expect from hydrogen's higher ionization potential and electron affinity, the HF product has less polarity than its LiF analog. The hydrogen and fluorine nuclei share the bonding electrons of hydrogen fluoride, even though they are on the average located nearer to the fluorine than the hydrogen nucleus. The figure 28 schematic formulas illustrate this sharing concept.

The high ionization potential of hydrogen makes it much stingier than the more massive group 1 elements when it comes to completely relinquishing its lone electron. However, when bargaining with an adversary like the group VII fluorine atom, which sports an impressive electron affinity, it will to agree to share its lone electron with fluorine's seven to satisfy the full outer shell ambitions of both.

Figure 28 Chemical bonding by Electron Sharing

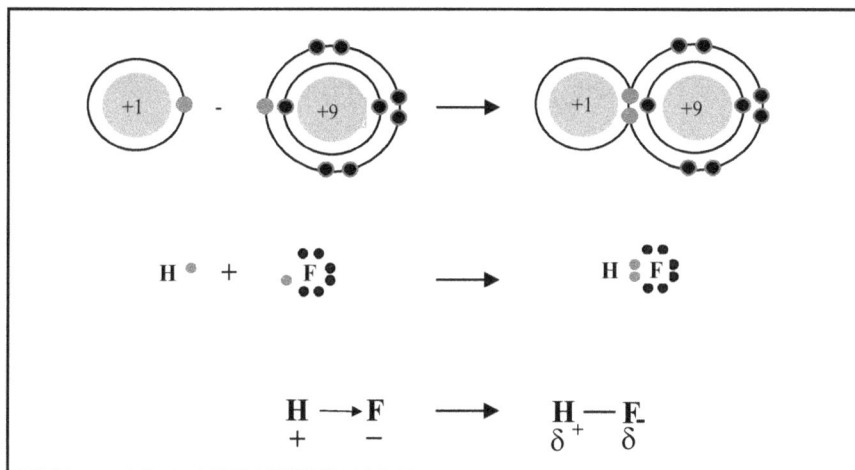

There are numerous constructions for demonstrating the polarity of hydrogen fluoride and similar compounds including the two shown above. In the lower left, an arrow shows the positive to negative direction of polarization while the one on the right uses the lower case Greek letter δ (delta) to indicate the partial nature of the electron sharing, or the fact that the bonding electrons will more likely locate nearer the fluorine nucleus than the hydrogen nucleus.

As shown in figure 29, two hydrogen atoms can also share their single 1s electrons to achieve a helium-like orbital configuration, however in this case the symmetry of molecular hydrogen means the bonding electrons have equal probability of being found near either nucleus making it a non-polar compound. Similarly, Group VII elements can form covalent bonds to assume inert gas-like configurations. Two fluorine atoms can share electrons to simulate a neon-like structure with eight electrons in the outer shell, yielding the non-polar molecule fluorine F_2. However, the 1s and 2s orbitals of the fluorine atoms are filled with electrons so an unpaired electron from their 2p orbitals serve as bonding electrons; being farther from the nucleus in the second shell makes the bond considerably weaker than the H_2 bond. This means that F_2 dissociates more readily than H_2 and is therefore more reactive with other materials; in fact, F_2 reacts explosively with most organic compounds.

Figure 29 Non-Polar Bonds

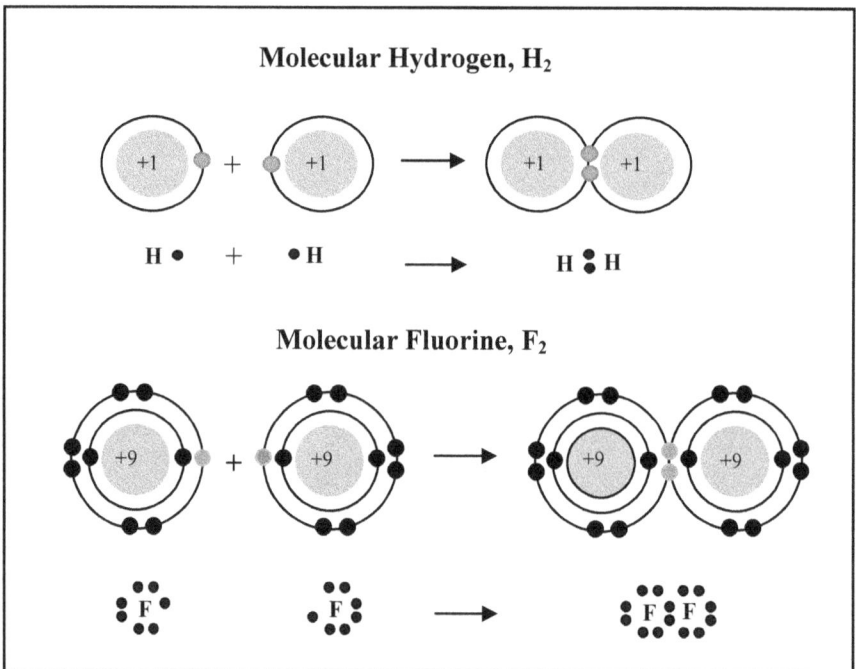

Moving toward the center of the periodic table to the elements of group III, we encounter a mixed bag of bonding behaviorists. Boron and aluminum each have 3 electrons in their outer shell whereas gallium and indium have 13, and thallium 27. However, all members of group III require 5 additional electrons in their outer shell to achieve inert gas configuration. The first electron of AN 5 boron has a rather high ionization potential at 8.3 e.v. while the second and third are 25 and 38 e.v. respectively This property makes it difficult for boron to lose electrons to form cations and limits its bond forming capabilities to the electron sharing, covalent type. The larger volumetric size of heavier members of group III and IIIb) makes their valence electrons farther from the nucleus causing them to have substantially lower first ionization potentials. These range from 5.98 for aluminum to 6.11 for thallium, providing them with some ionic bonding behavior.

Carbon

The elements of the periodic table that served as building blocks for all living things were a gift to the Earth from the stars. As seen in chapter 3, these great stellar atomic factories found ways to combine protons, neutrons and electrons that culminated in their manufacturing more than one hundred elements. From that point on, the construction of everything that makes Earth different from all other planets yet to be observed became the responsibility of chemistry—more specifically, the chemistry of carbon.

The high ionization potential of carbon combined with its unique electron configuration gives it nearly complete covalent bonding behavior. As a member of group IV, carbon, in addition to the two 1s electrons in its first shell, should have four electrons in its outer shell—two filling the 2s orbitals and the other two assigned to 2p orbitals. This outer shell electron configuration however, cannot explain the observed bonding characteristics of carbon containing molecules. For example, the four C−H bonds of the simplest organic compound, methane (CH_4) form a completely symmetrical structure, with the four hydrogen atoms located at the corners of a tetrahedron. The driving force of this tetrahedral configuration stems from the tendency of methane's positive charged hydrogen atoms to position themselves the maximum distance from one another.

If the carbon/hydrogen bonds had to rely on the 2s and 2p valence electrons of carbon, they could not achieve this symmetrical structure since the bond formed from the carbon 2s and hydrogen 1s electrons would differ in structure and energy from the three formed from carbon 2p and hydrogen 1s electrons. The symmetrical structure of methane requires that all four bonds be identical in energy and structure. The solution to the problem won multi-science practitioner Linus Pauling the 1954 Nobel Prize in chemistry. In addition to quantum chemistry, Pauling also made major contributions in such diverse areas as crystallography, molecular biology, and medical research; and even had time to win the 1962 Nobel Peace Prize for his efforts to affect a nuclear test-ban treaty during the cold war era. In order to achieve this carbon/hydrogen equivalence, Pauling introduced the concept of *hybridization* wherein carbon's single 2s orbital and the three 2p orbitals merge to form four identical orbitals that are symmetrical about four tetrahedral axes. One electron then occupies each of the four new hybrids designated as sp^3 orbitals indicating they each have 1 part s and 3 parts p character.

Figure 30 graphically illustrates the s and p orbital wave functions and the hybridized sp^3 products. The spheroids represent the space around the nucleus where there is high probability of the 2s electron being located. For the s orbital, the probability is highest near the center of the nucleus as indicated by the darker gray density, and decreases with distance. Electrons in p orbitals on the other hand have very low probability of being near the nucleus; the probability initially increases with distance along its axis and then decreases again, producing the dumbbell pattern.

166

Figure 30 Hybridization of Atomic Orbitals

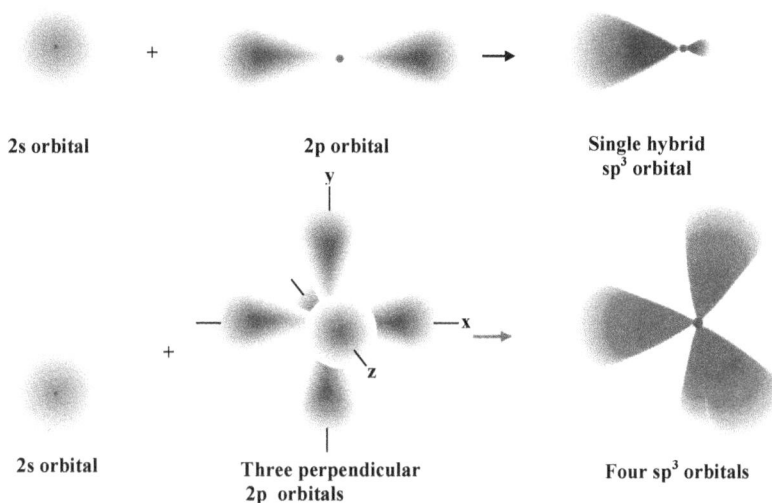

2s orbital   2p orbital   Single hybrid
                                               sp³ orbital

2s orbital   Three perpendicular   Four sp³ orbitals
                    2p orbitals

As shown at the top right of figure 30, the configuration of an individual hybrid sp³ orbital has most of the electron density on one side of the nucleus with only a small tail on the other side. This orientation places the bulk of the electron density in a position to overlap with electrons from other atoms giving carbon atoms their exceptional ability to form atomic bonds.

The lower drawing of figure 30 illustrates the configuring of the four sp³ orbitals along the tetrahedral axes to form the structure of carbon's bonding electrons. Figure 31 shows the 109° 28′ angle between each of the four axes. This angle, called the *tetrahedral angle* is common to all tetrahedrons.

The drawings of figures 30 and 31 represent the hybridized orbital configurations of an isolated carbon atom when it prepares to bond with other atoms. In the process of bonding however, carbon's sp³ orbitals must undergo further hybridization. In the case of methane shown in figure 32 below, four hydrogen 1s orbitals combine with carbon's four sp³ orbitals to form the tetrahedral structure of CH_4. Bonding results from the overlapping of hydrogen 1s and carbon sp³

167

orbitals to produce a type of covalent bonding called *sigma bonds*. Sigma bonds, the strongest of the covalent bonds, are characterized by their symmetry with respect to rotation about the bond axes.

Figure 31 Carbon Tetrahedral Configuration

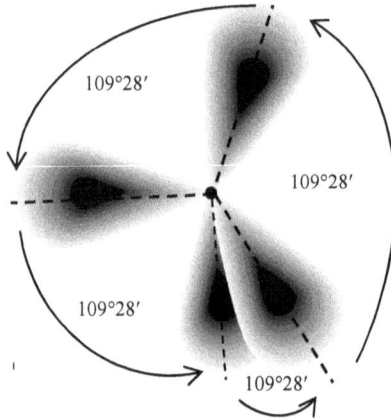

Molecular bonding

Figure 32 Molecular Orbitals of Methane

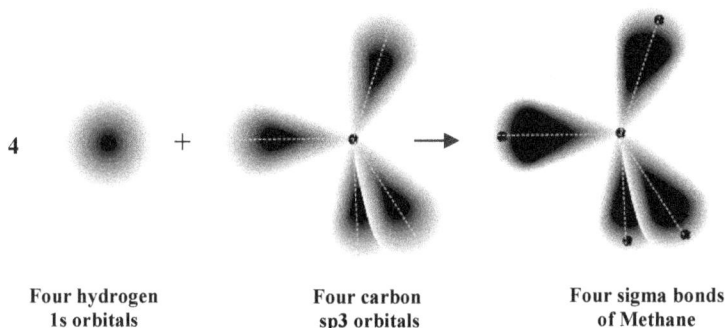

4   +   →

| Four hydrogen | Four carbon | Four sigma bonds |
| 1s orbitals | sp3 orbitals | of Methane |

Figure 33 isolates one of the hydrogen 1s and carbon sp^3 bonds to highlight the electron distribution about the bond. The small portion of electron density to the right side of the carbon nucleus points to the fact that the bond still retains some of the p character of the original unhybridized p orbital.

Figure 33 Carbon sp3 and Hydrogen Molecular Bond

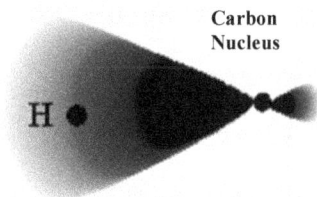

Carbon
Nucleus

H

Alkanes - Saturated Hydrocarbons (paraffins)

Cumbersome representations of molecular structure may be desirable or necessary for understanding bonding behavior; however, they become impractical in every day use, requiring a form of chemical shorthand. A two-dimensional drawing that identifies the bonds as sticks joining carbon and hydrogen as shown in figure 34 provides a less rigorous but far simpler description of methane. A carbon atom not only has the ability to form bonds with many other atoms but can also bond with other carbon atoms; a convenient property that makes possible the enormous variety of organic compounds that make up building blocks of life. Hydrocarbons are compounds composed solely of carbon and hydrogen. We refer to hydrocarbons as saturated if they contain the maximum number of hydrogen atoms consistent with the requirement that carbon have four bonds and hydrogen one. Unsaturated hydrocarbons, those containing fewer hydrogen atoms, will be discussed later.

Figure 34 presents the simplified stick structures of the first four saturated hydrocarbon compounds. Also shown are even less complicated drawings that eliminate the straight chain stick bonds and place the hydrogen atoms belonging to a given carbon to its immediate right. Vertical stick bonds can still be used to indicate placement of side chains (e.g., isobutane in figure 34). We should keep in mind however, that these structures are misleading representations of the molecule's geometry since they imply a flat linear configuration rather than twisting, three-dimensional chains that come from the tetrahedral arrangement of all carbon-hydrogen and carbon-carbon single bonds.

The chemical name for saturated hydrocarbons is *alkanes*, although commercial environments prefer the term *paraffins*. Alkanes follow the formula C_nH_{2n+2} where n can be any whole number. The series of stick figures in figure 34 illustrate the basis of the formula.

170

Regardless of the length of the hydrocarbon chain, the interior carbon atoms of the un-branched alkanes always have two hydrogen atoms attached while the terminal carbons have three. When a side chain attaches to an interior carbon of the alkane molecule, the attachment site carbon atom will have only one hydrogen; however, the terminal carbon of the saturated side chain will have a carbon with three thereby maintaining the 2n + 2 relationship. Once we understand this, we can identify a hydrocarbon with n equal to 1, 2, or 3 unambiguously by the formula alone, making the use of stick figures unnecessary. Methane can be CH_4, ethane CH_3CH_3, and propane $CH_3CH_2CH_3$; or, even simpler still by the number of carbon and hydrogen atoms as CH_4, C_2H_6 and C_3H_8.

Figure 34 Simplified Structures of C_1 C_4 Hydrocarbons

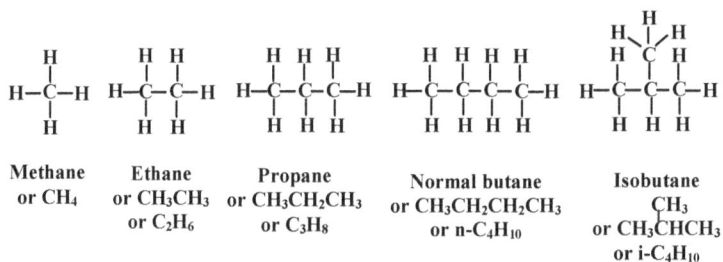

Methane	Ethane	Propane	Normal butane	Isobutane
or CH_4	or CH_3CH_3	or $CH_3CH_2CH_3$	or $CH_3CH_2CH_2CH_3$	CH_3
	or C_2H_6	or C_3H_8	or $n\text{-}C_4H_{10}$	or CH_3CHCH_3
				or $i\text{-}C_4H_{10}$

Alkanes with four or more carbon atoms (n equals 4 or greater) have bonding flexibility not available to the first three that permits molecules with identical atomic composition but different arrangements of atoms. While the chemical formula remains the same, this rearrangement of atoms produces compounds with different physical and chemical properties. The four carbon chain molecules of figure 34 provide the simplest example of this. One of the terminal CH_3 groups can switch places with a hydrogen atom from one of the two internal carbon atoms to give a new and different hydrocarbon. Hydrocarbons in which all the carbon atoms align in a continuous sequence with two hydrogen atoms attached to each of the internal carbon atoms are labeled *normal hydrocarbons*. Those with internal carbon atoms containing three (or as we will see later, four) carbon to carbon bonds, are called *isomers* of the parent normal hydrocarbon. Internal carbon atoms bonded to only two other carbons are *primary carbons*; those bonding with three, *secondary carbons*, and with four, *tertiary carbons*. Figure 34 above illustrates the simplified structures of butane's two isomeric hydrocarbons. The two internal carbons of normal butane are primary carbons while the single internal carbon of isobutane is secondary. Since butanes have only four carbons, there can be no tertiary carbons in their structure; however, the internal carbon of neopentane in figure 35 below is tertiary.

Saturated Hydrocarbon Groups

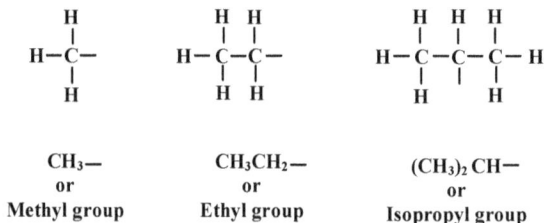

$$
\begin{array}{ccc}
& & \text{H} & & & \text{H } \text{ H} & & & \text{H } \text{ H } \text{ H} \\
& & | & & & | \;\; | & & & | \;\; | \;\; | \\
& \text{H}-\text{C}- & & \text{H}-\text{C}-\text{C}- & & \text{H}-\text{C}-\text{C}-\text{C}-\text{H} \\
& & | & & & | \;\; | & & & | \;\; | \;\; | \\
& & \text{H} & & & \text{H } \text{ H} & & & \text{H} \;\;\;\; \text{H}
\end{array}
$$

CH_3-	CH_3CH_2-	$(CH_3)_2\,CH-$
or	or	or
Methyl group	Ethyl group	Isopropyl group

Functional Groups

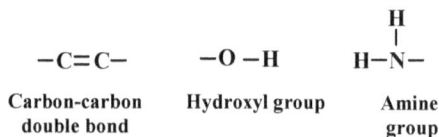

$$
-\text{C}=\text{C}- \qquad -\text{O}-\text{H} \qquad
\begin{array}{c}
\text{H} \\
| \\
\text{H}-\text{N}-
\end{array}
$$

Carbon-carbon double bond	Hydroxyl group	Amine group

When identifying differences in hydrocarbons, it helps to separate segments of the molecules into categories. These segments, a few of which we have listed above, are not viable entities on their own but serve only to identify parts of molecules. The bond missing an H, or any other atom, is its point of attachment to other molecules. These groups can be either saturated hydrocarbon groups or *functional groups*.

Functional groups derive their name from structural characteristics that give them a strong tendency to react with other elements or compounds to form new chemicals. Unsaturated fats for example digest more readily than saturated fats because they contain carbon-carbon double bond functional groups that provide reaction sites where stomach acids can begin the digestion process. The structural carbon atoms in saturated fats are all "hydrogen saturated" and, while not completely inert, are much less reactive, and therefore more difficult to digest.

We can draw three isomers with the formula C_5H_{12} as shown in figure 35 below. Like isobutane, isopentane, has a $-CH_3$, or methyl group branching from a secondary carbon atom. All four bonds of neopentane's single internal carbon attach to other carbon atoms, designating it a tertiary carbon.

Figure 35 Pentane and Hexane Isomers

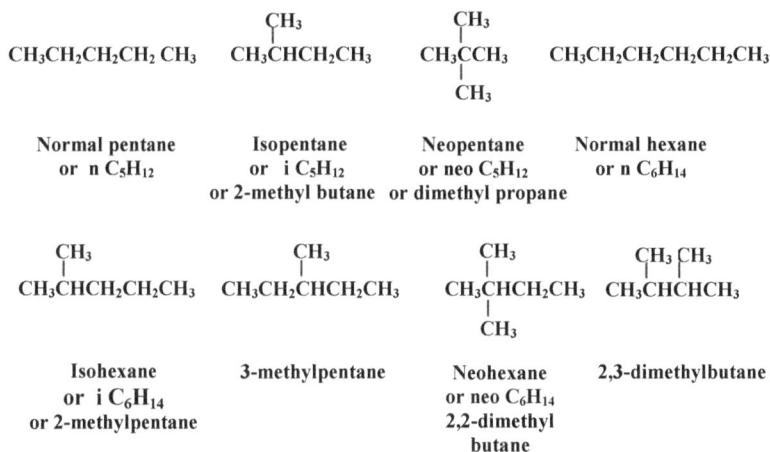

$CH_3CH_2CH_2CH_2\ CH_3$	$CH_3\overset{\overset{\textstyle CH_3}{\mid}}{C}HCH_2CH_3$	$CH_3\overset{\overset{\textstyle CH_3}{\mid}}{\underset{\underset{\textstyle CH_3}{\mid}}{C}}CH_3$	$CH_3CH_2CH_2CH_2CH_2CH_3$
Normal pentane or n C_5H_{12}	**Isopentane or i C_5H_{12} or 2-methyl butane**	**Neopentane or neo C_5H_{12} or dimethyl propane**	**Normal hexane or n C_6H_{14}**

$CH_3\overset{\overset{\textstyle CH_3}{\mid}}{C}HCH_2CH_2CH_3$	$CH_3CH_2\overset{\overset{\textstyle CH_3}{\mid}}{C}HCH_2CH_3$	$CH_3\overset{\overset{\textstyle CH_3}{\mid}}{\underset{\underset{\textstyle CH_3}{\mid}}{C}}HCH_2CH_3$	$CH_3\overset{\overset{\textstyle CH_3\ CH_3}{\mid\ \ \mid}}{C}HCHCH_3$
Isohexane or i C_6H_{14} or 2-methylpentane	**3-methylpentane**	**Neohexane or neo C_6H_{14} 2,2-dimethyl butane**	**2,3-dimethylbutane**

As the number of carbon atoms in saturated hydrocarbons increase, the possible configurations with the same chemical formula rise dramatically; with five C_6H_{14} isomers, ten C_7H_{16}, and eighteen isomers of C_8H_{18}; and the number continues to increase with chain length in this fashion. By the time we reach C_{40} (the straight chain parent hydrocarbon of which is n-Nonadecane)—as someone in bad need of a life has calculated—62,491,178,805,831 different structures can be drawn to represent its isomers. The overwhelming number of possible isomers makes the practice of assigning a unique name to each of these compounds impractical, and a systematic approach to allotting non-ambiguous identities becomes necessary. The method developed to accomplish this, called the *IUPAC* system (International Union of Pure and Applied Chemistry), uses derived names, wherein the longest continuous hydrocarbon sequence becomes the parent structure. Numbering of each carbon atom within the parent chain begins with the carbon at end nearest the location of the first attached group. Names of attached groups are listed in alphabetical order in front of the name of the parent structure prefixed with the number of its location on the parent chain.

Two of the five C_6H_{14} isomers, the second and the third in the figure 35 series, would qualify as isohexane if a single methyl group branching from an internal carbon were the only criteria. Using the conventions of the IUPAC system, chemists throughout the world can describe these two isomers with the same non-ambiguous names, 2-methylpentane and 3-methylpentane. In practice, the 2-methylpentane isomer has earned the isohexane designation due to its long usage prior to IUPAC, as well as its relative abundance.

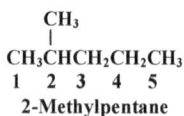

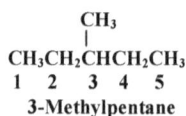

$$CH_3$$
$$|$$
$$CH_3CHCH_2CH_2CH_3$$
1 2 3 4 5
2-Methylpentane

$$CH_3$$
$$|$$
$$CH_3CH_2CHCH_2CH_3$$
1 2 3 4 5
3-Methylpentane

As shown below, the same system identifies complex isomers. In the first example, one of the eighteen isomers of n-octane illustrates the case where multiple groups attach to the parent chain, and more than one group attaches to the same carbon atom. The IUPAC name for the example with three methyl groups attached to a five-carbon pentane parent chain—two at the number 2 carbon and one at number 4—is 2,2,4-trimethylpentane. The second example shows more complex hydrocarbon groups, namely ethyl and isopropyl, attached at carbons 3 and 4 of an n-heptane parent chain. Although seldom used for simpler hydrocarbons, the system applies to them as well; for example, the IUPAC name for neopentane above is 2,2-dimethylpropane.

a C_8H_{18} Isomer   **a $C_{12}H_{26}$ Isomer**

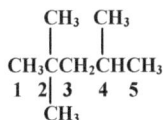

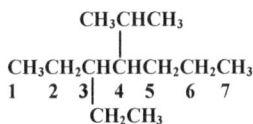

$$CH_3 \quad CH_3$$
$$| \qquad |$$
$$CH_3CCH_2CHCH_3$$
$$1 \quad 2| \ 3 \quad 4 \quad 5$$
$$CH_3$$

$$CH_3CHCH_3$$
$$|$$
$$CH_3CH_2CHCHCH_2CH_2CH_3$$
$$1 \quad 2 \quad 3| \ 4 \quad 5 \quad 6 \quad 7$$
$$CH_2CH_3$$

2,2,4-Trimethylpentane   **3-Ethyl-4-isopropylheptane**

While we illustrate these hydrocarbons as linear chains, remember they rotate freely along each carbon-carbon bond axis. Each of the four tetrahedral bonds of the individual carbon atoms has equal opportunity to bond with either carbon or hydrogen atoms, so we show the structure linear only for convenience; either of the two drawings below provides an equally valid representation of n-butane.

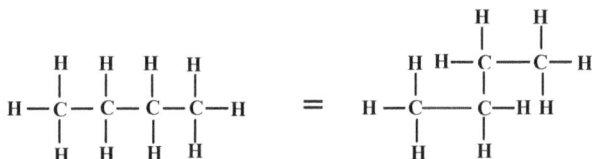

Alkenes - Unsaturated Hydrocarbons (olefins)

As mentioned earlier, hydrocarbons that have not used all of their available orbitals to bond with hydrogen or other carbon atoms are considered unsaturated. Their technical nomenclature is *alkenes*, although *olefins*, an old term derived from the Latin words oleum and facto—meaning oil making gas—is more common.

The absence of sufficient hydrogen to bond with all of carbon's

sp^3 orbitals forces the adjacent carbon atoms to form an extra carbon-carbon bond giving the configuration C=C instead of C–C.   As shown earlier, all alkanes, or saturated hydrocarbons, take the formula for C_nH_{2n+2}. However, for alkenes, the formulae vary according to the number of double bond, each of which reduces the hydrogen content, causing the formulae vary as follows:

Double Bonds	H Atoms Lost	Formula
0	0	C_nH_{2n+2}
1	2	C_nH_{2n}
2	4	C_nH_{2n-2}
3	6	C_nH_{2n-4}

Alkenes can be viewed as hydrogen deficient hydrocarbons and termed unsaturated since they do not have the maximum number of hydrogen atoms allowed by their bonding electrons. The absence of hydrogen means some carbon p orbitals have no s orbitals with which to bond forcing them to bond with each other. The result is a carbon-carbon double bond consisting of one *sigma* (σ) *bond* and another called a *pi* (π) *bond*. Since pi bonds are less stable than sigma bonds—a property that gives chemists, and Mother Nature, a valuable tool for synthesizing other compounds—they eagerly assume the stable sigma configuration and will readily react with H_2 or other chemicals to achieve that goal. This gives alkenes the quality of *reactivity*, a property absent from the more stable alkanes.

Another type of hybridization forms the double bonds in unsaturated compounds, in which only two of the three p orbitals from each carbon hybridize to produce sp^2 orbitals.   Figure 36 illustrates sp^2 bonding in the simplest olefin, ethylene, C_2H_4.   The upper drawing shows the formation of sp^2 hybrids for one of ethylene's two carbon atoms wherein two of the 2p orbitals combine with one 2s orbital to create three equivalent sp^2 hybrids.

The center drawing shows the third p orbital overlapping with the remaining un-hybridized p orbital of the other carbon creating a π (pi) molecular orbital.   Unlike the tetrahedral sp^3 configuration that generates sigma bonds, sp^2 orbitals lie in a plane with each axis separated by an angle of 120°, resulting in a *trigonal* state of hybridization.

The bottom of figure 36 shows bonding through the two types of orbital configurations to form C_2H_4. Two of the sp^2 orbitals from each carbon overlap with hydrogen 1s orbitals making σ bonds while the third forms a σ bond with the sp^2 orbital of the other carbon atom. The remaining un-hybridized p orbitals of each carbon atom connect making a π *molecular bond* arranged perpendicular to the molecular plane to complete the carbon-carbon double bond.

Figure 36 Hybridization in Unsaturated Hydrocarbons

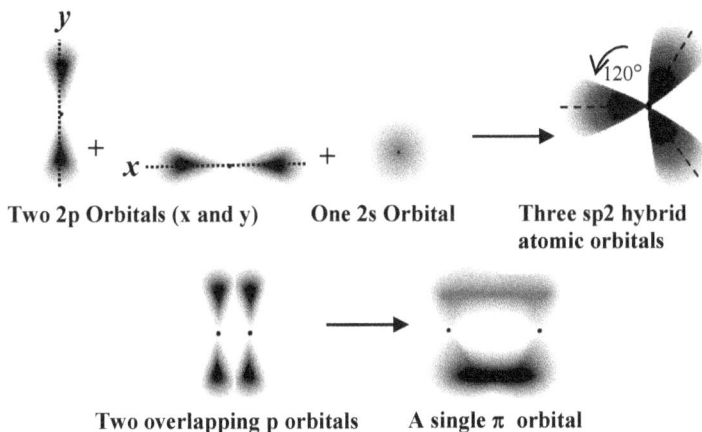

| Two 2p Orbitals (x and y) | One 2s Orbital | Three sp2 hybrid atomic orbitals |

Two overlapping p orbitals   A single π orbital

Molecular Orbitals of Ethylene

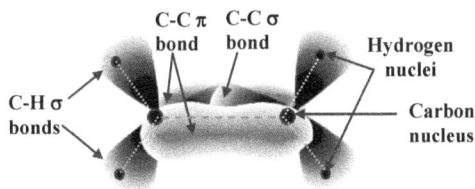

While the bond configurations in figure 36 represent the bond formation sequence, the final diagram does not give an accurate picture of ethylene's molecular structure. It implies the σ and π bonds making up the carbon-carbon double bond have different characteristics whereas in fact, once formed, they behave identically. Furthermore, while all six atoms of the ethylene molecule lie in a plane, the three bond angles are not 120° each as predicted; on either carbon atom the two =C—H bond angles measure122° and the —C—H bond angle 116°, fulfilling the 360° requirement:

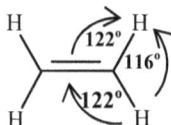

Like alkanes, alkenes also use simplified names. IUPAC nomenclature is similar to alkanes with the –ane ending of the corresponding alkane changed to –ene. However, –ylene is more commonly used in the names of the simpler C_1 to C_4 alkenes and their derivatives (e.g. ethylene for ethene, propylene for propene, and butylene for butene). When naming alkenes with branched chains, the longest sequence of atoms housing the double bond becomes the parent structure and numbering of individual carbon atoms begins at the end nearest the unsaturated—or double bond—linkage.

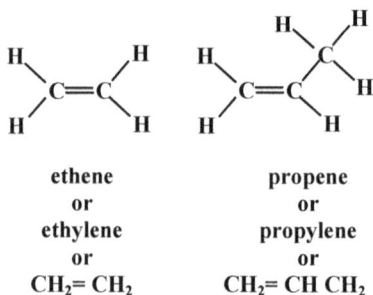

ethene
or
ethylene
or
$CH_2 = CH_2$

propene
or
propylene
or
$CH_2 = CH\ CH_2$

Alkenes of a given molecular formula have an even greater number of isomers than the corresponding hydrogen saturated alkanes. Obviously, no alkene fits n = 1 for the formula C_nH_{2n}, since

178

the double bond designating it an olefin needs two carbons. And, like alkanes, only one structure each fits n = 2 and 3, namely C_2H_4 ethene and C_3H_6 propene.

However for n = 4, four isomers have the formula C_4H_8 versus only two for alkanes. One source of additional isomers, compared to saturated hydrocarbons, stems from available options for location of the double bond. The double bond for C_4H_8 can be between the first and second carbons or between the interior carbons of the molecule. The first two C_4H_8 isomers, the olefin equivalents of normal butane and isobutane, namely butene and isobutene, have the double bond located between carbon atoms 1 and 2.

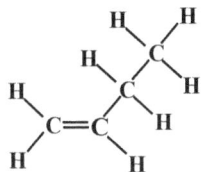

1-butene
$CH_2= CH\ CH_2\ CH_3$

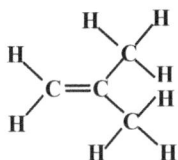

Isobutene
$(CH_3)_2CH_2= CHCH_3$

An additional source of alkene isomers results from the restriction of rotation around the bond axis due to the rigidity of the double bond. As shown below, we can draw two additional isomers when the double bond lies between carbons 2 and 3. Cis-butene has two methyl groups located on the same side of the double bond within molecule, while they lie on opposite sides in trans-butene.

Of course, cis and trans isomers don't exist for the alkenes with double bonds at the terminal (or end) carbons since the number 1 carbon atom has no hydrocarbon group to oppose the group on the number 2 carbon; only hydrogen atoms. Corresponding alkane isomers do not exist since unrestricted rotation about all carbon-carbon bonds makes these configurations equivalent.

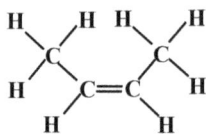 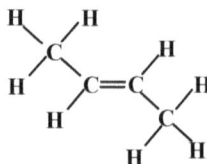

cis-butene

H_3C CH = CH CH_3

trans-butene

H_3C CH = CH CH_3

Like butanes, two possible locations of the double bond exist for pentenes, the C_5 olefin: and like the C_4s, they have one cis and one trans isomer. Note that the designation of carbon 1 or 2 is indifferent to numbering from either the left or the right side of the molecule. It always begins with the side nearest the double bond.

 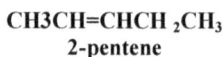

CH3CH2CH2CH=CH_2

1-pentene

CH3CH=CHCH$_2$CH$_3$

2-pentene

Hexenes have a second internal double bond site giving them an additional pair of cis-trans isomers; as the parent chain length increases, so does the number of possible double bond locations and isomers.

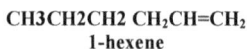 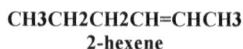 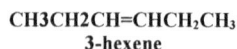

CH3CH2CH2 CH$_2$CH=CH$_2$

1-hexene

CH3CH2CH2CH=CHCH3

2-hexene

CH3CH2CH=CHCH$_2$CH$_3$

3-hexene

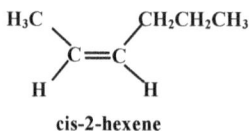 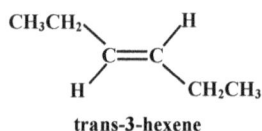

cis-2-hexene

trans-3-hexene

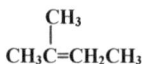 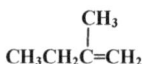 

CH$_3$CHCH=CH$_2$

3-methyl-1-butene

CH$_3$C=CH$_2$CH$_3$

2-methyl-2-butene

CH$_3$CH$_2$C=CH$_2$

2-methyl-1-butene

CH$_3$CH$_2$CHCH=CH$_2$

3-isopropyl-1pentene

180

Cyclic Hydrocarbons

The tetrahedral configuration of carbon's bonding electrons provides it with yet another source of flexibility for creating isomeric compounds. In addition to forming linear chains on hydrocarbon chains three or more carbons in length, two carbon atoms can each shed one of their hydrogen atoms and bond with the other to produce cyclic compounds. Cyclic hydrocarbons come in two varieties; the first, called *alicyclic*, are analogous to the straight chain alkanes and alkenes. Compounds of the second group have multiple double bonds and belong to a special group of unsaturates called *aromatics*.

Alicyclics

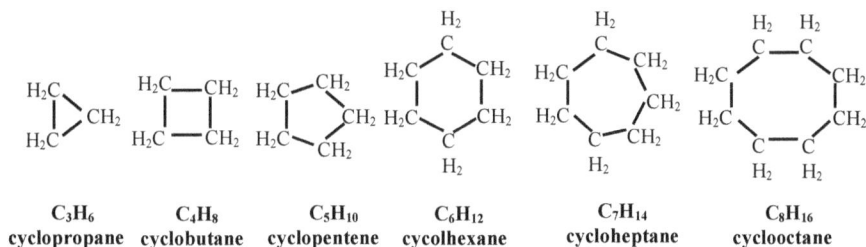

C_3H_6	C_4H_8	C_5H_{10}	C_6H_{12}	C_7H_{14}	C_8H_{16}
cyclopropane	cyclobutane	cyclopentene	cycolhexane	cycloheptane	cyclooctane

Names for the simple cyclic hydrocarbons derive by adding the prefix cyclo to their straight chain analogues as shown above for the series C_3 through C_8. The bond angles of cyclopropane, and to a lesser extent cyclobutane, fall well below the normal 109°28′ tetrahedral angle causing the bonds to be severely strained (called *steric strain*), making them unstable relative to their straight chain counterparts and the longer chain alicyclics. This instability however translates to higher reactivity making them particularly useful to the chemical industry.

Like their analogues, alicyclics can have carbon-carbon double bonds as well as attached hydrocarbon and functional groups. In molecules with more than one attached group, numbering begins with the carbon atom of the first functional group (as opposed to hydrocarbon group) with subsequent groups identified according to their attached site; e.g. methylcyclobutane and 1-amino-4-ethyl-1,3-cyclohexadiene below.

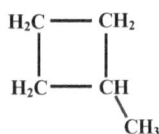

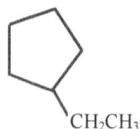

methylcyclobutane   cyclopentene   1-amino-4-ethyl-1,3-cyclohexadiene

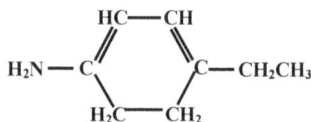

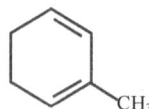

Since all carbon atoms must have four bonds, the number of hydrogen atoms attached at various sites on alicyclic rings becomes obvious, and in the interest of expediency are normally not shown. In the sampling of these structures below, a carbon atom sits at each corner of the simplified drawings and each uses two of its bonds to connect with adjacent carbon atoms. Therefore, since a saturated ring with no attached groups or double bonds (a cycloalkane) must have two hydrogen atoms at each corner to accommodate the other two bonds, we can ignore the hydrogens in the simplified structure. When the ring contains double bonds, the carbons at each end of the bond have only one hydrogen. Attached functional groups or alkyl groups also displace hydrogen atoms at the site of attachment (e.g. the ring carbon at the CH_3 group site in 2-methyl,1-3cyclohexadiene below has no attached hydrogen; two of its bonding electrons help form the double bond, one the single bond, and the other bonds with the CH_3).

cyclopropane   ethylcyclopentane   cyclohexenene   2-methyl,1-3cyclohexadiene

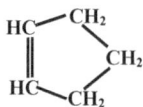

Cyclopentane is the most common naturally occurring alicyclic;

however, there is essentially no limit to alicyclycs' size. Chemists have synthesized compounds with rings containing more than thirty carbons.

Aromatics

Aromatics represent a very specialized category of cyclic hydrocarbons, limited to rings containing six carbon atoms. Benzene, discovered by Michael Faraday in 1825 is the simplest aromatic. It has the formula C_6H_6, which would imply an unsaturated hydrocarbon with multiple double or triple bonds. However, its chemical reactions differ greatly from those of known alkenes or *alkynes* (hydrocarbons with triple bonds). It was not until 1865 that Friedrich August Kekule' von Stradonitz proposed a cyclic structure like the one below, to account for some of the dissimilarities between benzene and other unsaturated compounds.

Benzene

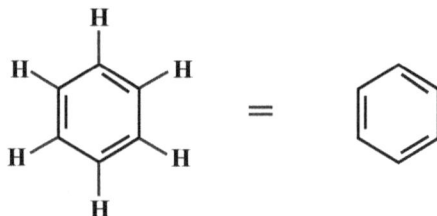

    Even though we continue to use Kekule's structure today for convenience, chemists recognized early on that it contained major inconsistencies. One of these involved location of the double bond in derivative compounds with groups attached to adjacent carbon atoms like 1,2-dimethyl benzene (orthoxylene) below. Literal interpretation of the Kekule' structure suggests there should be two orthoxylene isomers each with its own unique chemical properties. In one isomer, the two attached methyl groups would be located on either side of a double bond; and in the other, positioned across a single bond as shown below. Chemical researchers worked for many years attempting to either isolate the two compounds, or if only one of the structures proved feasible, establish the location of the double bond.

Orthoylene

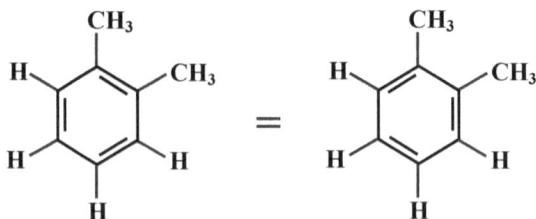

The puzzle did not resolve until much later, after development of more sophisticated analytical techniques such as X-ray diffraction and spectroscopic analysis. Researchers established there was indeed only one such compound and all the bonds within the ring were equivalent. Neither drawing rigorously defined the structure of orthoxylene, benzene, or any other aromatic compound.   Their description depended on a phenomenon called *resonance*.   Resonance usually occurs in molecules or ions that cannot be adequately described by any single structure. One of the simplest of these is the carbonate ion (CO_3=), an anion of carbonic acid (H_2CO_3). Carbonic acid, one of the acidic components of acid rain, forms by carbon dioxide reacting with water in the atmosphere.

$$CO_2 + H_2O \longrightarrow H_2CO_3$$

A carbonic acid molecule readily ionizes to form a carbonate ion (anion) with two negative charges and two singularly positive hydrogen ions (cations).

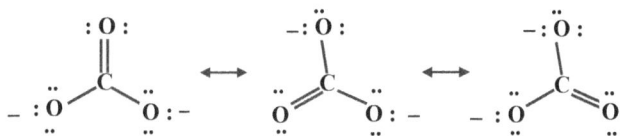

$$H_2CO_3 \longrightarrow CO_3= + 2H^+$$

The negative charge of an oxygenated ion typically resides on the oxygen atom due to its higher electron affinity compared to carbon. If the ion contains only 2 oxygen atoms, we would expect that one negative charge would be associated with each of the two single bonded oxygen atoms.   However, since the carbonate ion contains three oxygen atoms, each with an equal shot at hosting the two electrons imparting the negative charge, an apparent asymmetry results suggesting three different equally valid structures for the carbonate ion[1].

The double-headed arrow between the structures indicates all of the configurations have equal validity and suggests the combination more accurately describes the structures than either one. In fact, the carbonate ion is symmetrical and all three carbon-oxygen bonds equal. We can view the electrons comprising the double bond and the two negatively charge oxygen atoms as distributing among the three carbon-oxygen bonds by resonating between the structures.

We can describe the structures of benzene, orthoxylene, and other aromatics in the same manner. As is the case with alkenes, three of the four electrons in aromatic ring carbons occupy sp^2 orbitals; two of these form σ bonds with neighboring carbon atoms and the other with a hydrogen atom. The fourth electron remains in a p orbital and forms a π bond. However, in aromatics, instead of bonding only with the neighboring atom, the three π bonds distribute equally over all of the six carbon atoms in the ring to give the *resonance structure*. The diagram below illustrates the equivalency of the two possible Kekule' benzene drawings and another structure proposed in 1899 by German chemist Johannes Thiele. Chemists still use Thiele's representation today when they want to emphasize the resonating property of aromatics.

Numerous other representations, including the following, prove

---

[1] The electron configuration accounting for the negative charge on the oxygen atoms in the carbonate ion drawing is as follows: Oxygen has 6 outer shell, or valence electrons in its base state. Two of these combine with bonding electrons from the carbon atom to form the C=O bond, leaving 4 that are electronically neutral since they are balanced by positive charges in the oxygen nucleus. The 2 oxygen atoms attached to carbon by single bonds use one electron to form the single bond leaving 5 free electrons. Due to its higher electron affinity, the oxygen atoms take one electron from a bonding element such as hydrogen or sodium bringing its total to7 electrons, or a surplus of 1; hence the negative charge.

useful when we need to illustrate the three-dimensional aspects of benzene and more complex aromatics.

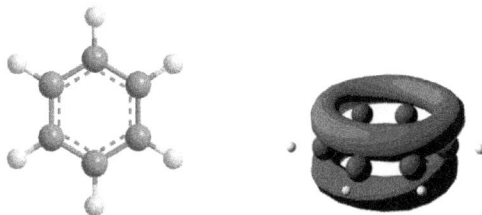

Heterocyclics, Molecules of Life

Heterocyclics, the core ingredients of all life processes, are organic ring compounds that contain other atoms (called *heteroatoms*) that have covalent characteristics such as oxygen, nitrogen, and sulfur within the ring structure. The four bases that make up the DNA of all living things are heterocyclic compounds that contain nitrogen as the heteroatom. Furthermore, deoxyribose, an oxygen containing heterocyclic sugar with an attached phosphate group, binds the bases

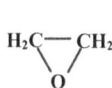

Ethylene oxide Propylene oxide   Trimethylene oxide   Tetrahydrofuran

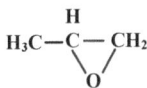

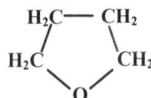

together in the ladder-like double helix matrix. The proteins that

187

make up our muscle, flesh, skin, and blood, as well as all plant tissue, are heterocyclic linkages, manufactured according to DNA blueprints.

Above are chemical structures of the first four saturated oxygen heterocycles. As in cyclopropane mentioned earlier, steric strain of ethylene oxide and propylene oxide make them valuable raw materials for production of chemical polymers used in the manufacture of plastics.

Below are five member ring unsaturated heterocycles of oxygen, nitrogen, and sulfur. Note that furan is a doubly unsaturated homologue of tetrahydrofuran above.

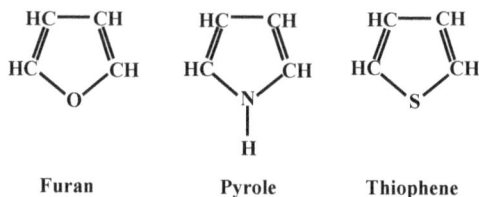

Furan   Pyrole   Thiophene

Of course, heterocyclics are not limited to five-member ring compounds nor to a single heteroatom in the ring. As illustrated below, they may contain two or more different heteroatoms as in thiazole and oxazole, or two or more of the same as in imidizole and pyrazole.

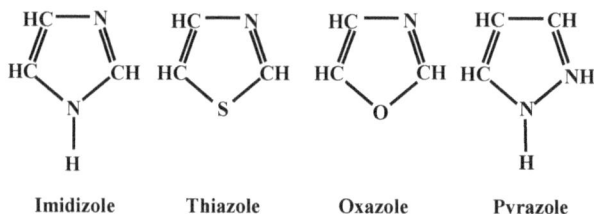

Imidizole   Thiazole   Oxazole   Pyrazole

Below, we show the six-member heterocyclic ring structure of a special heterocycle, pyrimidine, along with two of its derivative

188

compounds. As we shall see in the following chapter, cytosine and guanine are two of the four compounds that form the structure for the DNA of all living things. These same structures were likely involved in the origin of the first live creature on Earth, and its further evolvement into the millions of plant and animal species that inhabit our planet.

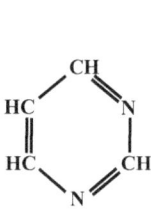

Pyrimidine

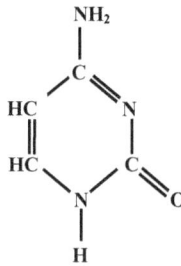

Cytocine

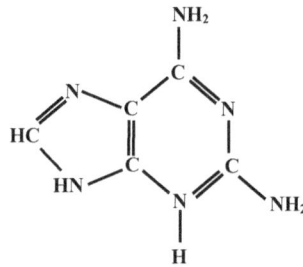

Guanine

189

5   Life

The great tragedy of science—the slaying of a beautiful hypothesis by an ugly fact.

T. H. Huxley, Biogenesis and Abiogenesis

Humble Birth

Probably the two most fundamental questions to challenge scientists and fascinate the rest of humankind since our species first developed the ability to consider such subjects are: 1) How did the world come into being; and 2) What is the origin of the creatures that inhabit it? These questions are so profound that all humans, from the early simple tribesmen to inhabitants of advanced civilizations, have hungered for answers—and there has always been someone willing to provide them. The answers have varied from one civilization to another but always based on the best—or the most believable—information available at the time. This still holds true

191

today; however, with the passage of time and the accumulation of knowledge, many aspects of these subjects are known. While science has solved many of the mysteries surrounding both questions, present knowledge fails to give a satisfactory answer when attempting to probe our ultimate origins. Just as relativity and quantum mechanics—the sciences governing the very large and the very small realms of our universe—fail to address the conditions of the Universe at its moment of birth (the period prior to 10^{-43} seconds from the Big Bang), science has yet to describe the synthesis of the first living entity. In the case of the evolving universe, science has provided consistent and provable information describing events beginning with the birth of the Universe 13.7 billion years ago through the creation of galaxies, our solar system, and our planet along with its oceans and atmosphere. In the case of living creatures, there is incontrovertible evidence that all of Earth's complex life forms evolved from a single microscopic being that lived more than 3.5 billion years ago. However, in both cases, researchers have met dead ends when attempting to probe the very beginning of these two phenomena. Just as the ability to describe the subatomic Universe has thus far eluded science, so too has the search to explain the first moment of life.

What we know with far more certainty however, is the mechanism that led from this first tiny organism to all living things that followed. The source of much of the evidence lies in the great quantity of fossilized remains of now extinct species that preceded us. We have voluminous data for the more recent species of our heritage—those living since the beginning of the Cambrian era one-half billion years ago. We credit this primarily to skeletal structures that fossilized well, and larger body sizes that make the fossils both easier to find and withstand the ravages of a turbulent Earth. Animals of the period prior to the Cambrian (the Precambrian) however, were considerably smaller and many did not leave hard evidence of their existence since they had not yet developed skeletal structures to fossilize.

Hundreds of millions of years of such fossilized evidence, combined with modern chemical and other scientific information, leads to the inescapable conclusion that all life forms change with time. Modern humans have been on Earth for about two hundred thousand years and have undergone significant changes in brain and body size during this brief ecological period. During the previous six million years, numerous human-like creatures preceded us, each different from her predecessor and exhibiting even less resemblance to modern humans. Earlier still, our ancestors and chimpanzee ancestors were the same. Living things have been around for three and a half billion years; however, the further back we go, the less the inhabitants resemble current ones. Finally, at a point three and a half billion years ago—just a half billion years after the formation of our molten planet—the only life forms were microscopic single cell entities, the creatures from which the rest evolved.

From this, we know that all forms of life change. DNA mutations occur in the genetic structure of living things that in some cases make the recipient of the mutation better adapted to its environment. We also know these newly mutated genes pass on to offspring, making them better adapted with a better chance of reaching childbearing age, thereby spreading the mutated gene throughout the population. The accumulation of such beneficial mutations over thousands to millions of generations leads to the creation of new and frequently better-adapted species. This simply stated pattern was responsible for the evolvement from that tiny unknown first life to the complex biosphere of Earth today. However, the origin of that first being remains cloaked in mystery. In our search for this final and critical piece of our ancestral puzzle, we can take heart in advances made in the last two hundred years. Within this mere tick of the human historical clock, we have made phenomenal progress toward this goal. However, at this early stage in our quest to understand the birth of that first live being, the best we can hope to do is propose scenarios consistent with the meager facts as to how it might have happened.

Until the middle of the seventeenth century, the prevailing view regarding the beginning of life on Earth was that God had created man and other complex creatures, while less significant life forms simply emerged spontaneously from earth or from decayed matter. Sixth century BCE philosopher Anaximander may have been the first to suggest that life originated from sources other than divine intervention in claiming that living creatures arose from the action of the Sun on wet substances; that eels and other aquatic forms spontaneously emerged from "lifeless matter". He may have also been the first to hint at evolution when he stated that humans were originally a kind of fish, and took a long time to mature to their present state.

Aristotle provided a more sophisticated view of the emergence of life two hundred years later in the fourth century BCE. He proposed that all life consisted of four terrestrial elements: earth, air, fire and water, along with a fifth essence called the *quintessence*, or *ether* that existed only in the heavens. Aristotle thought that all living organisms were mixtures of these factors and a nonphysical force he called "*anima*" the Latin word for "soul". All living things possessed aspects of this soul such as growth, motion, and sensation; however, only humans possessed the ability to reason. Aristotle believed that more advanced animals "...spring from parents according to their kind..." while insects and the like simply emerge spontaneously from decaying vegetation: aphids from dew deposited on plants, maggots and flies from spoiled flesh, and mice from stored grain. Thus these beliefs remained for nearly two millennia.

In 1648, Jan Baptista shocked the moral leadership of the world and initiated two centuries of heated debate, by proclaiming, "All life is chemistry." Nevertheless, the conventional view held sway since no one presented contradictory evidence strong enough to overcome centuries of established mythology. The nineteenth century however, brought three major breakthroughs that necessitated a rethinking of the origins of life.

In the first of these, Louis Pasteur proved that even the lowliest of bacterial life is born of parents much like itself. Pasteur's work discredited the concept of spontaneous generation, but did not address the essential question: How did the first generation of this species originate?

The second nineteenth century discovery, Charles Darwin's theory of natural selection, provided an answer. Darwin reasoned that some differences between members of a group are "heritable", and that subjected to a changing environment, those possessing the most adaptable inherited traits will better adjust to the new conditions, and therefore have more surviving offspring. As a result, the percentage of these traits will be higher in members of the next generation and will continue to increase in succeeding generations, resulting in a species with "improved" capacity to cope with the changed environment. The inescapable implication of natural selection is that through the continued accumulation and concentrating of such traits, combined with ever-changing environments, simple organisms evolve into more complex ones. This line of reasoning ultimately concludes that all life forms existing today could have evolved from a single simple organism.

The primary thrust of origin-of-life research in the twentieth century has been to shed light on how, in the absence of divine intervention, small molecules could spontaneously coalesce to become life. All of this however first begs the question: "What is life; what characteristics does an organism need in order to be defined as Alive?" One definition of life is anything that can reproduce and sustain itself by utilizing the resources available within its environment. Caus Emmeche and Alleta d'A. Gelin led a group of scientists who compiled a more specific list of the minimum requirements of a living being:

1) Auto conservation: The main function of every living organism is making sure that it can continue its existence.

2) Auto reproduction: Any living system can reproduce or proceeds from a reproduction.

3) Storage of Information: Each organism contains genetic information. This appears stored in DNA, and is read and translated by proteins according to a universal genetic code, which is common to all creatures.

4) Breathing-fermentation: Every living being must have a metabolism that will transform energy and matter taken from the environment

into energy and compounds that can be used by the different parts of the living organism.

5) Stability: Through the creation and control of its own internal environment, all creatures remain stable in front of the perturbations of the external world.

6) Control: The distinct parts of an organism contribute to the survival of a group and, therefore, to the conservation of its identity.

7) Evolution: The mutations in the hereditary material and natural selection permit the perfection, adaptation and complexity of living beings. For many, life is a mere product of evolution.

8) Death: It is determined by the genes and aggressions from the exterior, it marks the final phase of all living creatures.

While many continue to follow a faith-based view that the creation of man was a complete and rather instantaneous event that happened a few thousand years ago, scientists and other scholars, as well as a substantial portion of the informed public, have accepted the theory that complex life did evolve from very simple organisms. By the middle of the twentieth century, people believed that the first such organism lived about 540 million years ago at the start of the Cambrian period (see figure 49). This was an understandable assumption since scientists had found many fossilized shells and skeletons dating back to then, but at that point, hard evidence of earlier life dropped off sharply. However, in the 1950s paleontologists began to find very small fossils that predated the Cambrian period by 250 million years. They had not found such fossils earlier due to their small size, and the fact that they came from soft-bodied creatures without shells or skeletons meant they did not fossilize well. However, these discoveries of early animal life did not signal the beginning of life on Earth.

Over the following decades, the knowledge that such fossils existed motivated a host of researchers to take up the search for missing evolutionary links. Success came rapidly; paleontologists began uncovering microfossils of multicellular algae-like material dating back to 1.5 billion years. In 1954, researchers from the University of Wisconsin discovered primitive multicellular 2 billion year old organic matter on the north shore of Lake Superior. As techniques for identifying these life forms improved, the ages of discovered microfossils continued to move backward in time. As recently as April 2004, a team led by Norwegian scientist Harald Furnes found evidence of 3.5 billion year old bacteria etched in ancient volcanic lava in the Barberton Greenstone Belt near, Johannesburg, South Africa. The original site had formed beneath the ocean floor, but over the eons has uplifted, making it accessible. The deep subterranean location of the find provided a host of factors favorable to the origin of life. The site not only had access to seawater, hydrothermal vents, and a wide range of catalysts needed for the synthesis of life, the deep sea environment provided needed protection from the regular bombardment by large meteorites that characterized the period.

These factors suggest the Barberton fossils might have been among the earliest living creatures; however, they were not the first. It is not likely that we will ever find fossilized evidence leading us all the way back to the first creature that meets the Emmeche and Gelin criteria for life. Even if these early creatures were large enough to leave detectable fossils, it is likely their habitats were limited to a very small region of the ocean floor for a tens of millions of years, during which many species were evolving and becoming extinct. That such tiny fossils, or accumulations of them, would survive the turbulence of the evolving Earth's land masses, and at the same time be numerous enough for us to discover them, seems unlikely. Furthermore, the brief generation spans of these simple organisms would correspond to rapid evolution of species; therefore, many links between the oldest fossilized remains and that first live being will likely remain missing.

There are two schools of thought regarding the rarity of life and its beginning on Earth. The first suggests that the probability of chemical reactions needed to produce a living entity all coming together at one time is so extremely low that the event could have happened only once. This scenario holds that the Earth would be no more likely to have been the site of this miraculous event than many of the other trillions of planets in the universe. Therefore, this original and singular miracle of life likely occurred outside our solar system and reached Earth as microscopic passengers aboard meteorites born from the destruction of their parent solar systems. The second, while agreeing the odds against life "springing-up" are indeed enormous, counters that the opportunities in terms of time and locations where the event could happen are even greater; therefore life must have sprung forth many times throughout the myriad galaxies, one of these on our planet.

In private correspondence, Darwin expressed his feeling that the birth of this first living creature, while through the direction of God, did indeed occur on Earth via chemical processes "… in some warm little pond with all sorts of ammonia and phosphoric salts, light, heat, electricity, etc. present." On the other hand, being a religious man, he bowed to church pressures writing in the final paragraph of the Origin of the Species, "…The Creator originally breathed life into a few forms or into one." The rest however was evolution: "…There is grandeur in this view of life. Whilst this planet has gone cycling in accordance to the fixed law of gravity, from so simple a beginning endless forms most beautiful and most wonderful have been and are being evolved."

A convincing argument for the earthly origin of life is being made by a group of scientists who integrate biochemical information with aspects of the mathematical discipline known as *chaos theory*, a study of systems that exhibit chaotic behavior. Chaotic systems result from conditions where nonlinear factors (e.g. their mathematical description requires exponential variables, such as x^2) are at work that cause each point in the system to be closely approximated by other points that have different future trajectories (e.g. can lead the system to different results). These factors impact behavioral aspects of the systems, causing extreme sensitivity to initial conditions. This sensitivity permits an arbitrarily small change in the initial state of a system to lead to a vastly different future outcome, making the underlying process appear random, even though the system is well defined and contains no random parameters. The classical example of the sensitivity to the initial conditions factor of chaos is the well-known butterfly effect, which suggests that the tiny changes in the atmosphere initiated by the flapping of a butterfly's wings could cause a tornado. This of course implies that the wing-flapping could set off a chain of events, each of which would continue to magnify the effect, eventually resulting in a large-scale phenomenon such as a tornado.

The first true experimenter in chaos theory was in fact a meteorologist named Edward Lorenz. Lorenz came to a surprising conclusion through work on the problem of weather prediction in the early 1960s. He was experimenting with long-term weather prediction using a computer programmed with a set of twelve equations designed to predict weather factors such as atmospheric pressure and wind direction for various scenarios. One day, he decided to rerun one of his computer runs, but rather than starting from the beginning, in order to save time on the notoriously slow machines of the day, he began the calculation in the middle of the run instead of at the beginning. When the run finished, Lorenz was astonished to find that the sequence had produced an entirely different result.

As shown in figure 37, the pattern of the second run followed that of the original for some time, but it then gradually began to diverge, finally ending up with a wildly different outcome. Lorenz soon figured out what had happened. Normally the computer automatically rounded numbers stored in the program to six digits, whereas in order to save paper when printing the results, he had rounded them off to three digits. The results both amazed and confused Lorenz. By rounding a six decimal place number like 0.506127, to three places at 0.506, a difference of only 0.000127, he had obtained a totally different outcome.

On further analysis of the two runs, he found the tiny difference in the starting conditions, amounting to a mere 0.025%, was insignificant with respect to the accuracy of the information represented, yet it caused the difference in the results to double every four hours of the simulation. The discrepancy, though barely noticeable at first, grew exponentially into a completely different result within a surprisingly short period of time—a classical example of nonlinear growth—and a prerequisite for chaotic behavior. Since weather forecasters cannot estimate data input to the program to 0.025% accuracy, Lorenz—not appreciating the far-reaching implications of his result—simply concluded that accurate weather prediction beyond a period of a few days is impossible. However, it turned out his discovery had significance well beyond weather prediction and served as the beginning of what later became chaos theory.

Figure 37 Weather Prediction Divergences

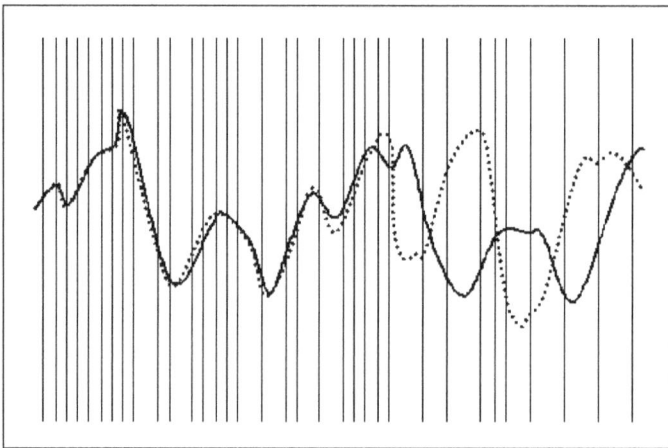

In addition to weather related phenomena, there are numerous other paths that lead to chaotic behavior, including branching of trees, growth of species populations, the atmosphere, solar system, plate tectonics, and the effects of turbulence on fluids. However, application of chaos theory concepts is not limited to scientific areas. The arts use chaos effectively as well. Tone-deaf musicians have created pleasant music through chaos principles, and by using relatively simple equations, anyone can design a beautiful and realistic tree. The one common theme to these systems however is self-similarity; that is, one or more of a system's parts has the same shape as the whole. Even so, the term chaos is somewhat misleading in describing these phenomena, since it implies completely random, complex, and unexplainable behavior. In fact, however, although systems described by the theory appear to be in a state of disorder, they are deterministic and randomness is not involved. The goal of chaos theory is to discover the underlying order within the apparent disorder and randomness of such systems.

All systems that exhibit chaotic behavior have the following in common:

They are *fractal*—a term derived from the word fraction and has come to be associated with images that display the attribute of *self-similarity*.

They are *nonlinear*—a property that leads to the exponential growth necessary for chaos.

They have an *attractor*—a mechanism that provides the feedback through which nonlinearity is manifested.

The *Koch curve*, named in 1904 for its originator, Swedish mathematician André Koch, supplies the most often used example of the fractal nature of chaos due to its simplicity and the interesting pattern that results. Although the epitome of simplicity, creation of a Koch curve can be extremely laborious (in fact infinitely laborious) since there is no end to the number of repetitions involved in its construction.

Figure 38 illustrates the steps in creating a Koch curve. In step 1, you draw a straight line. Step 2, divide the line into three equal parts, then attach an equilateral triangle to the middle third of the line and erase the base of the triangle that was previously the center third of the original line. In step 3, add four more equilateral triangles, one in the middle of each straight line created in step 2 and erase the bases of each of these as well. Keep adding new triangles and erasing bases ad infinitum, and the Koch curve emerges. The self-similarity, or fractal aspect of the Koch curve, is readily seen via an enlargement of any segment since it will look exactly the same as the original.

Figure 38 Koch Curve

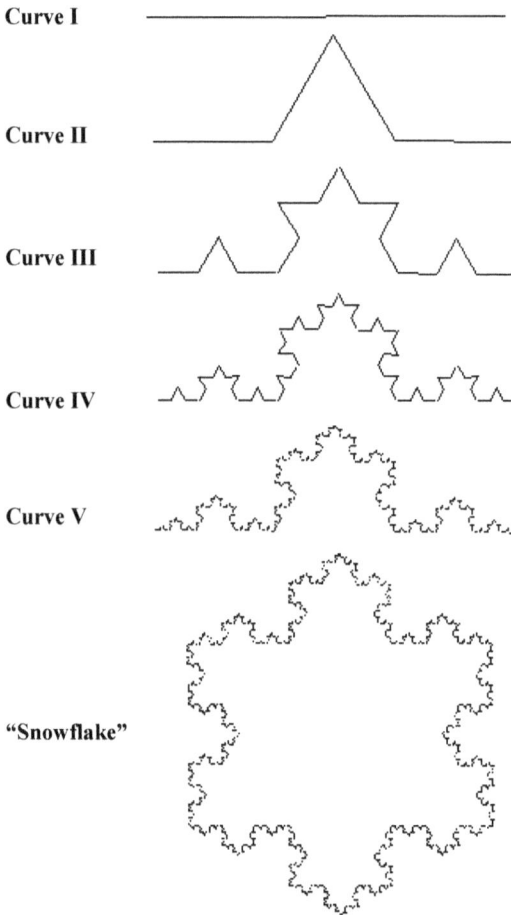

The bottom drawing of figure 38 is the more popular variation of the Koch curve, built using an equilateral triangle as the starting point and treating each side in the same manner as the straight line of curve

I. Addition of the first set of three triangles produces a six-pointed Star of David that morphs into the perfect little snowflake after repeating the process only four times. Each time we repeat this same process, we have performed iteration. In mathematics, iteration is the process in which an equation uses the result from one calculation as a factor used to calculate a subsequent result.

The Koch curve however is more than just an attractive curiosity; it also presents an interesting paradox. As we add new triangles to the figure, the line gets longer, and since the process of adding triangles and erasing bases can continue indefinitely, the Koch curve produces a line of infinite length. However, the end points of the figure 38 "snowflake" remain the same as the original starting line and covers an area less than that of a circle drawn around the original triangle; essentially producing a line of infinite length within a finite area.

To cope with this paradox, fractal dimensions were invented. To get a feel for fractal dimensions, first consider conventional geometric dimensions. A square has two dimensions and as such can occupy space while a line is one-dimensional and cannot. However, the infinite number of angles contained in a Koch curve gives it a crinkly aspect, requiring dimensional characteristics—with respect to space occupation—somewhere between a line and a square. It turns out that the Koch curve has a fractal dimension equal to 1.2619, a number that makes intuitive sense, albeit in a convoluted sort of way, since it lies between 2 and 1—the dimensions of a square and a straight line respectively. Furthermore, since it is somewhat more efficient at occupying space than a line but substantially less than a square, its value is nearer to 1 than to 2.

The fractal dimensional value of 1.2619 results from the self-similarity aspect of the Koch curve. A single segment of the Koch curve, represented by curve II of figure 38 is built from four equal length lines. If you multiply the size of any of the 4 self-similar segments of the curve by a factor of 3, you get a segment equal in size to the comparable segment from the previous iteration. For example, increasing the right side of curve III in figure 38 by a factor of 3 gives a structure identical to that of curve II. Now if we look carefully at this operation, what we have done is take one-fourth (1/4) of curve

III and scaled it by a factor of 3, or put another way, we divided the length by 4 and multiplied by 3. Using this relationship, we can calculate the fractal dimension of the Koch curve the same way we determine dimensions of conventional figures such as lines, squares, and cubes. A 3-inch one-dimensional line has a numerical value of 3^1 = 3, a 3-inch two-dimensional square is 3^2 = 9, and a 3-dimensional cube is 3^3 = 27. Just as the case with conventional figures, the fractal dimension of the Koch curve is the exponential value (n) to which the number 3 is raised to get 4; i.e. 3^n = 4. Solving for n to four decimal places gives $3^{1.2619}$ = 4, a fractal dimension of 1.2619.

Figure 39 Feigenbaum Diagram

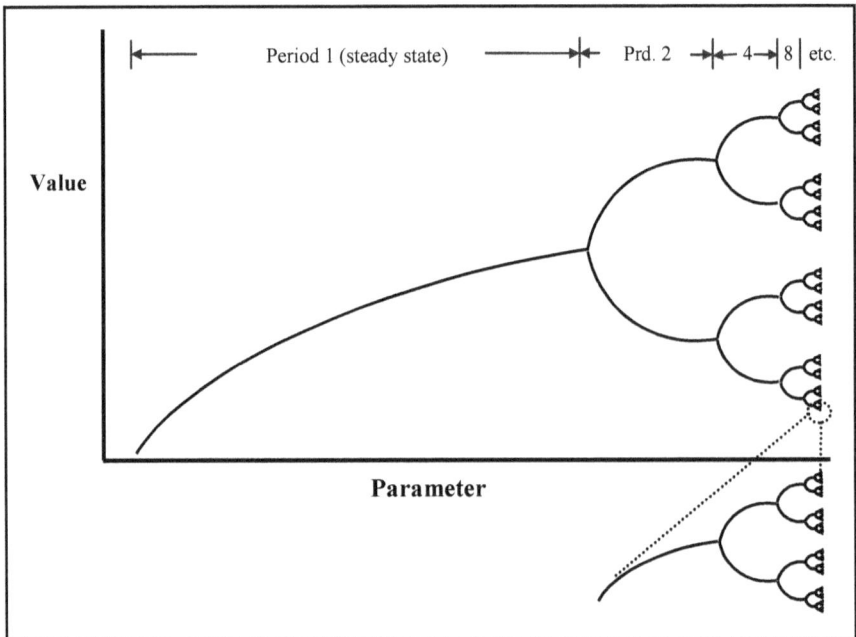

Measuring the length of coastlines with their many bays and other irregularities presents a similar phenomenon. However, coastline measurements taken from map information will invariably produce a greatly underestimated value since mapmakers must ignore the myriad minor bays and other shoreline irregularities due to their size and abundance. A walk around a costal bay would reveal many minor bays too small to include on the map, and countless more so tiny we would simply step across them. An even more intimate observation would reveal microscopic bays formed by individual grains of sand and if one cared to delve even deeper, molecular irregularities would stretch the length by additional orders of magnitude. No matter how much we magnify a coastline, further magnification will reveal more bays.

One of the most interesting mathematical examples of chaos, and fortunately one of the easiest to describe, is the period doubling route, the best known of which is the *logistics equation*—also called the population equation. The main aspect of the logistics equation, one that leads to chaos through period doubling, is the fact that it is self-referential; like the Koch snowflake, it feeds back on itself. The figure 39 Feigenbaum diagram, named for Mitchell Feigenbaum, illustrates the period doubling aspects of chaos theory. In 1945, while working at the Los Alamos Laboratories in New Mexico, Feigenbaum showed that period doubling is common to many systems—all of which result from an iterative process that feeds back on itself. These systems include such varied phenomena as electrical circuit oscillators, business cycles within the economy, and animal population growth. The most fascinating aspect of Feigenbaum's discovery is that all such self-referential period doubling phenomena do not just approach chaos in a similar manner but proceed in precisely the same manner.

The process begins in what we call a steady state, or period 1, characterized by a series of single values representing the system under consideration, that vary as another parameter is increased. Although a single number identifies each point on the steady state portion of the curve, it requires numerous calculations to reach it; the result from one calculation becomes a factor in the equation used to obtain the next calculation. Typically, in period doubling, results from the individual calculations vary wildly at first but in time begin to merge, eventually producing a single number that does not change with further iterations. When the variable parameter (the x-axis) reaches some critical value, the equation moves into period 2 where, as in period 1, the results fluctuate wildly at first but eventually settle down;

however, this time a single answer to the equation never emerges. At each point on the curve, the equation has bifurcated, alternating between two stable values and continues to do so regardless of how long the iterative process continues. The equation has entered the period doubling phase.

As the value of the parameter increases further, the equation enters the period 4 range. In this stage, each of the two stable values bifurcate as well, settling down to 4 stable values, while still further increases give 8, then 16, 32 and so on. The most fascinating aspect of the logistics equation, and the one responsible for sending the system into the chaotic region, then begins to emerge. Each successive bifurcation occurs with a smaller and smaller increase in the variable parameter. In fact, the bifurcation interval not only gets smaller but does so according to a constant factor; each successive bifurcation—or period doubling—results from a parameter increase precisely 4.669 times smaller than that of the previous doubling, a figure known as *Feigenbaum's number.*

The logistics equation is best known for its application in estimating fluctuations in animal populations from one generation to the next, and is in fact more often referred to as the *population growth equation.* We show it below in its most basic form; however, when used in practice, it usually takes a more robust form with additional terms that reflect factors such as predator/prey relationships and competition with other species for food supplies.

Logistics (or population) equation, $x_n = Rx(1 - x)$
  Where:
  x_n = population of next generation
  R = birth rate
  x = present population

A colony of mites confined to a small area, such as a single azalea bush, provides a simple example of the transition from order to chaos through the period doubling property of the population equation. Each year the adults lay their eggs, but do not survive the ensuing winter. Then the following spring, the eggs hatch to become the next generation. The birth rate R, expressed as offspring per individual per generation, is a function of the number of eggs produced by each individual mite, combined with a factor to account for the number that actually hatch. Since all chaotic systems result from conditions involving *fractals*—a term derived from the word fraction—we need to express the present population, x as a fraction. We do this through a procedure called *normalization* by first estimating the maximum number of mites that could possibly live on the azalea. Then we divide the number of mites in the present generation by this maximum value, thereby expressing the present population as a fractional value between 0 and 1.

Now to get the starting point for the size of the next generation, we multiply the fractional population value x by the birth rate R to get Rx, which in our example is the number of eggs hatched for the next generation. However, this alone does not take into account the mites that do not survive to adulthood due to competition for food. In a large population (e.g. one represented by x = 0.9), the azalea plant will be nearly saturated with infant mites, causing a high death rate due to starvation before having the opportunity to lay the eggs that would constitute the next generation. Conversely, a low value such as 0.1 would mean plenty of food for all and a high survival rate. To account for the premature death rate in the logistics equation, we multiply Rx by $(1 - x)$ to get $Rx - Rx^2$. Obviously, a high population value x means a low value of $(1 - x)$ which will offset the high value that x imparts to the Rx factor of the equation, whereas the reverse will occur for a low x.

For a given birth rate R, any single value of x will provide a legitimate value of the next generation population, x_n. However, since the answer is sensitive to the magnitude of the value for R as well as the present population x, the result from a single calculation does not answer the question: At what value of x would one expect the population to be stable? Between the extremes, we need to iterate— repeatedly calculating the equation, $x_n = Rx(1 - x)$, then substituting the result from one generation into the calculation for the next—until a value for x is reached that no longer changes in subsequent calculations. In other words, it is stable. During the process of

207

calculation, the value for x_n fluctuates at first but gradually moves toward a stable value. The iterative aspect of the calculation provides the feedback criteria necessary for chaotic behavior, whereas the x^2 factor obtained by multiplying Rx by $(1 - x)$ satisfies the nonlinearity requirement, and the fractional values of x supplies the fractal property.

Figure 40 Bifurcation of Population Growth

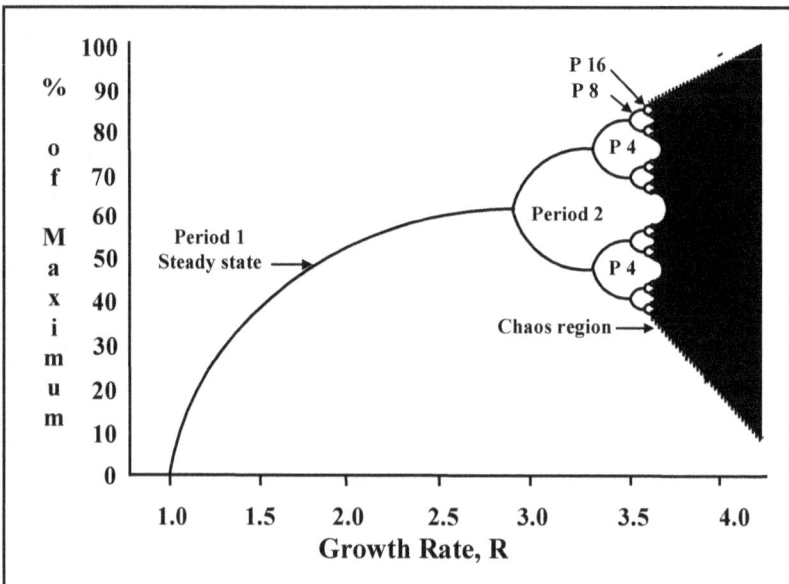

As shown in figure 40, for a birth rate R of less than 1, the population fails to reproduce itself and becomes extinct, regardless of the starting population. This is true since each adult on average will leave less than 1 offspring. At R-values above 1, an attractor kicks in. The attractor concept, first introduced by 19th-Century mathematician Jules-Henri Poincaré, represents the situation where an equation yields a single, constant solution for an entire range of numerical inputs. It contains a region of values that, given a sufficient number of iterations, will "attract" a range of numeric inputs into the result to affect a stable solution.

The concept transcends mathematical applications, and may even explain how brains function in memory processes. Our brains contain billions of neural networks representing all of the things we have learned and experienced. Attractors in neural networks provide an efficient storage system that makes it unnecessary for all events to be stored as separate entities in order for us to recall them. The networks store individual elements such as sights, sounds, and feelings in different regions of the brain associated with each particular sense.

Since all neural networks have similar activation patterns that bring about an attractor state, the system can see the commonality of the separately stored elements and produce the memory of a complete event when only a single element (e.g. an odor conspicuous to the event) has been triggered. You have probably experienced this process at some time when trying to recall the name of a casual acquaintance; the person suddenly enters the room, and upon seeing this singular face, you immediately know said person's name. Connecting the face with the name involves the use of information located in different segments of the brain, the temporal lobe for name and the cortical for visual information. The small subset of neurons from one brain area turns on the larger attractor fed by signals from a host of memories stored in various regions of the brain; and suddenly not only the name but a stream of other information is at your disposal.

In the population equation, the attractor is active within the range of R-values between 1 and 3. For birth rates above 1, and any starting population x greater than 0, values for the new population x_n settle into a steady state. At once, the rate of increase rises rapidly, reaching 0.17 at $R = 1.20$ and 0.29 at 1.40 (17% and 29% of maximum population respectively). As we further increase the growth rate, the population continues to grow but at a steadily decreasing rate, finally leveling at 0.66 (shown as %) when R becomes very close but not quite equal to 3.0.

Once the birth rate exceeds 3.0, first indications of chaos begin to emerge. Instead of stabilizing at a constant population, it begins to vacillate from one generation to the next, one settling well above the 0.66 value and the other below. At $R = 3.1$, the values alternate between 0.56 and 0.76, and diverge gradually to 0.45 and 0.84 at R = 3.40. This scenario is not too surprising and is consistent with expected results for a system involving limited resources coupled with rapid population growth, which in turn leads to over-population. An over-populated generation will deplete the food supply, causing many of its members to starve before reproducing. The next generation, being smaller and therefore owning a plentiful food supply, will all survive and produce many offspring for the following generation, and the cycle will continue to repeat as long as the birth rate remains constant.

As R increases further, the chaotic character of the equation begins to manifest in accordance with Feigenbaum's number (4.669). Increasing R by ever-smaller increments causes continued doubling of the number of attractors and population variants. At R = 3.45, each of the two branches of the curve bifurcate, leading to populations that oscillate between four different levels, each with highs above the previous sequence and lows beneath the previous lows. When increased to R = 3.54, stable populations alternate in the same manner on an eight-generation pattern; and at 3.56, sixteen.

The bifurcations continue doubling with ever-smaller increases in the birth rate until chaos emerges and population prediction becomes impossible. When R = 3.5699, the number of bifurcations in the population curve reaches infinity. However, this does not signal an end to the process. There are still ranges of values for R within the chaotic region where order reappears. One such region lies between R = 3.8 and 3.9 (indicated by the first white band in figure 40) where, just as in the region below R = 3, a single line represents population.

The chemical aspect of the effort to explain the earthly origin of life almost certainly involves *autocatalysis*, or more specifically, *autocatalytic loops*. Autocatalytic loops are essentially attractor or feedback phenomena wherein a molecule produced in a reaction also catalyzes the reaction. Equations that describe reactions involving feedback are necessarily nonlinear; the rate of reaction is proportional to the square of the concentration of the product.

Stuart Kauffman, a theoretical biologist at the Santa Fe Institute in New Mexico, first connected the world of chaos theory with autocatalysis in life synthesis. Kauffman worked from the belief that at its most basic level, a living organism is merely a system of chemicals with the capacity to catalyze its own reproduction, and as such, is a classic example of a chaotic process operating through a feedback mechanism. In studying systems where different autocatalytic sets compete for the same raw materials, Kauffman found that autocatalysis tended to happen at what he termed "the edge of chaos" and concluded that life could have sprung from this process rather than through the chance of the instantaneous gathering of just the right combination of ingredients. Instead of a wedding of all the elements of life through one highly improbable accident, Kauffman believed that with the essential ingredients in place, the autocatalytic process functioned as a kind of search engine, systematically assembling myriads of chemical combinations until eventually the right formula emerged, and so life emerged. In fact Kauffman, and other scientists involved, concluded that given the presence of the basic chemical construction materials of living things, and operating through autocatalytic networks, the evolution of life was not only possible, but given the proper conditions and adequate time, was inevitable.

Work in the early 1950s by Alan Turing, the multi-disciplined English scientist, provided one of the earliest insights into the feasibility of life emerging from an autocatalytic system. Turing is best known for breaking the German Enigma Code that led to the Allies intercepting a wealth of critical intelligence regarding Nazi military operations and hastened the end of World War II. While his interests did not involve questions regarding the origin of life, they did initiate a chain of discoveries that eventually led to connecting chaos theory concepts with chemistry that culminated in Kaufman's hypothesis. Through mathematical principles discovered while pursuing these efforts, related to a phenomenon called *symmetry breaking*, Turing found it possible for complex systems to "learn" through experience gained within the system itself.

In a paper published in 1936, "On Computable Numbers," Turing described what he considered a hypothetical "imaginary device" involving self-organizing complexes. He visualized a machine that he could feed a strip of paper—also imaginary since in terms of calculation techniques available at the time, it would need an almost infinitely long piece of paper—divided into squares on which he would write numbers or symbols. These included symbols for add, subtract, multiply or divide, which the machine could read, erase, change or rewrite as needed. The machine could interpret numbers or symbols within these squares as instructions to move back and forth along the tape—performing calculations as it did so; then write the results into other squares for use in subsequent calculations. Alan Turing foresaw that by using this machine in this manner, writing the answers, then recalculating using the revised answer, the machine would have the ability to perform any calculation if given the proper instructions. The great mathematician's imaginary invention, known today as the universal Turing machine, required only the simplest mathematics but nevertheless—as you might have already guessed—became the inspiration for the modern computer.

Following the end of WW II, Turing began to study the application of chaotic systems to morphogenesis, the science of biological growth processes. Turing's specific interest concerned the question of how a fully structured embryo could develop out of the featureless conglomeration of cells called the *blastocyst* that result from egg fertilization. Turing believed the process had to involve *symmetry breaking*, a phenomenon related to several rather diverse areas of science that range from magnetism to particle physics. However, in each case a transition from an orderly state to a chaotic state occurs. Symmetry breaking occurs during phase transitions such as when a magnetized piece if iron is heated to its Curie point (for iron, 760°C), where heat energy overcomes the orderly alignment of dipoles responsible for the magnetic field. Iron loses its magnetic properties until it again cools below the Curie point where the north and south ends of the individual dipoles realign and magnetism resumes.

In view of these principles, Turing proposed that certain selected chemicals could diffuse through a mixture of other chemicals initiating patterns of localized concentration variants, effectively breaking the symmetry of an otherwise uniform mixture. He further suggested that a similar process might be involved in breaking the symmetry of the featureless blastocyst to form and develop a highly structured embryo, thus creating pattern where none had existed before.

Although catalysts were the heart of Turing's proposal, not being a chemist, he did not specify the individual chemicals and catalysts to achieve the results he envisioned, leaving that chore to laboratory grunts. Instead, he simply referred to his materials as chemical A, B, C, etc. and assigned them properties such as reactant, product, catalyst, inhibitor, and autocatalyst. Reactants are of course the starting materials consumed to form the desired products of a reaction. Catalysts are chemicals not consumed in chemical reactions, but promote or speed up the reaction rates of other chemicals. Inhibitors on the other hand, discourage or slow reactions. Some chemicals called *autocatalysts* even serve to catalyze reactions that produce themselves. Autocatalysts drive the process since they set up a condition in which the more a particular chemical is produced, the faster the reaction proceeds to make even more of itself. This makes the process nonlinear and introduces the feedback aspect needed for chaotic behavior.

Another factor in Turing's proposal is the presence of chemicals that behave as catalysts for two or more reactions at the same time. He calculated that patterns could emerge from a chemical mixture if such a dual catalyst—call it chemical A—not only increased the production of A but also catalyzed the formation of a second chemical B, which served as an inhibitor that slowed the A producing reaction. Key to Turing's scheme was the diffusion rates of A and B through the mixture once they had formed. Since the dual catalytic nature of A would set up a competition for the production of both A and B, it was necessary that as they formed, they would diffuse through the liquid at different rates, thus generating small pockets of slightly different composition, some richer in chemical A and others in chemical B. In order to prevent the autocatalytic feedback characteristic of A from causing runaway production of itself, chemical B had to diffuse through the liquid at a greater speed. Being an inhibitor, the faster diffusion of B would not only prevent it from totally shutting down the production of A, it would also make its presence in the liquid a widespread phenomenon occupying the bulk of the mixture volume. In other words, the slower diffusion of A would make it a localized phenomenon, producing small regions with high concentrations of A.

To picture the results of Turing's brainchild, first consider a freshly prepared beaker containing a quiet mixture of A and B plus the other reactant chemicals. The different diffusion rates of chemicals A and B would guarantee the presence of random concentration fluctuations throughout the vessel. This would generate small regions where the concentration of A slightly exceeds average, causing the formation of more A as well as B. B however, tends to diffuse away from these regions, allowing more A to form there, but the inhibitor characteristic of B will discourage A formation in the intervening space. Now, if chemical A just happens to have a bright red color and B green, what you will see is an initially colorless liquid spontaneously begin to form red spots throughout the beaker while the surrounding liquid becomes green. The resulting pattern eventually exhausts the supply of original chemicals but can continue indefinitely by adding new reactants and removing products without disturbing the calm of the vessel. While Turing's discoveries fell short of shedding light on embryo development, it did suggest that the symmetry of a chemical system could be broken to create patterns where none had existed before by maintaining the system in a non-equilibrium state.

Oscillating chemical systems resurfaced in the early 1950s through research by Soviet biophysicist Boris Belousov aimed at determining how enzymes (protein molecules that act as catalysts in biochemical reactions) break down organic foods to produce energy. He prepared a cocktail of chemicals composed of malonic acid ($CH_2(COOH)_2$), potassium bromate ($KBrO_3$), sulfuric acid (H_2SO_4), and a catalyst of ceric ions in the form of ceric ammonium nitrate ($Ce(NH_4)_2(NO_3)_5$), included to simulate the catalytic behavior of enzymes. After preparing the mixture, Belousov was shocked to see its color spontaneously oscillating in clock-like fashion. First, it would appear completely colorless for about thirty seconds before changing to yellow and back again—colors representing two different valence states of the element cerium, namely cerous and ceric ions (Ce^{+3} and

Ce^{+4}). Belousov could not have been more surprised had he been dining and found that between sips, his wine vacillated back and forth; first a rich red burgundy, then a clear dry chardonnay—perhaps a pleasant experience if he happened to have ordered surf and turf. However, upon submitting his results for publication in 1951, Belousov's colleagues, certain his results were at odds with the second law of thermodynamics, did not take his work seriously. The second law states that chemical reactions always head for degenerate equilibrium with "order decaying uniformly to disorder"; all systems should proceed to a low energy state and remain there, not jump back and forth between high and low states. "The second law is inviolable", therefore Belousov must have erred in his experiment.

However, the interpretation of the second law at that time was overly simplistic since it had not been reconciled with non-equilibrium systems. Most chemical reactions achieve a state of homogeneity and equilibrium quickly; however, Belousov's reaction was unique. The competition among catalyst, autocatalyst, and inhibitor could maintain the system in a non-equilibrium state for days.

Later, in the 1960s, a professor at Moscow State University in Russia and former colleague of Belousov convinced a young postgraduate student under his tutelage named Anatoly Zhabotinsky to resurrect Belousov's work. Zhabotinsky not only repeated the results, but he continued to experiment with other chemical formulations that produced even more dramatic results, creating mixtures that alternated between bright shades of blue and red. When Zhabotinsky presented the subject to an international conference in Prague, Czechoslovakia in 1968, his results were too convincing to be immediately rejected.

Further investigation into the phenomena revealed that the experiments did not violate the second law of thermodynamics since the second law dealt with equilibrium states, and these were non-equilibrium systems. The second law of thermodynamics requires that a decrease in energy should accompany all spontaneous processes; therefore, a reacting chemical system must move steadily toward equilibrium. This means that during the course of a spontaneous chemical reaction, the amount of reactants must always steadily decrease, while products increase. Until the 1960s, the conventional wisdom of the scientific community was that this monotonic aspect of equilibrium prohibited such oscillations in the concentrations of chemical species during a reaction. They finally realized however, that the second law only requires that the reacting materials and final products change uniformly.   Oscillation of low concentration intermediate species resulting from feedback loops were not in violation.

In recognition of the contributions of both chemists, the class of reactions that exhibit such non-equilibrium behavior was dubbed the Belousov/Zhabotinski, or the BZ reaction.   The first step in a simplified version of the BZ reaction converts material A into B followed by a second autocatalytic reaction:

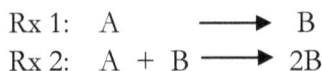

$$Rx\ 1: \quad A \longrightarrow B$$
$$Rx\ 2: \quad A + B \longrightarrow 2B$$

Rx 2 does not consume chemical B, therefore, it acts as a catalyst; each molecule of B on the left side also appears on the right side of the equation.   But since the reaction also produces B, it acts as an autocatalyst by generating positive feedback.   In reaction 3, another chemical C functions as an autocatalyst in a competing reaction that consumes B and produces more C.   Then in reaction 4, C converts to a non-reactive chemical D.

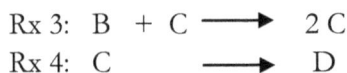

$$Rx\ 3: \quad B + C \longrightarrow 2C$$
$$Rx\ 4: \quad C \longrightarrow D$$

Now all we need is an *indicator* (a neutral substance that changes color or some other characteristic) that turns red in the presence of B and blue in the presence of C. The reaction can start with an excess of either A or C and the system will eventually reach the same state. But let us assume it begins with an excess of A and small quantities of C. Since in autocatalytic reactions, rates (speed) are proportional to reactant concentrations; and, since there is much more A than C, reaction 2 will initially dominate over 4, causing high concentrations of B that give a red color to the mixture. The rapid increase in the concentration of B causes the second autocatalytic reaction 3 to kick in to produce chemical C, which further builds up, thus reducing B, and the mixture turns blue. However, reaction 4 diminishes the concentration of chemical C by converting it to a non-reactive product D, allowing reaction 2 to dominate again, bringing the color back to red. The system will continue oscillating in this manner so long as there is a continued supply of chemical A, and waste product D leaves the system.

The BZ Reaction remained little more than a curiosity until the early 1970s when a group led by Richard Noyes of the University of Oregon developed a quantitative model connecting the chemistry of the BZ Reaction with the mathematics of nonlinear dynamics. Noyes, and his colleagues, E. Kőrös, and R.M. Field, proposed a complex system of reactions, now known as the FKN mechanism. The system contained an array of eighteen reactions involving twenty-one different chemical species, plus an assortment of equations to describe the rates of the individual reactions. The model covered all of the aspects of Turing's system, except that the Oregon team dealt with real rather than hypothetical chemicals.

The complexity of the model (nicknamed the *Oregonator* in honor of the patron institution) prohibits its use here, although a simplified version is included in the appendix. The Oregonator represented a significant step in understanding chemical processes that may have been involved in the synthesis of the first life on Earth. Using autocatalytic mixtures of chemicals, the University of Oregon researchers found they could set up a system that would begin in an unchanging mono-color state, and then by adding reactants, initiate oscillating behavior. When the reaction takes place in a stirred beaker, the entire contents regularly change color from red to clear and back again. However, the truly fascinating aspects of the Oregonator occur in a quiet environment. Unstirred, the reactions generate traveling waves of color looking much like a pond into which stones are dropped; but, instead of generating water waves, waves of alternating color propagate outward from the point of entry. Even more dazzling color displays come from boiling away volatile portions of the mixture and placing the concentrated reactants between two glass plates. This causes beautiful patterns of concentric or spiral waves of color to pass through the liquid. Of course, the Oregonator is more than an intriguing and visually appealing diversion. It proved that complex patterns of self-organization can arise through simple interactions involving autocatalysis and feedback mechanisms.

Many prominent researchers have concluded that synthesis of the first life form involved autocatalytic feed back mechanisms. Stuart Kauffman believes that life "bootstrapped itself into existence" in such a manner through autocatalytic sets of chemicals. He suggests that if the primordial chemical soup of the young Earth had enough different compounds, that such compounds would have acted in this manner to reproduce and evolve. In an article published in the August 25, 2000 issue of *Science Magazine*, German scientist Günter Wächtershäuser described results from experiments involving negative feedback in autocatalytic systems that produced an active form of acetic acid, thought to mirror an ancient metabolic pathway in bacteria. He argues that such a process might have taken place on metal sulfide surfaces located in hot deep ocean vents billions of years ago, and could have been one of the first steps in the evolution of life.

Existing life forms provide endless examples of biomorphesis feedback mechanisms, so it seems likely that such processes played a major role in the synthesis of the very first life form. In his book *Emergence*, which we will discuss later, Steven Johnson used the existence of positive feedback loops within single cell individual slime molds to serve as an ideal example of emergent behavior within species. Without any central command, millions of tiny individual slime molds (cells) will "self organize" into a single mobile blob, often several inches in diameter, designed to protect the community in times of stress. Individual cells emit a chemical called *acrisin*, and have a built in attraction to acrisin secretions left by other cells. When threatened, the individual secretions increase, causing slime molds to begin following acrisin trails left by other cells, producing an ever-increasing feedback loop that causes cells to merge toward one another, or "join a cluster."

In *Deep Simplicity – Bringing Order to Chaos and Complexity*, a highly recommended book on chaos theory and self-organizing behavior, author John Gribbin expressed his belief in the importance of the chaos scenario to emergence of the first life. Gribbin believed the event happened in a "…primordial chemical broth…" containing chemicals that acted as catalysts for the formation of other chemicals as in the BZ reaction where "…chemical A catalyzes the formation of chemical B…" along with an energy source to make it happen.

Many of the starting chemicals that must have been involved in first life synthesis may well have arrived from the interstellar cloud that condensed to form our planet 5 billion years ago. In the 2002 volume 416 issue of *Nature Magazine*, two groups of scientists reported having synthesized compounds required in all life process by subjecting chemicals known to exist in interstellar space to conditions that exist there as well. After exposing mixtures of water, methyl alcohol, ammonia, and hydrogen cyanide stored in a closed container, to ultraviolet radiation at temperatures below minus 258° C, one team produced three of the twenty amino acids involved in the protein synthesis of all living things, namely glycene, serine, and alanine.

The second group, using similar conditions but slightly different chemical mixtures, succeeded in synthesizing a variety of sixteen separate amino acids. The relative ease of these experiments leads one to believe that, given the hundreds of millions of years for random experimentation, the BZ reaction scenario may well have been responsible for forming the necessary chemicals of life. It is quite conceivable that such a chemical B emerged from the primordial brew that served as a catalyst for making a third chemical C, and C catalyzed a reaction to make a chemical D, and so on. Eventually within this progression, a chemical formed that catalyzed the formation of the starting chemical A; and then another that just happened to be a catalyst for producing chemical E, an inhibitor for chemical A production thereby closing the loop. With all of the catalysts in place, a self-sustaining network of connections between chemicals that could lead to life only needed a continuing supply of reactants and an energy source such as light or heat from the many hypothermal vents of the young oceans.

The Earth 3.5 billion years ago had all the facilities of a great laboratory for producing the chemical components of living organisms, although it is still not completely clear that the first ingredients of life originated then. Some scientists, including Carl Sagan, believe the precursors of life arrived already assembled from outer space via meteorites. The Murchison meteorite, discovered 1969 in Australia, contained—in addition to all five of the nitrogenated bases of DNA—two-hundred and fifty different hydrocarbons and seventy-four different amino acids. Many held symmetry properties peculiar to those produced in life processes. As noted in chapter 4, two molecules can have identical compositions and configurations, while orientated differently as to their placement on one of the carbon atoms in the molecule. This kind of isomerization occurs when a molecule contains an *asymmetric carbon atom*—a saturated carbon with four different groups attached. Such asymmetric molecules always have two possible structures that are mirror image of each other, and therefore, not superimposable. You can demonstrate this property of asymmetry by simply comparing your right and left hands. When placed palm to palm they appear mirror images, but placed one on top of the other, they are opposite and therefore different.

These compounds, called *stereoisomers* or *enatiomers*, have identical physical properties such as boiling and melting points; however, they may react differently with other asymmetric compounds and respond differently to certain asymmetric physical probes. Polarized light is one such disturbance, which chemists use as an analytical tool to identify stereoisomers. When a polarized light beam passes through an asymmetric material, its plane of polarization rotates. When the rotation is clockwise, the compound is *dextrarotatory* (rotation to the right); when it rotates counterclockwise, it is *levorotatory* (rotation to the left). As shown in figure 41, these terms shorten to dextro and levo, or d and l, when used as prefixes for identifying enantiomers.

Figure 41 Stereoisomers or Enantiomers

Asymmetric Carbon atoms

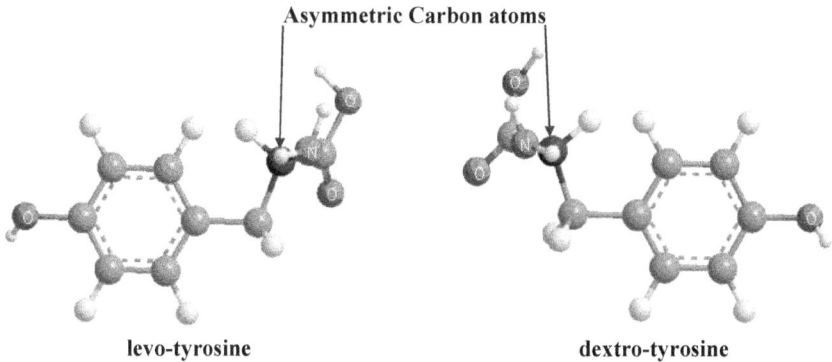

levo-tyrosine   dextro-tyrosine

Laboratory synthesized amino acids contain equal amounts of right and left hand (or d & l) enantiomers, while life forms use only the levo, or left handed ones. The Murchison meteorite contained both types, but a definite excess of left-handed amino acids suggesting they originated from extraterrestrial life. Figure 41 shows ball and stick models of the enantiomers of tyrosine, one of the twenty amino acids used in protein synthesis. The asymmetric carbon atoms of the two enantiomers are the black balls, and both have a hydrogen atom, an amine (NH_2) group, a carboxyl (COOH) group, and a 4-hydroxyphenyl group attached (in figure 41, oxygen atoms are labeled O and nitrogen atoms N. Hydrogen atoms are pictured as small light gray spheres, and carbon atoms large darker gray ones. The asymmetric carbon atom is black). We can see the mirror image aspect by comparing the placement of the amine and carboxyl groups. Furthermore, we cannot make them identical by rotating the groups along a carbon-carbon bond axis; continued rotation would return them to their original mirror image. However, switching locations of any two of the groups on one of the molecules would make them identical compounds.

Figure 42 shows simplified structures of the 20 amino acids encoded by DNA to synthesize protein, with asymmetric carbon atoms in larger bold type. Solid carbon-carbon bond lines connecting

221

the asymmetric carbon mean they align with the plane of the page, whereas wedge-shaped bonds place the attached group in front of the page and the dotted line behind the page. Note that all carry the l designation, identifying the levorotatory configuration unique to life processes. l-tyrosine above appears in the fourth position from the left in the bottom row.

Figure 42 Twenty Amino Acids of Protein Synthesis

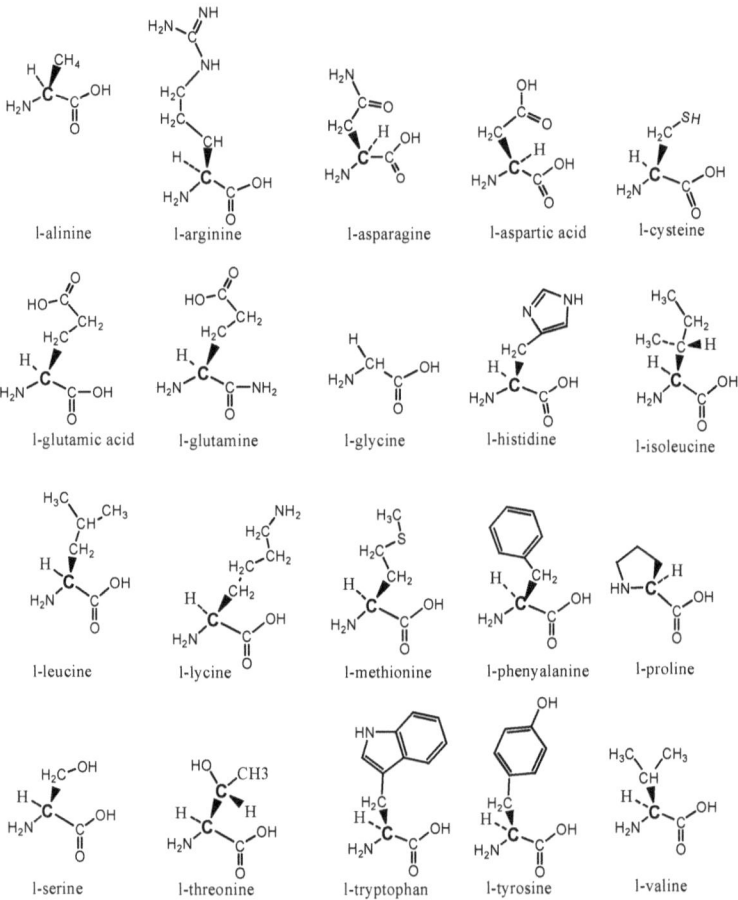

l-alinine	l-arginine	l-asparagine	l-aspartic acid	l-cysteine
l-glutamic acid	l-glutamine	l-glycine	l-histidine	l-isoleucine
l-leucine	l-lycine	l-methionine	l-phenylalanine	l-proline
l-serine	l-threonine	l-tryptophan	l-tyrosine	l-valine

Even if comets did send organic materials useful to life, the early

Earth still contained the basic ingredients of life. It included all the elements of the periodic table, a huge inventory of chemicals and an endless supply of natural catalysts. It contained ocean-size reaction flasks with energy from the Sun or hypothermal vents to warm them, and hundreds of millions of years to perform Edisonian style experiments. As early as 1951, Harold Urey and his student Stanley L. Miller, working out of the University of Chicago, synthesized complex molecules that can readily convert to the amino acids used in protein synthesis. Their technique involved sending electrical current through a flask containing methane, ammonia, hydrogen, and other gasses known to exist in the atmosphere 4 billion years ago. This and a host of other experiments that followed showed the early Earth possessed all of the building blocks of life; the only thing missing was instructions.

The Cell

The first "live" organism may not have been a cellular creature but simply the first in a number of evolutionary steps involving self-reproducing entities that eventually culminated in the cell. This viewpoint is a tempting alternative to the tremendous complexity necessary for a theory based on spontaneous fusing of all the elements of cellular life. At the same time, we must keep in mind that no one has yet presented a concrete proposal that accounts for all of life's processes except through the cell.

Researchers have described cellular life as simply a mass of chemical reactions surrounded by a skin or membrane that controls rates of diffusion through its surface in order to manage much of the life of the cell. Energy and organic matter enter the cell through the membrane, causing the cell to split and double, followed by more energy input and more splitting and doubling, allowing the organism to grow.

Cells of all living things have nuclei containing DNA (deoxyribonucleic acid), which guide their reproductive processes to produce proteins that form the cell structure. DNA is the long double-helix molecule, brought to public attention by the project for mapping the human genome, in which the arrangement of nucleic acid pairs dictates the genetic information for cell manufacture and determines every property of the resultant life form. This happens through a process called *replication*, or DNA synthesis, and is responsible for copying the double-stranded DNA molecule, the heart of cellular growth.

We often call DNA the instruction manual for the construction and maintenance of life. Four nitrogen containing organic bases, collectively referred to as nucleotides, are the key ingredients in the DNA of all living things; they are cytosine, guanine, thymine, and adenine—shortened to C, G, T, and A for illustrating DNA structures.

Figure 43 lists the common and chemical names along with the IUPAC chemical formulae for the DNA components. In the DNA molecule, the four bases always combine in threes, or triplets (e.g. ATG, CGA, etc.), to form words called *codons*. In keeping with the instruction manual analogy, instead of pages codons are written on long continuous chains of a sugar compound called *deoxyribose*, and phosphate molecules (shown in figures 48 and 50). The individual bases form the four letters of the DNA alphabet, while codons, three letter words with no spaces between, combine to supply instructions for replication based on their order of sequencing. Starting with this four-letter alphabet, mathematically they can be arranged sixty-four possible ways to make three letter codons, or words (4 letters used 3 at a time means $4^3 = 64$ different possibilities).

Figure 43 DNA Bases

Letter	Common Name	Chemical Name	Formula	Structure
C	Cytosine	4-amino-2-hydroxypyrimidine	$C_4H_2N_2NH_2OH$	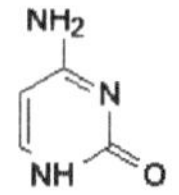
G	Guanine	2-amino-6-hydroxypurine	$C_5HN_4H_2NHOH$	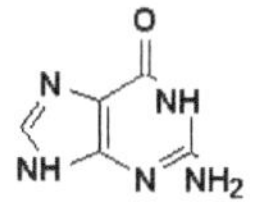
T	Thymine	2,6-dihydroxy-5-methylpyrimidine	$C_5HN_2CH_3(OH)_2$	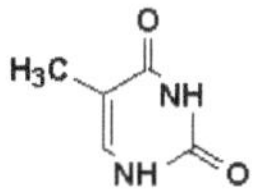
A	Adenine	6-aminopurine	$C_5H_3N_4NH_2$	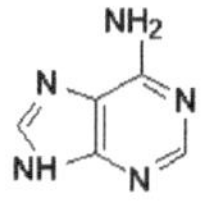
U	Uracil	2,4-dihydroxypyrimidine	$C_4H_2N_2(OH)_2$	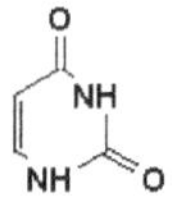
	Deoxyribose	2,4-dihydroxypyrimidine	$C_4H_5(OH)_3$	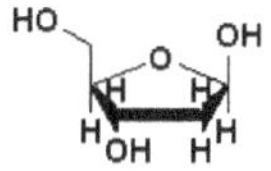
	Ribose		$C_4H_4(OH)_4$	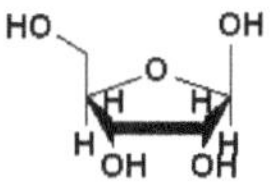
	Phosphate group		$PO_4^=$	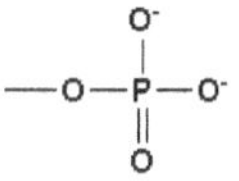

Even though the manufacture of proteins requires only 20 of the amino acids, the process uses all 64 codon combinations. As shown in Table 6, several of the possible codon pairings apply to the same amino acid, e.g. there are six different "spellings" for serine: TCT, TCC, TCA, TCG, AGT, and AGC, while tryptophan and methionine have only one each, TGG and ATG respectively.

Table 6 Codon Table

T = Thymine
C = Cytosine
A = Adenine
G = Guanine

		T	C	A	G
T		TTT phenylalanine	TCT serine	TAT tyrosine	TGT cysteine
		TTC phenylalanine	TCC serine	TAC tyrosine	TGC cysteine
		TTA leucine	TCA serine	TAA Stop	TGA Stop
		TTG leucine	TCG serine	TAG Stop	TGG tryptophan
C		CTT leucine	CCT proline	CAT histidine	CGT arginine
		CTC leucine	CCC proline	CAC histidine	CGC arginine
		CTA leucine	CCA proline	CAA glutamine	CGA arginine
		CTG leucine	CCG proline	CAG glutamine	CGG arginine
A		ATT isoleucine	ACT threonine	AAT aspargine	AGT serine
		ATC isoleucine	ACC threonine	AAC aspargine	AGC serine
		ATA isoleucine	ACA threonine	AAA lycine	AGA arginine
		ATG ethionine, Start	ACG threonine	AAG lycine	AGG arginine
G		GTT valine	GCT alanine	GAT aspartic acid	GGT glycine
		GTC valine	GCC alanine	GAC aspartic acid	GGC glycine
		GTA valine	GCA alanine	GAA glutamic acid	GGA glycine
		GTG valine	GCG alanine	GAG glutamic acid	GGG glycine

Four of the sixty-four codons serve as a kind of punctuation for reading the genetic code. Since there are no spaces between the codon words, START and STOP indicators point to where reading of a sequence begins and where it concludes. ATG is always the START signal, whereas TAG, TGA, or TAA will signal STOP. Therefore the sequence: ATGAAATTCGATCCCTAA would signal: start (ATG), lysine (AAA), phenylalanine (TTC), aspartic acid (GAT), proline (CCC), serine and stop (TAA). Table 6, called the forward codon table displays the amino acid specified by each of the 64 codons while Table 7, the reverse codon table is its opposite in that it shows the codons that belong to each of the 20 amino acids.

Table 7 Reverse Codon Table

Alanine	GCT, GCC, GCA, GCG
Leucine	TTA, TTG, CTT, CTC, CTA, CTG
Arginine	CGT, CGC, CGA, CGG, AGA, AGG
Lycine	AAA, AAG
Aspargine	GAT, GAC
Methionine	ATG
Aspartic Acid	GAT, GAC
Phenylalanine	TTT, TTC
Cysteine	TGT, TGC
Proline	CCT, CCC, CCA, CCG
Glutamine	CAA, CAG
Serine	TCT, TCC, TCA, TCG, AGT, AGC
Glutamic Acid	GAA, GAG
Threonine	ACT, ACC, ACA, ACG
Glycine	GGT, GGC, GGA, GGG
Tryptophane	TGG
Histidine	CAT, CAC
Tyrosine	TAT, TAC
Isoleucine	ATT, ATC, ATA
Valine	GTT, GTC, GTA, GTG
START	ATG
STOP	TAG, TGA, TAA

Series of codons form "paragraphs" that are steps in the instructions called *exons*, frequently separated by long meaningless letters of "chatter" called *introns*. Exons and introns combine to form a strand of completed "instructions" in the manual called *genes*. Thousands of genes, each carrying different instructions, make up "chapters" called *chromosomes*, and the sum of all the chapters is the manual—the machine's *genome*.

The human body contains 100 trillion (10^{14}) cells, each measuring about one tenth of a millimeter across, that house a black amorphous mass called a *nucleus*, which contains copies of its owner's unique genome. With the exceptions of blood, egg, and sperm cells, each of the trillions of cells carries two complete sets of the genome, one from each parent. Egg and sperm cells carry only one set of the genome—the mother's for egg cells, and the father's for sperm cells—while blood cells have none. Each parent's genome has twenty-three chromosomes, each containing approximately 20,000 different genes with perhaps 100 more yet to be discovered. While we inherit nearly identical genes from our mother and father, small differences occur in the chemical arrangement that can alter cell characteristics. When our offspring later breed, they pass on their own unique genome containing information selected from both parents with these new characteristics intact.

Each of the four bases has a property that allows DNA to build new cells identical to the original ones via the replication process. This property comes from very specific bonding attractions; attractions that the bases have for each other in which base A pairs only with base T while G pairs only with C—reminiscent of the Boston social propriety wherein Lodges talk only to Cabots and Cabots talk only to God. In the presence of nucleotide mixtures, this selective attraction allows the single DNA strand to make copies of itself by individual bases on the strand pairing with their respective mates from the mixture. The result is an adjacent complimentary strand in which T is always opposite A, and C opposite G; conversely, A opposes T, and G opposes C. For example, the sequence GCTTAG on the original strand would become CGAATC on the *complimentary pair* (keep in mind that there are no spaces between the three letter words in the genome).

Pairing of the complimentary strand takes place through a relatively weak and non-permanent form of bonding called *hydrogen bonding*, where a nitrogen or oxygen atom of the pairing nucleotide that has negative character, attracts certain loosely held hydrogen atoms, possessing partial positive charges. The dashed lines of figure 44 represent the hydrogen bonds for the adenine-thymine and the guanine-cytosine pairings.

228

Figure 44 Hydrogen Bonding in DNA Replication

Adanine-Thymine Pair

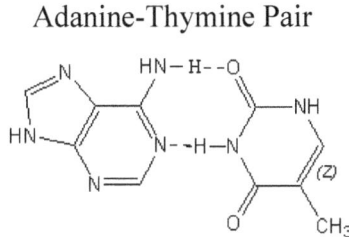

Guanine-Cytosine Pair

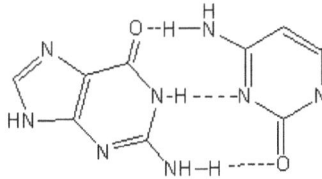

These paired configurations form the basis for the DNA double helix with its intertwined complimentary pairs. Figure 45 shows three rungs (or nucleotide pairs) of the ladder. As in figure 44, dash-lined crossbars are hydrogen bonded attachment points for the nucleotide pairs of the four bases while the long chain sugar and phosphate structures on each side hold them together. To appreciate the complexity of DNA from a human genome, imagine this chain continuing for another million rungs.

After pairing, the complimentary pair will in turn be copied, thereby returning it to the original configuration. Those of adult age prior to the digital revolution can appreciate the comparison with printing a photographic negative to get the original positive image.

Figure 45 DNA Structure

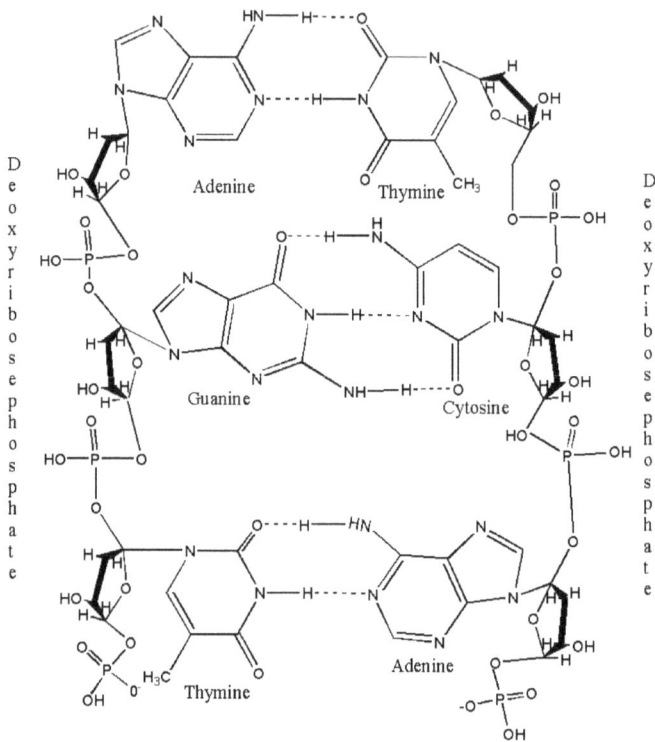

DNA to Protein

A *ribosome* molecule is an enzyme frequently described as a factory designed to build protein from a set of genetic instructions found in the genome manual. Ribosome is a complex structure made from both protein itself and from RNA (ribonucleic acid). The ribosome copies RNA from the DNA molecule to serve as a template to produce a specific protein. However, it copies only one strand of the double helix DNA and changes all of the Ts (thymine) to Us (uracil). In addition, ribose replaces the sugar deoxyribose (see figures 48 and 50), adding an extra oxygen atom to the helix structure. Every strand of DNA contains the instructions for manufacturing all of the proteins as well as the other constituents of the life form in which it resides. It is the job of the ribosome to access the instruction for a particular protein, copy its information, gather the raw materials needed to build it, and arrange said materials in the precise order necessary for the protein to function.

The manufacture of protein within the cell is a multi-step process. The first step gathers—and in some cases produces—the amino acid building blocks. The cell itself produces twelve of the twenty amino acids needed for human protein biosynthesis from glucose and other compounds through a series of chemical reactions. The other eight (called *essential amino acids*) must be ingested directly from the diet. The *transcription process* is the second step, in which genetic information needed to make a functional protein transfers from DNA through a complementary RNA (called *messenger RNA* or mRNA). Transcription occurs in three sub-steps as follows:

The Transcription Process

Initiation – The stretch of double-stranded DNA containing the instructions for the particular protein temporarily separates to provide access to a single strand, which bonds with RNA in the next step. An enzyme called *polymerase* assists in the bonding by attaching to the RNA forming a structure called *RNA polymerase*.

Elongation – The individual amino acid nucleotides

bond covalently one-at-a-time beginning at one end of the growing polynucleotide chain. The appearance of the START codon sequence ATG signals the beginning of the sequence. In this step, all of the nucleotides on the DNA chain except thymine bond with their complimentary nucleotides: cytosine (C) on the DNA chain bonds to guanine (G), adenine (A) to thymine (T), and guanine (G) to cytosine (C). Uracil (U) however replaces thymine in bonding with DNA chain adanine (A).

Termination – A transcription termination sequence, or STOP, is recognized (TGA, TAA, or TAG) and a newly formed messenger RNA strand, which is a complementary copy of the DNA message releases from RNA polymerase.

Figure 46 DNA Transcription

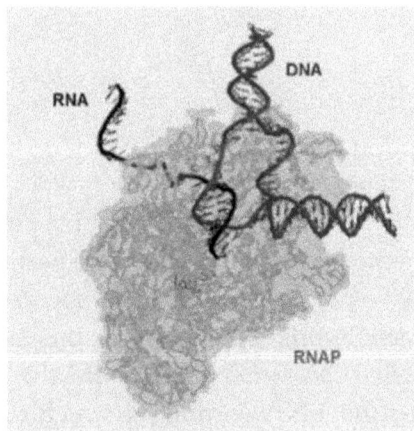

The process illustrated in figure 46 below, shows the RNA

polymerase, making a complimentary RNA copy of the separated section of DNA in the presence of RNA polymerase. The polymerase transfers the nucleotide that corresponds to the complimentary pairing nucleotide, to its appropriate DNA site. It then continues down the DNA chain, one nucleotide at a time until it reaches a STOP sequence, at which time it separates from the RNA polymerase.

The final step in protein biosynthesis is the translation phase where the mRNA prepared in transcription is decoded to produce the specific protein. This task goes to *transfer RNA* (tRNA), a group consisting of 74 to 93 compounds each designed to attach to one, and only one of the twenty amino acid building blocks (figure 43) of protein manufacture. An anti-codon triplet sequence, the precise compliment to the three-codon mRNA sequence that specifies the amino acid to which the tRNA is dedicated, sits on another site of the tRNA molecule. The tRNA loaded with its anti-codon triplet travels inside the ribosome, where it combines through complementary base pairing with the corresponding mRNA codons. The amino acid dangling from the other end of the t-RNA is then placed in proper alignment—according to the genetic code read from DNA—with those from other t-RNA, to play its part in the assembly of a protein. Having deposited its cargo, the t-RNA goes off in search of another of its own special amino acid molecules in order to continue the process.

Figure 47 illustrates the complexity of protein biosynthesis through a ball and stick model of the enzyme *hexokinase*, a protein molecule that acts as a catalysts in biochemical reactions. This complex structure is built from the 20 amino acids according to specific DNA instructions for hexokinase protein found in the human genome. Corresponding models of ATP (adenosine triphosphate) and glucose molecules in the top right-hand corner, illustrate the scale of the molecule.

233

Figure 47 Protein Molecule – Hexokinase

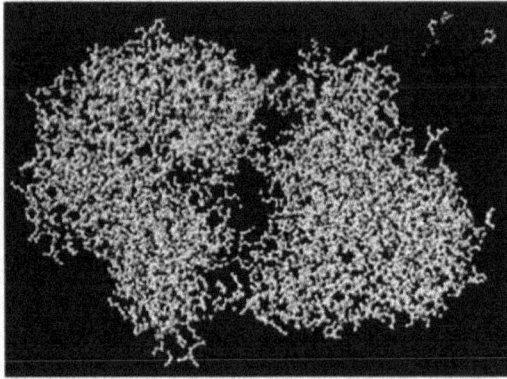

Early Life

Many of the doubts among scientists surrounding the origins of life do not carry beyond the initial organism. Some believe the first life was born from a chance occurrence through chemical processes occurring on our planet, some that it arrived in part or in total from outside our solar system, while others believe God had a personal hands-on role in the event. Regardless of how it first arrived, overwhelming evidence has brought nearly all serious scientists to agree that all subsequent complex forms, from bacteria to plants and animals, evolved from this tiny primitive organism.

If the hexokinase molecule above appears complex, consider that even the simplest single cell bacteria are still orders of magnitude more so. Still, when compared to multicellular life, they appear simplistic. The earliest ancestor of all living things, often called *Luca* (for Last Universal Common Ancestor) must have been DNA based since DNA is involved in the fundamental processes of all living organisms. However, all attempts to establish a theory that the first life happened through a series of spontaneous chemical reactions terminating in a living being complete with DNA, invariably meet with a paradox. This stems from the fact that in all current life forms, the nucleic acid building blocks of DNA (C, G, T, A, plus deoxyribose and a phosphate group) require proteins for their synthesis; further, proteins are only produced when their corresponding nucleotide sequence is present (e.g. CGA, ATG, etc.). The fact that it would appear unlikely in the extreme that the highly complex structures of both the proteins and nucleic acids could come together at the same instant, has led some to suspect that life could not have originated by purely chemical means.

In the late 1960s, Francis Crick, Carl R. Woese and Leslie Orgel, proposed that a life form utilizing a variation of current RNA could have preceded the DNA Luca. Several aspects of RNA strongly suggest that it predated both DNA and protein. In fact, there is reason to believe that an "echo" of the original life form still resides within the human genome as well as that of all other living things. Matt Ridley reports in his book *Genome*, about a short sequence of codons in the human genome located near the center of chromosome 1 that "...repeats itself over and over again...like a familiar theme tune..." with each sequence separated by random text. While only 120 codons long, this exon (or paragraph) is constantly being copied into segments of RNA called 5S RNA, housed in a ribosome along with other RNAs and proteins dedicated to the most basic of all life functions, translating DNA information into proteins. These facts: the commonality to all current life forms, repetitiveness within the DNA strand, and its performance of such a basic life function, lead many to conclude that 5S RNA was part of the very first living entity.

As we have seen, all life from algae to humans use RNA to build proteins and transmit the genetic information copied by DNA from one place in the cell to another. Since RNA is simpler than DNA, it can make copies of itself without outside assistance, and has the ability to catalyze reactions that splice and rebuild DNA. In 1995, two University of California scientists, Jack W. Szostak and Charles Wilson, proved that certain types of RNA could perform functions beyond their accepted role in DNA transcription. While testing properties of a wide range of RNA compounds, they discovered that some behaved as enzymes for catalyzing the production of chemicals other than the RNA components. Scientists had previously thought the compounds, called *ribozymes* (from **ribo**nucleic acid and en**zyme**), served the sole purpose of catalyzing reactions involving other RNA molecules. However, Wilson and Szostak found they could catalyze a wide range of reactions, including ones involved in protein production.

Wilson not only proved that RNA could have been the first life form; he performed experiments that demonstrated the feasibility of RNA evolution. He started by adding amino acids to a random sampling of billions of messenger RNA and found most of the mRNA incapable of reacting. A few however, did react and could make copies of themselves but like DNA replication, some of the copies contained errors. Furthermore, just as in DNA replication, some of the errors hindered the copying procedure while others enhanced it. Then, rather than allowing time and chance to be the evolution driving force, Wilson selected only the speedier mutated mRNAs and fed them more amino acids. After repeating this procedure over many generations (very brief generations in the case of RNA), each time selecting only the fastest reacting mRNAs, a greatly improved version emerged.

The fact that the ribonucleotides of RNA synthesize more readily than the deoxyribonucleotides of DNA, and since it is easier to imagine possible pathways for the evolution of RNA to DNA further enhances the plausibility that RNA life could have preceded DNA. However, laboratory attempts to synthesize RNA from basic chemicals have disappointed chemists, leading some to believe that an even simpler self-replicating system must have preceded RNA.

Even if the first life form were RNA based, a biosphere based solely on RNA would have been very limited. Its instability would subject all but extremely small RNA life to what geneticists call *error catastrophe*. Error catastrophe happens when copying errors (mutations) occur too frequently in the replication process, thus destroying its functionality before the normal culling effect of natural selection takes place. When isolated, RNA rapidly loses its structural integrity along with the genetic information it carries. Extremely small organisms might survive a few hours, but since the effect magnifies with size of the organism, there could be no opportunity to evolve beyond the simplest life forms. Furthermore, the instability would also limit RNA life to very moderate temperature environments. A living RNA entity would also have needed properties not seen in current life forms, such as the ability to replicate without the presence of proteins, and it would need to serve as the catalyst for the synthesis of proteins. However, invoking the anthropic principle, the fact that we are here, may well mean an error eventually occurred in one such RNA life form that led to the more stable DNA, which then co-opted a ribosome machine and used RNA to make copies of itself.

The cell membrane is another element critical to cell functioning, and therefore to life itself since it controls the flow of energy and the organic compounds necessary for growth, through the membrane surface. While the development of the first earthly cellular membrane is a far simpler task than the evolvement of the RNA/DNA segments of the life process, there is considerable doubt as to its source. One theory credits inorganic clays. Inorganic clays contain atoms that can readily lose electrons to become cations. Within the clays, cations align in loosely held sheets that serve as templates to attract selective molecules to their positive charged sites. Once attached, the new molecules can arrange in a particular order that aligns their reactive sites (functional groups), enhancing their reactivity.

Like many of the chemical ingredients of life, cell membranes could have come from extraterrestrial sources as well. Chemicals found in meteorites called *carbonaceous chondrites* naturally congeal into membranes and may have produced a kind of oil slick at the edges of the early seas similar to the foam that accumulates at the water's edge today.

Acceptance of the possibility that materials unique to living things on Earth arrived from outer space does not imply they came from planets with earth-like complex life forms. Based on the extraterrestrial origin of Life hypothesis, for every planet that advanced to higher life forms, billions may have produced some building blocks of life yet never evolved to become even the simplest bacteria. However, considering the enormous meteorite inventory wrought by the many billions of super novae star deaths, and the accompanying destruction of their planets throughout the eons, it seems reasonable to believe that some of their remnants arrived here and played a role in Earth life.

Energy

All life forms use ATP (adenosine triphosphate) to transport energy-producing fuel (food) to the sites where it converts to bodily tissue. ATP consists of three phosphate groups attached to a heterocyclic carbon/nitrogen ring that release energy when broken from the ring. Although all surviving species synthesize their own ATP, very early life forms must have harvested it from their environment.

Cells of different life forms generate energy for driving its division processes in various ways; however, each involves a chemical reaction that converts one compound to other compounds, liberating energy as a byproduct. One form of bacteria, called *methanogens*, use hydrogen and carbon dioxide to produce water and methane, commonly known as swamp gas, while others reduce sulfates to hydrogen sulfide (H_2S) and water plus energy.

$$CO_2 + 4H_2 \longrightarrow CH_4 + 2H_2O + e-$$
$$SO_4 + 5H_2 \longrightarrow H_2S + 4H_2O + e-$$

Oxygen is a deadly poison to methanogens that limits their habitats to reducing environments (absence of oxygen) such as bogs, marine sediments, and deep soils. Nevertheless, these tiny creatures have served humankind well, producing most of the natural gas that heats our houses. Methanogens also present a potentially attractive avenue for a more controlled natural gas manufacturing process through conversion of sewage to methane. Methanogens have a symbiotic relationship with cows and other rudiment animals by taking residence in their rumen, or fore-stomachs, where they assist the host in chewing her cud. The major byproduct of this relationship is about 50 liters per day of natural gas (methane), which finds its way into the atmosphere. This might suggest that, when the hills of West Virginia become oceanfront property, Elsie the cow will share some of the blame with gasoline guzzling SUVs.

Figure 48 Early Life Phylogenetic Tree

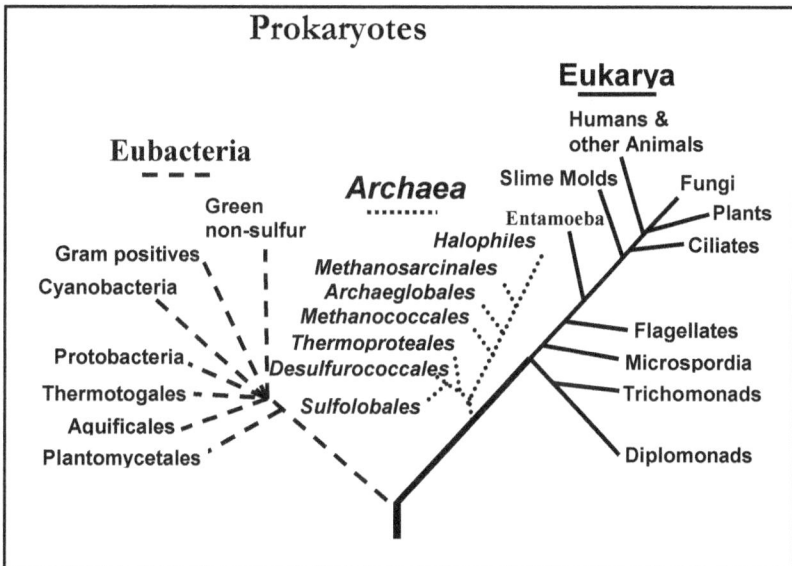

239

Another form of bacteria, called *hyperthermophiles* (also called *archaea*), derive energy from direct reduction of sulfur through a process called *chemosynthesis*. Their name comes from the extreme heat they require to function, and cannot survive what we call ambient temperatures. Hyperthermophiles may be the oldest bacteria, and still exist near volcanic vents and hot seeps on the ocean floor. Archaea also have the simplest structures of all living things. A single cell houses their entire bodies, and unlike other bacterial species that evolved later, they do not have a nucleus to house their DNA. Many cannot reproduce at temperatures below 80° Centigrade (176° Fahrenheit), yet they thrive in boiling water conditions.

Scientists have long suspected that the warm oceans created the first raw materials of life. However, it wasn't until the early 1980s that the submersible craft Alvin found what turned out to be the most likely site when they discovered "black smokers" 2.5 kilometers below the surface of the Pacific Ocean. Black smokers, alternately referred to as "The Garden of Eden" and the "Chimneys of Hell," are hydrothermal vents that emit dark sulfurous fumes and other minerals contained in 400 degrees Celsius water that is cycled from the molten mantle below the ocean floor. One theory is that the first living cell originated under such conditions by utilizing energy from the vents to break up carbon dioxide and convert its carbon to the organic compounds essential for life. Energy was abundant near the "smokers" that regularly spewed enormous quantities of hot metallic and sulfurous compounds. One of these, iron pyrites, formed from the reaction of hydrogen sulfide and iron sulfide, was the energy source.

$$H_2S + FeS \longrightarrow FeS_2 \text{ (pyrite)} + H_2 + 2 \text{ e- (electrons for energy)}$$

This theory also accounts for the cell membrane-like structures since they can self-assemble at the junctures where iron pyrites contact the surrounding water. Furthermore, the regular structure of iron, with positive-charged sites where complex organic molecules can accumulate to form bonds, could have served as templates to manufacture large complex organic compounds like RNA.

The fact that the early atmosphere contained very little molecular oxygen (O_2) adds further emphasis to the theory that the first life relied on sulfur rather than oxygen as the source of electrons for energy. The early Earth contained much higher quantities of metals in their base, un-oxidized states. While some oxygen formed continuously through the impact of solar UV radiation on atmospheric water vapor, converting H_2O to O_2 and H_2, the quantities were relatively small and oxidation (rusting) of iron and other minerals depleted oxygen as fast as it formed. Since plants had not yet arrived, photosynthesis, the main source today for recycling oxygen back to the atmosphere had not yet begun.

Preparing the Atmosphere

If not for the imperfect nature of DNA, Earth's population would have continued to consist of nothing but colored smears of simple bacteria living on the ocean floor near hot sulfurous vents. Without photosynthesizing bacteria to generate oxygen from CO_2, evolution might never have advanced beyond the bacterial stage; and even with photosynthesis, absent errors in DNA copying, it would have faltered when the supply of carbon dioxide depleted. Fortunately, simple bacteria, like human beings, are not all alike. Occasionally the DNA replication process produces an error or mutation. If the error occurs in an important bond sequence, it will usually produce a nonviable entity; however, it sometimes results in an improved model. Human beings accumulate about one hundred mutations per generation in our million-codon genome. Most of these prove benign but can be fatal if the error occurs at a critical location in the sequence. Of course, they can also be beneficial and occasionally an individual within a species is born with a mutation that provides a reproductive advantage over its peers. If great enough, that advantage can provide the recipient an inordinate number of progenitors and, over enough generations, will spread throughout the species. Natural selection dictates how fast the mutation propagates, but a series of favorable ones occurring over time can produce a new, more complex species, better equipped to survive and thrive in its particular environment.

Over the next 100 million years or so, countless genetic mutations occurred in the DNA of archaea. Most mutations were terminal, but others had the effect of producing organisms more efficient in competing for life's resources. Chief among these is the life energy source. Reliance on sulfur from H_2S for energy to power Earth's inhabitants would have confined life to hydrothermal vents and greatly limited the opportunities for evolution of complex life. Furthermore, fermentation (breaking down of complex organic molecules into simpler ones in the absence of oxygen) is not a very efficient method of producing energy. The inefficiency lies in the fact that only about one-third of the carbon atoms in the organic compounds involved in the process (e.g. glucose), actually give up their electrons for energy. The process yields acetic acid or ethanol and carbon dioxide, while releasing energy to build other organic materials such as ATP.

Dissolved oxygen, on the other hand was abundant in the oceans that covered the Earth by this time. Initially, bacteria that could exist in the presence of oxygen without actually using it for energy, evolved from this setting by feeding on anaerobe's (an organism that does not need oxygen for growth) fermentation products. Cyanobacteria (blue-green algae) that derived their energy from the Sun and food from carbon dioxide came next. Cyanobacteria could not only survive in an oxygen atmosphere, they could use it to their benefit and ultimately to the benefit of all life forms that followed.

The arrival of these creatures proved a key step in the history of life. Paleontologist Andrew H. Knoll heaped the highest praise on cyanobacteria calling them "the most important organisms ever to appear on Earth." Cyanobacteria contain the green pigment chlorophyll deposited in efficiently arranged plates that function as solar panels, allowing them to absorb light that breaks down carbon dioxide into carbon for producing glucose and other organic compounds for growth, and oxygen for release to the atmosphere.

Cyanobacteria proliferated over the next billion years along shorelines, turning much of its surface blue-green. It also did something far more important for all subsequent life forms; it changed the composition of the atmosphere. Cyanobacteria existed as the top layer of stromatolite mats (from Greek strōma, meaning bed or stratum, and lithos meaning rock), just above anaerobic bacteria that don't require oxygen, but feed off the remains of the cyanobacteria covering them. Even though the Sun at that time was smaller and not as hot as today, the high concentration of atmospheric carbon dioxide produced a hot "greenhouse world," that accelerated their growth. The CO_2 from the atmosphere, absorbed in rainwater, became the cyanobacteria food source, and the O_2 waste product of photosynthesis returned to the atmosphere. It was no easy task since the Earth environment attempted to impede this progress by removing O_2 as fast as produced through oxidation reactions with iron, silica and other materials. However, the vast numbers of cyanobacteria, exhaling oxygen for the next thousand millennia, still supplied the oxidation reactions and at the same time added free oxygen to the atmosphere.

Eventually, atmospheric oxygen from photosynthesis, and carbon dioxide recycled back to the atmosphere via volcanic activity reached equilibrium. The movements of the Earth's continental plates subducted some of the collected carbonated minerals like silica and calcium carbonates into the hot mantle under very high pressure. Subsequent volcanic eruptions returned the hot carbonates to the lower pressure surface where they decomposed restoring CO_2 to the atmosphere. However, the cyanobacteria worked harder than the continental plates. The interaction between these processes over the next billion years lowered CO_2 concentrations and changed the atmosphere to one similar to that we enjoy today thus making possible the emergence of the complex life forms that followed.

Life Expands

The first *eukaryotes* represented a major evolutionary step and were the most advanced life on Earth for about one billion years. While only single cell creatures, they were larger and more complex than archaea. Furthermore, housing their genetic material in a nucleus sheltered by an internal membrane allowed more efficient conversion of ingested materials into the energy and chemicals needed to support growth. The cell contained structures called *mitochondria* that use oxygen to burn fuel for energy, and *chloroplasts* that perform functions such as photosynthesis and enzyme production. The first eukaryotes probably evolved in stromatolite mats where symbiotic relationships between cyanobacteria and aerobic bacteria already existed. Most biologists believe the first eukaryote originated within these mats when one of the bacterial types internalized another without actually digesting it, producing a hybrid that could perform the functions of both within one individual. The result was a new level of complexity in life forms.

Combining small but different simple organisms presented a method for forming more complex and effective ones. After capture of the appropriate cyanobacterium by the eukaryote, the bacterium became the *chloroplast organelle* (a structure for carrying out a specialized function within eukaryote cells) responsible for photosynthesis. A single cell may house thousands of mitochondria and chloroplasts needed to perform other functions. This became the next miracle of evolution, without which, bacterial mats would have represented the peak of life's evolvement on Earth.

Nucleated cells presented the potential for building large multicellular organisms with complexes of cells dedicated to different functions. Divisions into sexes occurred allowing natural genetic variation for greater adaptability; by offspring receiving genetic material from two parents, new and useful DNA mutations could spread far more rapidly through the species population.

Micropaleontologists have found fossil evidence of multicellular life dating back to about 2 billion years ago. This coincides with the point when photosynthesis by bacterial mats had produced enough atmospheric oxygen to sustain microbial development. In general use, the word fossil describes anything old, generally exceeding 10,000 years. To scientists, fossil has a more specific meaning—an object, in which minerals have replaced the host's original composition. Fossilization minerals are usually crystalline compounds of silica, calcium, and aluminum oxides, or carbonates such as quartz, feldspars and calcite. Crystals are giant molecules with atoms arranged in an orderly lattice pattern that repeats over-and-over again billions of times. Fossil formation usually occurs when minerals dissolved in a liquid crystallize in the porous structure of the bones, or on rare occasions, the tissue fabric of a creature buried in an oxygen-free atmosphere. Over the millennia, as the animal's original body structure disappears, minerals dissolved in water continue to deposit as it passes through the porous structure gradually replacing bone and tissue and assuming the form of the host; and if undisturbed, remaining true to it for many millions of years after the original remains are gone. Fossilization is not limited to animals; in fact, entire forests fossilize. Arizona's Painted Desert fossilized 200 million years ago when minerals leached from ground water displaced the entire bodies of trees leaving microscopically detailed evidence of their cellular structure.

While ancient fossil discoveries are exciting, they provide limited information if their age is unknown. While scientists cannot date fossils themselves, igneous rock—usually crystalline minerals that started as lava—found near the fossil can be. Volcanic rocks contain a variety of elements with unstable isotopes that over time convert to more stable isotopes via radioactive decay. As we saw in Chapter 5, such elements decay at predictable rates called *half-lives*, which can be used to measure the age of the material. Half-lives of these elements vary greatly from as little as a few seconds to billions of years, providing accurate yardsticks for rocks of any age beginning with the Earth's formation.

Geologists estimate the age of potassium-containing rock formations by measuring the quantity of a potassium isotope, ^{40}K, or potassium 40 (atomic number, AN 40). Fifty percent of ^{40}K will decay to ^{40}Ar (argon 40) in 1.3 billion years. One needs only to compare the quantity of ^{40}K with that of ^{40}Ar trapped as a gas in the rock crystal. Equal amounts of ^{40}K and ^{40}Ar would date the formation of the crystals within the rocks at 1.2 to 1.4 billion years ago, while twice as much ^{40}K as ^{40}Ar would make them 600 to 700 million years old.

Most fossilization takes place in sedimentary rock, which builds from tiny fragments of limestone (mostly calcium carbonate) and sandstone products of the water and wind erosion of larger rocks. The fragments become suspensions in water or air as sand, silt, or dust. They then deposit elsewhere where they settle and compact to form new layers of sedimentary rock.

The discovery of Precambrian fossils is a rare event due to their extremely small size, and to the billions of years of volcanic and tectonic plate activity that destroyed the overwhelming majority of them. Most of the eukaryotes from the Precambrian (see figure 49), for which there is fossil evidence, were tiny threadlike plants that became part of the *stromatolite* communities by feeding on waste products of other bacteria in the mats. Some spiny spherical plants measuring about 0.2 millimeter with tough organic walls that aided fossilization have been found, as well as large single celled ones—1 centimeter across—dating 900 million years ago. Despite their small size, the development of these aerobic spheres was extremely important as they allowed life to expand its habitat. With oxygen their fuel source, they did not need to be near the hot sulfurous confines of hydrothermal vents, freeing them to colonize the entire ocean. This greatly increased the opportunities for genetic design of new cells with new functions to produce a greater variety of life forms.

246

Although multicellular plant evidence dates back nearly 2 billion years, archaic animal fossils date to only about 700 million years, near the end of the Precambrian. Paradoxically however, the *amoeba*, generally accepted to be an animal, lies below plants in the evolutionary chart; so if the amoeba is an animal, then some plants must have sprung from animals. *Slime molds* are surprisingly interesting unicellular amoeba like creatures that occupy a place where higher plants split off from fungi and multicellular animals (see entamoeba, figure 48). Slime molds spend most of their lives as individual single cells, but when a colony is threatened, they can come together, forming a globular slug-like structure. Such slime mold conglomerations usually inhabit shaded areas on rotted wood during hot damp weather, where they move about like a gelatinous blob, growing by extracting nutrients from vegetation. However, when the weather cools, it stops and hardens into a definite shape. At this point, it is fruiting, or producing spores to form new slime molds, a distant joint between mobile and rooted species.

Grypania, a curled up spiral shaped creature measuring one centimeter across, is the oldest known multicellular organism dating back 2,000 million years; however from that point forward until the end of Precambrian 650 million years ago, very few such species have been found. The few included a variety of microscopic shell fossils collectively referred to as "the small shelly fauna". The relatively low levels of atmospheric oxygen during the Precambrian, is probably responsible for the shortage of larger multicellular animals. While adequate for small aerobes, photosynthesis had not increased oxygen to the levels needed by larger animals.

Differentiation between plant and animal life of the Precambrian is somewhat fuzzy. Plants manufacture their own growth and substance through photosynthesis, while animals must exploit plants directly by feeding on them, or indirectly by eating animals that eat plants. Animals use oxygen for respiration fuel and are at least in some sense mobile. Nevertheless, plants must have preceded animals in both single celled and multicellular organisms because where plants led, animals followed.

Until only recently, biologists considered *sponges* plants; however,

247

they are in fact the oldest known multicellular animals, and probably the first non-plant ancestors of humans and all other animals. Our common ancestor with today's sponges lived about 800 million years ago. He (or more accurately it, since sponges are asexual) probably looked and acted very much like current varieties. These early sponges fed by passing water of ancient oceans through the millions of passages on the sides into a hollow interior where it extracted microscopic plant life, then expelled the food-free water through a large hole at the top.

Like plants, sponges are stationary creatures whose body structures do not include muscle fiber, nervous systems, stomachs, kidneys, livers, or other organs normally associated with animal life. The cells that make up the sponge are also simple. The body structure of most animals contains both *germ-line cells* and *soma cells*. Germ-line cells, usually found in testes or ovaries, function exclusively in the reproductive functions of most animals while soma cells constitute the remainder of the body. Sponges however, are made entirely of germ-line cells. Each cell has the ability to combine with other cells allowing them to reproduce by self-assembling. Researchers have separated cells of living sponges in the laboratory by forcing the sponge body through a fine sieve. Examined under a microscope, an individual sponge cell behaves very much like a single cell amoebae; however, in time it will begin to reconstitute itself, agglomerating to form a new sponge. The asexual nature of sponges explains their low progress throughout nearly a billion years of evolutionary history. Rather than inheriting the combined beneficial genetic mutation history of two parents, the sponge lineage must depend on a single parental line. There is an upside to this limited evolutionary progress however; the continued repetition of the same simple cells within the sponge has provided insight to the mystery of how complex multicellular creatures evolved from single cell protozoa.

We also share a Precambrian ancestor of 780 million years million years ago with another animal found in today's oceans called *Trichoplax*. Like sponges, Trichoplax has plant-like properties, but he is multicellular—albeit with the smallest genome of any animal ever measured—, the entire body composed of only four different cell types. This aspect leads many to believe Trichoplax may even predate the sponge as our earliest animal ancestor.

Figure 49 Geologic Time Scale

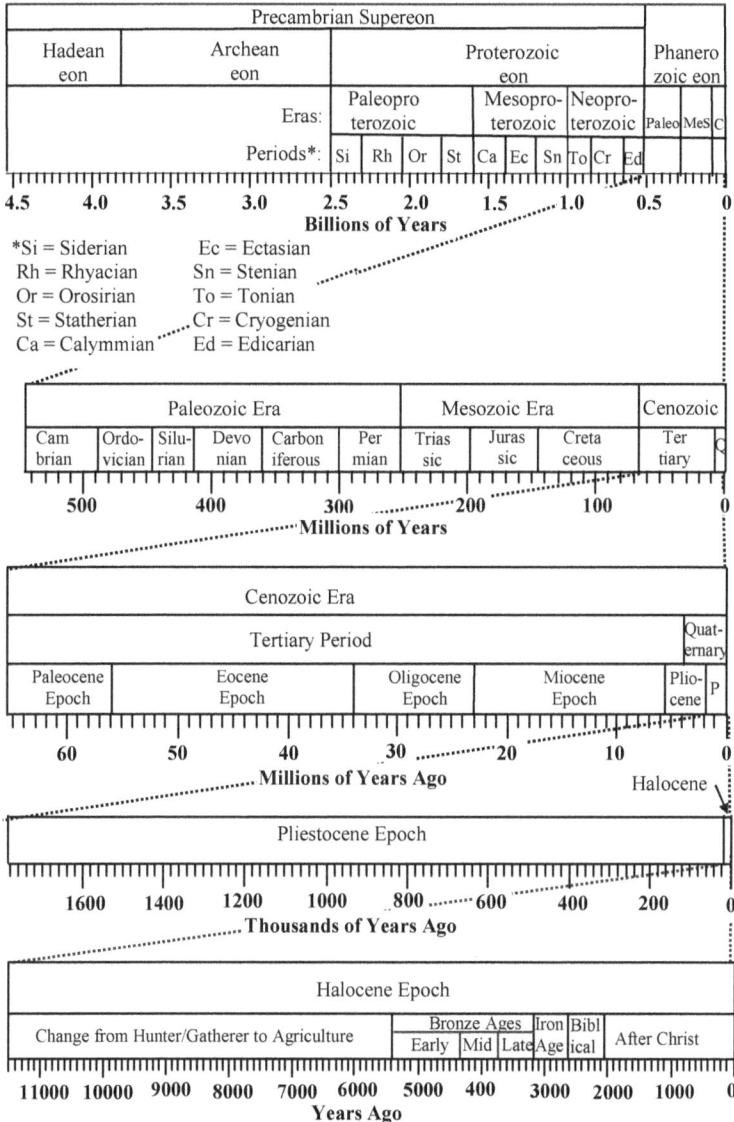

Precambrian Supereon															
Hadean eon	Archean eon		Proterozoic eon											Phanero zoic eon	
		Eras:	Paleoproterozoic			Mesoproterozoic			Neoproterozoic			Paleo	MeS	C	
		Periods*:	Si	Rh	Or	St	Ca	Ec	Sn	To	Cr	Ed			

4.5   4.0   3.5   3.0   2.5   2.0   1.5   ....1.0   0.5   0
Billions of Years

*Si = Siderian   Ec = Ectasian
Rh = Rhyacian   Sn = Stenian
Or = Orosirian   To = Tonian
St = Statherian Cr = Cryogenian
Ca = Calymmian Ed = Edicarian

Paleozoic Era						Mesozoic Era			Cenozoic	
Cam brian	Ordo- vician	Silu- rian	Devo nian	Carbon iferous	Per mian	Trias sic	Juras sic	Creta ceous	Ter tiary	C

500   400   300 200   100   0
Millions of Years

Cenozoic Era					
Tertiary Period					Quat- ernary
Paleocene Epoch	Eocene Epoch		Oligocene Epoch	Miocene Epoch	Plio- cene	P

60   50   40   30....   20   10   0
Millions of Years Ago

Halocene

Pliestocene Epoch

1600   1400   1200   1000   800 600   400   200   0
Thousands of Years Ago

Halocene Epoch					
Change from Hunter/Gatherer to Agriculture	Bronze Ages			Iron	Bibl	After Christ
	Early	Mid	Late	Age	ical	

11000 10000 9000 8000 7000 6000 5000 400 3000 2000 1000 0
Years Ago

249

The first hard evidence of Precambrian life came in 1957 from the fossil of a small fern-like plant given the name *Charnia masoni* for the young schoolboy John Mason, who found it while wandering through the Charnwood Forest of Leicestershire, England. While the Charnwood find did not result in the wealth of Precambrian species fossils the young man's find had implied, it did spark the interest of biologists, initiating a worldwide search for more such Precambrian antiquities. In 1946, an Australian mining geologist named Reginald C. Sprigg found what would be the mother lode of late Precambrian animal fossils in the Ediacara Hills of South Australia. One of the most prolific fossils of the Ediacara came from an animal named *Spriggina*. Named for its discoverer, Spriggina was a bilaterally symmetrical animal a few centimeters long with a crescent shaped head and body divided into numerous segments. Biologists assigned Spriggina to the group called *arthropods* (animals with jointed legs), although some scientists remain doubtful since none of the fossils revealed the presence of the necessary appendages; possibly from having been too small to fossilize.

Other Ediacara animals included a small disc-shaped animal named *Arkarua*, which is probably the oldest; and *Tribrachidium*, an unusual appearing creature with tri-radial (three-part) symmetry. Tribrachidium measured about five centimeters in diameter with three curved "arms" extending from the center of his disc-shaped body. All were below rocks of the Cambrian age, proving they had to predate the Cambrian.

A little later in the Precambrian, perhaps 730 million years ago, our ancestors had begun to appear more animal-like. Animals resembling current day jellyfish were among the Ediacara fossils, which is not too surprising considering the low level of evolutionary development in jellyfish. This is not only due to their lack of substance, but also to the simple arrangement of their body tissue. The jellyfish body is just two layers of cells separated by "jelly"; they do not have blood and their nervous system is very primitive and not organized into a central brain, ganglia, or major nerve trunk; their entire digestive system is a simple cavity with only one opening that serves as both mouth and anus.

The Ediacara also yielded the first Precambrian fossil evidence of larger animals such as *Dickinsonia*, a strange platter shaped creature that grew by adding segments, each measuring two centimeters across, until it reached about twenty centimeters. Others included *acoel*, a probable ancestor of modern day flatworms as well a host of *bilateral* animals (animals with left side to right side symmetry), that were still around 630 million years ago. Some of the living *acoel* share a symbiotic relationship with plant life by ingesting them and benefiting from their photosynthesis. The plants benefit from acoel's migrations to shoreline areas where they receive maximum Sun exposure.

These animals of the late Precambrian became victims of the Earth's first great species extinctions, and many other such events would follow. Their discovery represented a transition between kingdoms, a time between the fall of inactive animals with no predators, and the emergence of "true" predatory species. While most Precambrian animals did not survive, these simple creatures signaled that life had crossed a new evolutionary threshold and animals would continue to inhabit the Earth, even though future reversals would alter their evolvement.

Life Explodes

The Cambrian period that began 540 million years ago, and concluded 50 million years later, marks the beginning of the geological time scale measurement. British geologist Arthur Holmes, who developed the original scale in 1913, chose this point since at that time, it coincided with a total absence of fossil evidence; early 20th Century scientists had found many fossils dating from that time but none before. They therefore concluded that life itself began approximately at the Cambrian boundary and simply lumped the previous 3.4 billion years of Earth history together as Precambrian. However, new fossil information began emerging around the middle of the 20th Century that dated far beyond the 540 million year mark. This did not mean the end of the Precambrian classification however; as shown in figure 49, it was simply subdivided into eons—the Proterozoic, the Archaen, and the Hadean eons.

A mass extinction of species accompanied the end of the Precambrian. One of the most important theories—but not the only one—explaining this extinction, called "Snowball Earth", suggests that a major reduction in the Earth's surface temperature simply froze-out the animal population. The prevailing temperature of Earth's atmosphere at any point in time depends on a balance between heat from the Sun and the entrapment of radiation heat by greenhouse gasses like carbon dioxide, methane, and water. The Cambrian Sun was slightly smaller and less luminous than today and required more greenhouse heating to maintain temperatures hospitable to aerobic life (life that uses oxygen to provide energy needs). About 750 million years ago, the major landmass on Earth, a super-continent called

Rodinia, began to break and drift apart. The result was smaller subcontinents located near the equator with much greater coastal land area and tropical climates. The warm beachfronts proved temporary however. The new conditions brought more rainwater to what had been distant interior land regions and scrubbed more greenhouse carbon dioxide from the atmosphere causing global temperatures to fall.

Normally, when such cold conditions exist, temperature balance stabilizes through increases in atmospheric CO_2. Ice accumulation on rocks interrupts the normal CO_2 cycle balance by preventing its reacting with minerals within the covered rocks to make carbonates. Having checked the loss rate, CO_2 content can be reconstituted by CO_2 emerging from volcanic activity. This ice protection was absent at the end of the Precambrian due to the equatorial location of the subcontinents, so removal of carbon dioxide from the atmosphere continued, dropping concentrations below levels needed for normal greenhouse heating. Further aggravating the problem, brightness of the growing ice surface in non-equatorial regions reflected more Sun radiation back into space, preventing it from being absorbed thereby reducing the temperature further. The buildup of ice reached a critical level when global coverage spread to latitudes within 30 degrees on either side of the equator, and runaway freezing began. Within about 1000 years of this point, ice and snow covered the entire Earth and most of the larger life forms had become extinct.

While larger species suffered disproportionally, the oceans did not freeze to the bottom, primarily due to heat supplied from the Earth's core. Life for sulfur bacteria and other hyperthermophiles that lived near hydrothermal vents continued uninterrupted; and some cyanobacteria survived the icy conditions away from the vents.

The totally ice and snow covered Earth was—pardon the pun—the high water mark of the extinction, but it set the stage for a comeback. There could be no further leaching of CO_2 from the atmosphere since the oceans that supply the rainwater had frozen and the thick ice blanket covering all landmasses denied access to mineral-containing rocks. But the Earth's core remained hot and tectonic plates continued moving, allowing volcanic activity to continue unchecked. Over the span of about a million years, volcanoes belched enough CO_2 into the atmosphere from earlier mineral carbonate deposits to resume warming the planet.

Since snow and ice blocked CO_2 access to carbonate forming metals, over the next 10 million years carbon dioxide climbed to 350 times present levels. This brought about another huge greenhouse condition causing massive melting of ice, particularly near the equator. Evaporation of seawater plus carbon dioxide buildup combined to drive temperatures at the equatorial coastal regions above 120 degrees F for several centuries.

However, with the land again exposed, carbonate formation once again resumed, thus lowering atmospheric carbon dioxide and presenting the opportunity for the entire process to repeat. And it did. In fact, it happened as many as four times between 750 and 580 million years ago. Evidence for this comes from unique marks made by glacial ice makes as it moves over rocks. When lava-like material first hardens into rock, engrained magnetic materials become frozen, aligned like compass needles in the direction of the Earth's magnetic field, usually north to south. However, geologists have uncovered some glacier-marked rock formations with magnetic materials aligned parallel to the equator indicating they came from glaciations in tropical regions (this happens at the equator since magnetic north and magnetic south are balanced causing flux lines to run parallel to the equator).

Hard evidence has not established that the timing of these freeze/thaw cycles coincided exactly with the mass extinction of Precambrian life. However, the sudden nature of the extinction does suggest a global catastrophe preceded it and the Snowball Earth scenario qualifies as a major suspect.

In the history of life on Earth, massive genetic changes that increased life form diversity have always accompanied such great environmental stresses. The "explosion" of new species that followed the Cambrian extinction proved no exception. In the wake of such events, opportunities for communities of various life forms to become isolated from each other increase, which—as we shall see

254

later—results in generation of new species. The absence of now extinct predators and the reduced competition for available resources provided the ideal environment for life and species proliferation.

And proliferate they did. Within just a few million years such a plethora of new life forms emerged that the term "Cambrian Explosion" described the period. Animals differed greatly from their Precambrian predecessors in many fundamental ways. Since life forms could no longer live in the absence of predators, new species evolved equipped to succeed in the changed environment. New survival equipment included things such as legs for mobility, eyes to locate and to avoid becoming prey, and nervous systems to coordinate these activities. Sex and the competitive impulse to propagate came next.

The absence of a skeleton for mechanical support in Precambrian animals had limited their attainable body size, but those of the Cambrian learned to secrete skeletons. The first skeletons appeared early in the Cambrian, and with them, animals of much greater size. We do not know the precise ancestors of these animals but they were probably small, non-skeletal Precambrians that left no fossilized clues. The earliest known skeletal animal, a 2 mm long creature called *Cloudina*, had a tube shaped shell, likely composed of calcium carbonate (calcite), the same material constituting most current shelled animals. Other animals of the period had silica-based shells, like the skeletons of current day sponges; still others were phosphatic, like human skeletons. The tube-like structures provided them many evolutionary advantages that a host of subsequent life forms would copy.

Larger skeletized arthropods such as spiders, flies, crabs and beetles came next. *Trilobites*, with fully developed nervous systems, including a simple brain, as well as eyes, well-developed segmented limbs, lungs, gills and antennae were among the first. Like flies and lobsters, trilobites had compound eyes containing as many as 3,000 lenses made of calcite rods, the same material that makes up limestone. These eyes, the oldest known visual system, allowed trilobites to be hunters making them one of the most prolific species of the early Cambrian. The Trilobite presence brought on a marine ecology similar to that of modern oceans and established the first food chain, signaling another major step in the history of life.

The anthropological advancement that shed the most light on diversity of Cambrian life occurred at the beginning of the 20th Century when Charles Doolittle Walton happened upon the Burgess Shale deposits in the British Columbia Canadian Rockies. The discovery not only revealed a huge number of life forms, but also provided fossils with exceptional detail that included soft tissues. Burial of the animals in sediment happened so fast it preserved the soft tissue before having a chance to decay. Paleontologists, believing trilobites the sole animal occupants, had to reassess the early Cambrian since the fossils revealed an entire range of arthropods.

The shale yielded twenty-six different species of arthropods, each varying greatly in body size and appearance. While some were blind, others had well developed eyes. The length as well as the number of legs of some species varied greatly, and some of them terminated with claws. Curiously, these arthropods did not seem anatomically consistent with surviving arthropods thought to have been their descendants. Only one or two exhibited characteristics appropriate to ancestors, leading biologists to believe that Burgess Shale animals might not be ancestors after all but life designs that had vanished long ago leaving no descendents. In addition to arthropods, great variety flourished in other phyla (see phylum in Taxonomic Classification below), including mollusks, jellyfish, and brachiopods, all of which appeared suddenly near the beginning of the Cambrian, but with few bearing resemblance to surviving members of their phyla.

In 1989, these observations led Stephen Jay Gould to rethink the evolutionary tree of life concept, which paleontologists viewed as a kind of inverted bush that branched outward as it grew in height. This model implied that new life forms evolve continually from old ones, and that species extinction was the exception rather than the rule. Gould revised the concept by flipping the tree, changing it to a Christmas tree shape—wider at the base and thinning upwards, showing more species in earlier ages. The sudden bushing out above the base represents the early predator-free Cambrian while the narrowing at higher levels shows the effect of thinning of species with non-adapting, non-competitive designs, or those falling by the wayside due to chance. Gould's speculation as to the impact, should chance have chosen a different evolutionary path by selecting one of these early creatures over another was in his words "...the consequence of an early branch would have been not only a different tree but a different forest. It seems inconceivable that history would slavishly replay itself if it could somehow be restarted and no doubt a different turn would have had a very different outcome..."

Taxonomic Classifications

	Humans	Gorillas	Dogs	Bacteria
Kingdom	Animalia	Animalia	Animalia	Prokaryote
Phylum	Chordata	Chordata	Chordata	Omnibacteria
Class	Mammalia	Mammalia	Mammalia	Enterobacteria
Order	Primates	Primates	Carnivora	Eubacteria
Family	Hominidea	Poingidae	Canidae	
Genus	Homo	Gorilla	Canis	Escheria
Species	H. sapiens	Gorilla	C. familiars	E. Coli

In the early 1980s a study of soft shales deposited during the late Cambrian revealed the fossil of a creature thought to be an ancestor of all living vertebrates (animals with internal skeletons made of bone) from fish to birds to humans. Researchers had been finding tiny tooth like objects called *conodonts*, measuring a few millimeters in length, in Cambrian rocks for many years but their source remained a mystery since they always appeared alone with no animal attached. Euan Clarkson and Derek Briggs found an impression in shale of a slender worm-like object about two inches long that contained an arrangement of conodonts at one end. The impression, looking somewhat like a sand eel with fin-like apparatuses on each side, was of a creature related to an animal living today called *lancelet*, better known by its Latin name *amphioxus*.

Lancelets belong to a phylum class called *Agnatha* that consists of marine animals similar in appearance to fish but lacking both lower jaws and paired fins. They feed on other fish using a circular toothed outgrowth to bore into the side of a fish and suck its blood. Instead of a backbone made of bony discs common to vertebrates, lancelets have a *notochord*, a stiff rod of tissue running the length of their body that gives them a swimming action. By alternately tensing and relaxing muscles attached on opposite sides, lancelet propels forward in a somewhat controlled motion. Attached to the dorsal side of the notochord is a nerve tube with a small swelling near the front end that performs double duty serving as both an eye and the closest thing to a brain the poor creature has at its disposal. In later models, a backbone encircles both the notochord and nerve tube.

Lancelets belong to a grouping called *protochordates*. While protochordates are not vertebrates themselves, one member of their group living in the middle of the Cambrian some 550 million years ago is an ancestor to all current vertebrates. This does not say that lancelets evolved to become humans; only that a creature of the Cambrian—probably far more similar to a lancelet than a human or any other current vertebrate, spawned an offspring that is our ancestor. The creatures' siblings on the other hand could only claim lancelets, eels, and hagfish as their grandchildren several hundred million greats removed.

In the *Ordovician*, from 490 – 445 million years ago, the odd creatures of the Cambrian began to give way to more familiar looking species of today, some of which still exist. Geologists named the era for the Ordovices tribe, early inhabitants of the Welsh hills where they had found the plant and animal fossils of the period. Clams, sea urchins, starfish, sea cucumbers, and sea snails suddenly came into existence and the reefs contained corals.

The nearest human ancestors of 460 million years ago were *cartilaginous fish* that had begun to develop lower jaw structures. If current day sharks traced their ancestry to this period, they would find they share their two hundred million-greats-grandparents with humans—a fact you might want to remind one of should you meet him at the beach. A branching of our evolutionary tree occurred wherein one branch remained in the sea and became current day sharks and ray fish while the other developed a more rigid skeletal structure and evolved to become bony fish and the ancestors to all vertebrates. The embryos of current vertebrates reflect our evolvement from these ancestors since all skeletons begin as cartilage in the embryonic stage. With the exception of small amounts used as connective tissue for lining our joints and outer ears, all of the cartilage transforms to bone through ossification with calcium phosphate crystals. Except for teeth, this process does not occur in sharks leaving them with a skeletal structure composed entirely of cartilage. The hinged jaws, common to sharks and all other classes of vertebrates, probably evolved from the gill slit support structures.

Life Explores

Evidence linking the first self-reproducing life form to the varied landscape of flora and fauna in the world today is far from a complete unbroken record. Movements of the Earth's crust likely destroyed many of the clues linking prokaryotes to humans. Glibly stated, continental plates are islands of dirt floating on the great molten core of lava that makes up the bulk of the Earth's mass. Movement of continental plates formed many of the great mountain ranges, such as the Asian Urals, which resulted when two continents merged, pushing the crust upward. It is the release of this tension, built up by merging tectonic plates when they move under one another that causes earthquakes. Tectonic plate movements also cause volcanoes when crust melts from temperatures generated by friction at points where plates merge. The continuing effect of this reshuffling of the Earth's surface over hundreds of millions of years has served to destroy and/or hide the evidence of existence for many of our ancestors. Even with fossils that have survived, we can only credit the good fortune of their discovery to rivers and oceans that exposed them by carving away the native rock that had for so long hidden them. Therefore, it is likely that undiscovered fossils of species from this early period greatly outnumber those for which we have fossil evidence since vastly more fossil-containing rock remains hidden than exposed.

Another great ice age marked the end of the Ordovician era, killing off more than half of its species and ending the first great phase of marine life diversification and reorganization. The animals emerging from the Ordovician had special, useful adaptations, allowing them to survive and go on to populate the modern world. One of these would give rise to animals that walked on land with the ability to communicate and reason, and ultimately be a direct ancestor of humankind.

Except for a few bacteria living on rocks near streams constantly washed with water, life throughout the Cambrian period remained in the oceans and immediate shoreline areas. Most of the early terrestrial plant fossils date to the Silurian (445 to 415 million years ago). However, biologists have recently found older spores from the Ordovician, indicating that the greening of the Earth took tens of millions of years. The evolvement of spores with adaptations that allowed their dispersal through air instead of water proved a key step in development of land-based vegetation. Next came the development of a waxy coat on the outside of their green fronds that reduced water evaporation losses, protecting them from desiccation during their journey. Prevention of evaporation losses, however, also greatly reduces the plants breathing ability. Nature solved this problem by evolving stomata, very small holes in the plant surface protected by an aperture that permits air to enter, yet blocks water from escaping during dry periods.

Before the Silurian period, all plant life had been intimately tied to water, not just for their roots, but for their entire structure, which needed frequent water contact. Some of the earliest plants of the Silurian were small creeping ones with scale-like leaves covering upright shoots that bifurcated only once or twice. They probably evolved in lakes or rivers and represented the crossing of another threshold when their first shoots emerged above water to extract oxygen from the atmosphere. From an evolutionary viewpoint, breaking free from total reliance on a water habitat provided major advantages since some lakes and ponds dry out at times and overflow at others. In addition, reproduction by dispersing spores through the air provided a wide range of new habitats to evolve greater varieties of plant life.

Isolation is a key factor in the evolution of new species. Collisions of continents during the Ordovician produced the Appalachian and the Caledonia mountains, providing a host of isolated habitats that led to the breeding of diversified land life forms during the Silurian. As the early plants established themselves on land, animals also began leaving the water and a mutual dependency developed among them. An ecosystem emerged that returned organic matter from decomposed animals back to the soil. Fish of the Silurian were larger than those of the Ordovician, but they still had a simple opening for a mouth without a hinged jaw. However, some had developed the ability to tolerate fresh as well as brackish water. This was an early step in the sea to land transition, allowing them to swim upstream in rivers to take advantage of a fresh variety of nutrients previously enjoyed only by certain bacteria. Joint legged arthropods may have been the first animals to make the water to land transition in the Silurian. Some, including *Slimonia*, were very large, about the size of a man. Slimonia had spindly legs that carried a flat body, with eyes and a pair of pincers at the front and a paddle at the back. Although inefficient by later standards, Slimonia had lungs that required rationing its energy. Like current crocodiles, it probably lay motionless for long periods; then with a sudden lunge, would grasp its prey. The Silurian also produced ancient relatives of scorpions with stinger like spikes at their rear, probably used to disable prey.

Having their skeletons on the outside limited both the size and the efficiency of arthropods, particularly those that chose land as their venue. As arthropod legs increase in size, the thickness of the shell

261

covering the legs must become much thicker to provide support. This translates to increased difficulty in operating the leg joints, and greatly reduces the limb's efficiency. By the Devonian (360 – 410 million years ago), animals with backbones and jaws existed along with their jawless predecessors. One of these jawed fish was the ancestor of all future backboned land animals including birds, dinosaurs and humans. All land-living animals with backbones have genetic similarities that make it very likely this evolutionary step happened but once.

All current land vertebrates fall into a class called *tetrapods*, meaning they have four feet or limbs. Tetrapods did not suddenly appear on the early Devonian scene, sporting fully formed arms and legs designed for terrestrial needs. Limb evolution more likely began with poorly functioning mechanisms, transformed from fins or flippers that had refined through hundreds of millions of years of natural selection. Of course, limb evolution for all Devonian creatures did not follow a path that led to arms and legs designed for terrestrial use. In some animals such as birds, two of the limbs evolved for flying rather than walking, or in the case of snakes, limbs were completely lost; all however, evolved from four-limbed ancestors.

It is also interesting that nearly all tetrapods have five fingers and/or toes. Exceptions such as the horse, which has but one, resulted from multi-digit species having lost toes. There is no apparent evolutionary advantage associated with the number five when it comes to toes: in fact, today entire villages of human inhabitants have six toes and exhibit no known disadvantage. It is definitely not a lethal mutation and may well present some interesting advantages in some activities such as playing the piano. The number has not always been five however. On limbs of early land dwelling tetrapods, the number of toes varied from five to eight. At some point, a five-toed species advanced beyond the rest, illustrating that the selection of one species over another during the Devonian was often due, not to a fitness criterion, but merely to simple chance.

The arrival of birds obviously did not occur by a single genetic mutation wherein a finned mother gave birth to a winged daughter. Many intermediate steps must have marked the way. It would be extraordinarily unlikely that every one of the thousands of mutations necessary to cross the new phylum line would be either helpful or neutral, and none harmful—that a single evolutionary *"macromutation"* would allow an animal of one phylum to give birth to an offspring sufficiently different from its parent to be classified as another phylum. The number of DNA mutations needed for such an event is enormous. Furthermore, every one of them carries a certain probability of helping or harming the recipient. In fact, most mutations are neither harmful nor useful, but neutral. They have no impact on natural selection or on the functioning of the host, so we can remove them from the argument. Nevertheless, neutral gene mutations remain in the genomes of future generations. Also keep in mind that the genes of the parent had to function quite well in order for the individual to have survived long enough to reproduce. So any chance mutation is more likely to hurt than to help. But on the positive side of evolution, as we discuss later, natural selection prevents accumulation of harmful genes by inhibiting the reproductive success of the affected individual.

Richard Dawkins, in his book *The Ancestors Tale* described the likelihood of a single-phylum producing event, akin to the creationist's analogy likening Darwinian evolution to a tornado creating a Boeing 747 as it passes through a junk pile. Dawkins argued that the metaphor applied to scenarios like the overnight transition of an earthworm to a snail, but not to the "…gradual cumulative nature of natural selection…" As the string of mutations leading to new phylum progresses, every individual that contributes a new mutation that becomes part of the genome, at that point, becomes a most recent common ancestor of all future species bearing that gene.

Even so, based on fossil evidence alone, the earliest winged creatures were insects sporting fully developed wings. While there must have been intermediate stages, no one has found fossils representing the transition. This is not terribly surprising since size is very important in insuring the survival as well as the detection of fossils; and four hundred million year-old small, delicate winged structures don't make good candidates for museum displays.

Another method for identifying ancestries relies on the degree of evolutionary relatedness among species rather than attempting to identify a line of descent. The method, called *phylogenetics*, establishes ancestry by using diagrams, which map out species relationships as a series of regularly splitting branches, each based on particular characteristics of the animals. Branching points occur where species acquire new characteristics and pass them on to succeeding species or group of species while length of the branches establishes the time when their most recent common ancestor existed. Groups of animal or plant species descended from a common ancestor, all of whom share the new characteristic form a "clade" of that characteristic. Clades can range in size from a few species to one representing most of Earth's animal kingdom. Tetrapods exemplify a large clade. As clades go, the one representing mammals is small including cats and rats, and elephants as well as all other animals held together by the common characteristics of warm blood, four limbs, ear structure, and the manner in which they give birth to and suckle their young. In addition, there are clades within clades. The genus *Homo* is an even smaller clade within tetrapods and mammals that include humans and some apes that share the ability to make tools, walk upright, and copulate at all times.

Figure 50 shows a phylogenetic tree for the taxon *Eutheria*, which represents all placental mammals. Eutherians differ from other mammals in that during gestation, the fetus receives nourishment within a placenta and the mother carries the offspring until fully developed. Marsupials, the counterpart to placental mammals belong to *Metatheria*, the sister group of Eutheria. The phylogenetic tree begins with its root located on the far left side, which represents the common ancestor of all the species to the right that in this case lived about 100 million years ago. Each bifurcation of the tree corresponds to a *speciation event* where all species to the right descend from those to the left, but none of the ones above and below. For example, members of the order Primates, including humans, chimpanzees, gorillas, and monkeys relate more closely to each other than to mammals of all the other orders of the tree.

Figure 50 Phylogenetic Tree of Placental Mammals

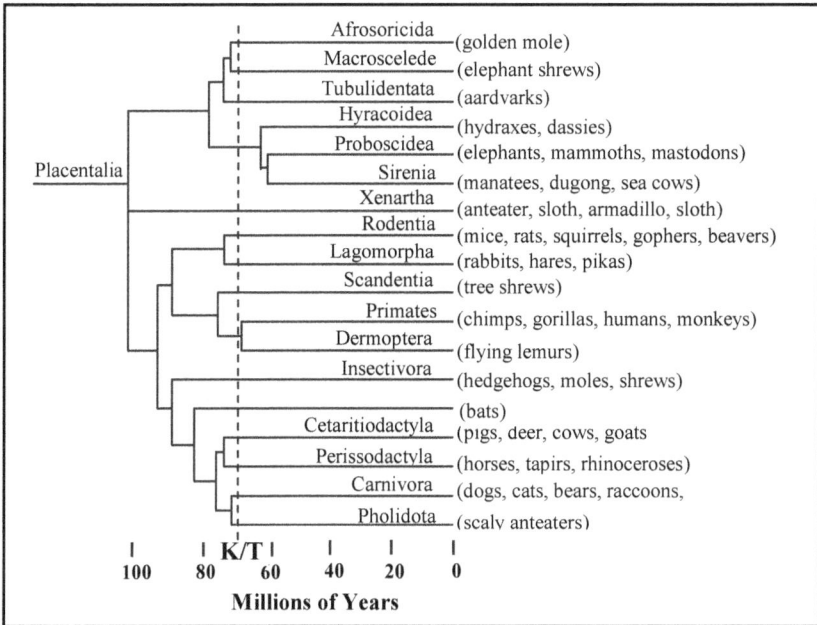

By observing the branching order, it is obvious that flying lemurs and other members of the *Dermoptera* order are our closest non-primate relatives. Humans and all of the other primates share a common ancestor with these tiny wide-eyed creatures that lived 65 million years ago near the end of the Cretaceous period. We share our second closest non-primate ancestor with tree shrews of the *Scandentia* order that lived 75 million years ago. An eighty-five million year old common ancestor of mice, cats, and beavers of the *Rodentia* order, and the hares, pikas, and rabbits of *Lagomorphia*, is also an ancestor of primates and shares the third spot. Ninety million years ago, a common ancestor to primates and all of the orders mentioned above, plus the six orders at the bottom of the phylogenic tree, hid from Cretaceous period dinosaurs. The most recent common ancestor of all placental mammals lived 100 million years ago.

Of course, our relationship to every species within a given order is not the same. There is also a phylogenetic tree for every group with

265

branching points placing some nearer the left side (closer to the primate order), and others to the right (further away). For example, a further branching of the *Cetarodactyla* order would show pigs closer to humans than cows, whales, or hippopotamuses.

Human and all other tetrapod lineages may have passed through a group of *lobe-finned fish* (so named because of the leg-like appearance of their fins) that evolved about 420 million years ago. Paleontologists thought lobefin lineages had been extinct for 70 million years since the only evidence of their existence came from fossils of that age or older. However, in 1938, a strange looking fish showed up in the nets of surprised fishermen operating off the coast of South Africa. It measured about five feet in length and sported very prominent fin-like appendages on the lower part of its sides, approximately a third of the way from either end of the body. The lucky find turned out to be a species of supposedly extinct lobefins called *coelacanth*, which later earned the nickname "old four legs". Like humans, old four legs—given a second moniker, "the living fossil"—is a living scion of our ancestral lobefin and still inhabits the Indian Ocean at depths of 600 to 2,000 feet. Coelacanths grow to more than seven feet in length and give birth to live young "pups" immediately capable of surviving on their own. Hans Brigan of the Max Planck Institute found more convincing—if less technical—evidence linking fish to land animals while aboard the submersible craft JAGO, where he filmed the swimming motion of a coelacanth that revealed fin movements identical to the crawling and walking motion of tetrapods.

Neoceratodus forsteri, a slow moving freshwater fish native to the Burnett and Mary Rivers of Queensland Australian, is the most ancient living member of a taxon, collectively called *lungfish*, from the *Devonian* period. *Neoceratodus* has no limb-like fins at all, but as his name implies, breaths through lungs like a tetrapod. The presence of two such marine creatures, each sharing an important trait with land animals presented biologists with a choice: Which of the two species is closer to humans and other tetrapods on the evolutionary scale? Biologists declared lungfish the winner after a phylogenetic analysis showed they exhibited more characteristics of tetrapods than do the lobefins.

If scientists were to evaluate the evolutionary progress of our species based only on the uncovered fossil evidence of our human ancestors, they would conclude that our advancement happened as a series of rare but major events. They would have to conclude that parents from a species that for tens of millions of years walked on all fours gave birth to a child with a body perfectly engineered for upright walking; or two very hairy individuals spawned an offspring with hair covering only a few discrete areas. The fact however, is that evolution involves slow but continuous change, propelled in a particular direction by natural selection. It involves many intermediates that are not sufficiently different to be new species but have experienced significant genetic divergence. Small genetic mutations that for any reason enhance survival or mating opportunities spread throughout the group, and thus become more pronounced as further mutations reinforce the original one in succeeding generations. Taken over the six million year span between current humans and the common ancestor we share with chimpanzees, there have been many intermediates; however, the intermediates are extinct. It is very likely that a male and female representing intermediates separated by time spans of not more than, say 10 thousand years could—if transported by time machine into one another's presence—mate successfully, the litmus test for belonging to the same species. However, intermediates separated by a million years would very likely not be capable of breeding.

We cannot prove that evolution progresses in tiny steps from accumulated fossil evidence. This is because fossilization requires very rare conditions in the first place, complicated by the fact that only a tiny fraction survives geological disturbances such as movement of tectonic plates, and an even smaller likelihood of finding those that might exist. As a result, many millions of years often separate fossils belonging to a particular species, leaving large information gaps.

This paucity of fossils also leaves evidential gaps at branching points in the evolutionary tree where a species splits and the two groups proceed down different evolutionary paths. New species can result when members of a group become separated geographically, placing the two segments in different environments. The new environment will then subject subsequent mutations within the two groups to different natural selection criteria, thereby imposing different rewards and penalties. Since the two groups can no longer share mutations through interbreeding, over time they will become incapable of producing offspring even if barriers separating them disappear. They will belong to different species.

Icthyostega, a 360 million year old fossil from the upper Devonian of Greenland (upper denotes the fossils position in the rock formation of the period; therefore upper signifies later or more recent) is one of a series of fossils that represents a rare exception to the absence of transitional fossil evidence. Icthyostega, thought to be a descendent of fresh water lobe fish and an ancestor to humans and all living land vertebrates, may be the earliest known pioneer in the emergence of terrestrial life from the sea. He resembled fish in many skeletal features such as a large fish-like rayed fin along with bony supports along the rear edges of his tail. Small scales imbedded his skin, a characteristic of lobefins but not amphibians, which have smooth un-scaled skin. He had however, lost his predecessors gills, retaining only some of the bone structure that had served as gill supports.

The major Icthyostega bone structure is nearly identical to that of land vertebrates; it includes well-developed five-toed limbs along with hip and shoulder girdles capable of supporting the body for movement on land. The skeletal limb structures consist of single large upper bones with the humerus in front limbs and femur in the rear, along with two long lower bones, radius and ulna in the front, and tibia and fibula in the rear. Like other land vertebrates, Icthyostega's wrists and digits consist of numerous smaller bones.

In 2004, paleontologists working Devonian period sediments on Ellesmere Island in Canada, discovered several incredibly well preserved fossils belonging to a new species given the name *Tiktaalik roseae*, that may well be a link in the evolutionary continuum leading from fish to land animals. Although limbs of the 375 million year old fossils had originally evolved as fins, they included bones that served as a shoulder, upper arm, elbow, forearm, and wrist that allowed him to move about on land or in shallow water with a seal-like gait. Tiktaalik had a crocodile shaped head and a neck that appeared to be designed for independent head and shoulder movement common to land animals. There was evidence that Tiktaalik had gills, but he probably had lungs as well, along with an overlapping rib cage to protect his inner organs from gravity.

Figure 51 Limb Structure of Devonian Tetrapods

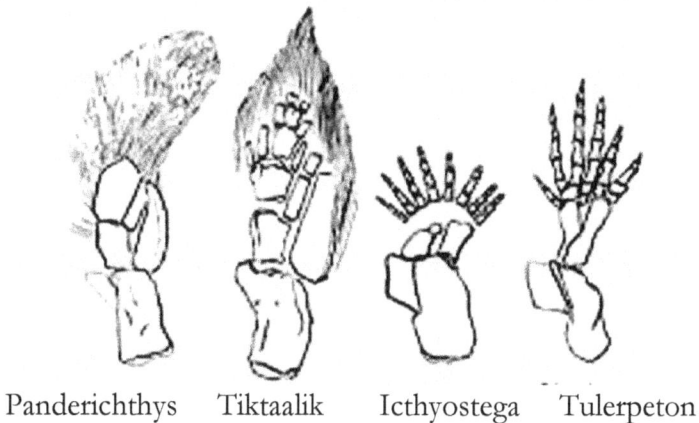

Panderichthys   Tiktaalik   Icthyostega   Tulerpeton

As shown in figure 51, it was Tiktaalik's and his Devonian relatives' limbs that provided the most interesting information regarding evolvement of animals from sea to land existence. The bone structure of the front limbs, when compared to his predecessor *Panderichthys*, suggests an intermediate in the transition from fins designed for movement through water to appendages suited for land locomotion. While still short of the well-developed digits of *Ichyostega* and *Tulerpeton*, Tiktaalik was definitely moving in the right direction.

Size Matters

Large, lush plant life characterized the Carboniferous (280–345 million years ago). The period derived its name from the great coal deposits formed from trees, some as large as oaks, growing at that time. Coal is primarily carbon that originated from ancient trees preserved in the anaerobic environment of bogs, which prevented them from oxidizing to dust. The period consists of two separate eras based on North American (or America's) geology: the *Mississippian*, named after deposits from the period found in the Mississippi Valley, and the *Pennsylvanian*, named for Pennsylvania's great coal deposits.

The Carboniferous was a very hot, humid, period during which plants grew in size, complexity, and density creating great forests. The Carboniferous landscape would at first glance appear not too different from current tropical rain forests except for one aspect, the absence of color; plants had not yet developed flowering mechanisms. Furthermore, Carboniferous forests were silent. While it included many varieties of insects, including cockroaches and some creatures akin to current dragonflies, animal life had not yet acquired the ability to make noise.

Flying insects abounded. Although poorly developed wings that jutted straight out from their bodies limited their speed and maneuverability compared to current aviators, they nevertheless provided a huge advantage over earth-bound cousins while searching for food and mates. The very earliest wings probably resulted from a series of genetic mutations that resulted in membranous outgrowths from the body that originally served other purposes, but proved useful in gliding from one plant to another. Once established, natural selection favored creatures with larger membranes, and subsequent mutations changed the membranous material to a more rigid structure, articulated in a manner to provide aerodynamic features.

The density of lush foliage created a competition among plant species for light, which favored the taller herbs that could spread their fans to the Sun, thereby shading rival vegetation below. Trees were the evolutionary answer to this new environment; however, plants of

270

the day proved inadequate and attaining the necessary height required major fundamental structural changes. Wide, thick, heavy leaves require greater support from the trunk. Dividing the leaf into thinner, sometimes feathery, leaflets solved this. Bundles of lignin added strength to the walls of tubular cells (tracheids) that transport water to the leaves, thereby bolstering the trunk and providing support to the tree crown. This allowed the roots to collect water and other nutrients and transport them through tubes to feed leaves located on higher branches. Capillary action, brought on by energy supplied by water evaporation at the leaf surfaces drawn water upward through the *tracheids*, supplied the "pumping power" for the water. The system was an engineering marvel; as weather became hotter, the rate of evaporation increased, causing even more water to be "pumped" upward. While trees are indeed among natures masterpieces, in the context of natural selection working through a sufficient number of experiments in a proper environment, it may boil down to a simple concept. As Richard Fortey said in his book, *LIFE, A Natural History of the First Four Billion Years of Life on Earth*, the tree is "…a logical consequence of striving for light, coupled with the evolution of leaves equipped with stomata."

The size of animal life also increased greatly in the Carboniferous era including giant scorpions and sharks, which began to look very similar to those of today. Carboniferous deposits yielded fossils of a close relative of today's horseshoe crab (Limulus) along with pigeon-sized dragonflies and millipedes measuring six feet in length. The large plant and animal sizes of the period is attributable to high levels of life's fuel, atmospheric oxygen, brought about by the larger mass of land flora photosynthesizing CO_2 to O_2, and outstripping oxygen consumption by oxidizing sources.

Plant and animal reproductive systems underwent major changes. Seeds, the plant female reproductive entity, became much larger and contained their own food. This allowed them to endure cold or dry periods—thereby increasing their opportunities for germination. Pollen, the male counterpart, remained small, preserving the advantage of wind dispersal for cross-fertilization. These were major advances in the movement toward our current civilization, since agriculture, one

271

of its vital components, would not have happened without them.

A similar revolution occurred in animal reproduction in the area of egg design. Amphibians needed water for hatching their small, numerous eggs and—like tadpoles—their young needed a water habitat during the early periods of their lives. Humans and all other mammals as well as birds, turtles, and reptiles share a common ancestor from the Carboniferous called *amniotes*. Amniotes are a group of animals who give birth from large sized, nutritious eggs, housed in a tough waterproof, yet breathable membranous skin called *amnion*, designed to protect against desiccation and allow offspring to achieve nearly full development prior to hatching. This major procreation breakthrough effectively substituted a system of producing many offspring, each with a poor chance of survival, with one involving fewer young but vastly improved survival chances. It signaled another significant event in life's evolvement since it allowed animals to move further from their marine habitats and reduced the immediacy of their water dependency.

Extinction

In 1915, Alfred Wegener of the University of Marburg, shocked even the scientists of the day, when he published his book, *Our Wandering Continents* proposing that the continents of the world were not static entities but wandering landmasses, alternately converging upon one another and drifting apart. This became the first serious work suggesting the position and size of Earth's continents, as well as the location of the North and South Poles, had ever been any different than today. The concept that continents actually floated about the Earth was so radical that geologists had not considered it even though they had amassed a wealth of information that was inconsistent with the Earth's present configuration. Much of this came from evidence of Permian era glacial flows in tropical areas of Africa, Arabia, and South America, that only an Earth with all of the continents connected could explain. In retrospect, it does not appear to require a great leap of the imagination to see the jigsaw-like pieces of western Africa fitting into the eastern coast of South America but they considered this coincidental since the immutability of the shape of the world had long been unassailable.

Furthermore, physicists, mathematicians and other scientists could find no acceptable mechanism for movement of continents— merging on one hand, then tearing apart again to drift away. One could reasonably ask why it should take so long to become aware of anything as enormous as floating continents. One answer is that we do notice it every time there is an earthquake or volcanic eruption. This of course presupposes that we know that continental drift is the cause. Without this knowledge, it is not surprising that the drift has escaped attention since, even though the continents have separated by some six thousand miles over the last 180 million years, the average drift rate calculates out to be less than two inches per year, or about the growth rate of your fingernails.

Of course, when we speak of converging and separating continents, we also talk about new and disappearing oceans; they don't come and go but are simply rearranged. Lake Baikal nicknamed "The Pearl of Siberia" in southeastern Siberia, which holds the record for the world's deepest lake at 5,315 feet may be the precursor of a future ocean. Baikal is a normal sized lake at about four hundred miles in length and varies from eighteen to fifty miles in width, but it sits atop a deep fault that may be in the process of splitting Asia apart. Geologists have found hydrothermal vents normally associated with mid-ocean rifts on Baikal's floor, along with bacterial mats and other life common to ocean environments. Another 180 million years of plate movement may well produce two separate continents and a new Baikal ocean.

Geologists initially proposed that India, Africa, South America,

273

and Australia were all part of a super-continent called *Gondwanaland* (or more commonly, *Gondwana*), named for the Gond inhabitants of central India. Later evidence added Eurasia and North America to the assemblage to complete the single continent *Pangaea* (meaning "all the Earth"). In the 1950s, work using a new technique called *palaeomagnetics* provided convincing evidence of the wandering pole positions. Geologists established that the poles had been located at various times, on what are now separate continents, a situation that could only have happened if they were once a single continent. However, proof of migratory landmasses did not surface until the 1960s when maps of the midocean floor revealed huge trenches and ridges that had resulted from volcanic activity. Geologists concluded that volcanic activity continually forces magma to the surface at the ridges, creating new ocean crust. The new material exerts pressure on the existing crust, pushing it away from the magma source and slowly changing the location of continents. This phenomenon, later called *continental drift*, explains both the existence and the subsequent breakup of Pangaea.

There were still individual continents during the *Carboniferous* (240–360 million years ago) providing hundreds of thousands of miles of coastal environment, producing weather patterns friendly to the development of a lush highly forested Earth. However, continental drift was already in the merging process, and by the beginning of the Permian era, about 240 million years ago, Pangaea had formed.

The convergence of the continents produced a land mass that extended from pole to pole assuring that pole positions would be located inland. Thus free from the moderating influence of the sea, polar ice began growing at a greatly accelerated rate. This brought on another period of great glaciations and accompanying colder climates, providing an environment for evolving the first animal with a metabolism that could maintain its own temperature. There is no direct evidence of such mammals until after passage of the glaciers in the Triassic some 50 million years later, but many paleontologists believe they will eventually discover a cold-tolerant animal fossil from the Permian that gave birth to live offspring.

A land of vast deserts best describes Permian Pangaea. A barren desolate landscape replaced the lush forests and swamps of the Carboniferous period throughout what is now North America and much of Europe. While all of the major landmasses were contiguous, Pangaea did contain vast interior seaways. One called the *Zechstein* that stretched from Russia through current Europe experienced alternating periods of drying out and refilling with seawater. Another, the *Tethys* named for the wife of the sea god Oceanus, meaning "mother of everything", teemed with many forms of Permian life. The Tethys separated the hot equatorial region of Gondwana from Eurasia and extended to the west from the Mediterranean to what is now Texas. Rock deposits from the Tethys contained unprecedented varieties of *corals, bryozoans, brachiopods, snails,* and *ammonites.* The last trilobites in the fossil record came from the Tethys deposits along with fossils of single cell *foraminiferans* called *fusulines* that measured several centimeters across.

The trend toward larger more sophisticated animals continued in the Tethys during the Pangaea episode of the Permian. Some reptiles were on the scene at its beginning and dinosaurs, sea lizards, and mammal-like reptiles made their appearance before its end. The Permian began with very small mammal-like reptiles but some reached lengths of up to 10 feet during the Triassic. Fossil evidence showed that many had the small heads and blunt teeth of herbivores, but others had large heads and muscles with sharp fang-like teeth for tearing and chewing meat typical of carnivores. The sail-backed reptile *Dimetrodon*, often wrongly depicted as a dinosaur, appeared in the Permian. The sail may have acted as a kind of solar panel storing heat from the Sun to provide the cold-blooded creature morning warmth to get her going early, thus providing a competitive advantage. *Anteosaurus*, a smaller more muscular mammal-like reptile, preyed on a more abundant hippopotamus-like herbivore called *Ducynodont* who sported long tusks and a beak-like jaw good for shredding the tough-leaved plants of the Permian. Legs of the mammal-like reptiles also underwent evolutionary design. They moved from positions on their sides to below their bodies providing an upright gait with greater speed and load bearing capacity, ultimately expanding their diet options and enhancing their defense mechanisms.

The Permian era ended about the same time Pangaea was beginning to breakup with an extinction event that eliminated 96% of the species, and 60% of all families of animals alive at the time. Scientist cannot identify any single meteoric or atmospheric event that was responsible for the Permian extinction but rather lean to a series of catastrophes, none of which has received universal agreement as the primary cause. Unlike earlier extinctions, a series of independent events separated by millions of years appears responsible for such universal annihilation of species. Some evidence points to the seas as having experienced reductions in oxygen and salinity content that would have impeded reproduction of some species and suffocated others. Biologists have proposed both global warming and cooling as probable culprits, and each might well have contributed at different times during the transition. The extinction affected some groups of animals more than others with Marine animals of the Tethys that accounted for much of the diversity suffering more than land dwellers. Trilobites that had been around for 100 million years, disappeared along with brachiopods and many of the coral that built ocean reefs. Fortunately for humankind, some of the mammal-like reptiles survived even though the majority did not.

Jennifer McElwain, while still a postgraduate student at Sheffield University in England in 1999, was studying fossilized leaves growing before, during and after the die-off of Triassic plants. These included rare leaf fossils with cuticles still in tact, taken from a rich deposit discovered by British paleontologist Thomas Harris in the 1920s at Scoresby Sund, Greenland. The cuticle is a waxy polymer coating on the leaves of all plants that contain stomata, the tiny pores that absorb CO_2 from the atmosphere via photosynthesis to provide plant food and produce oxygen. Since the number of stomata on leaves varies with the amount of atmospheric CO_2 available for removal, McElwain realized that she could track CO_2 concentrations from the three periods by comparing the number of stomata on the fossilized leaves. A greater number of stomata would mean higher atmospheric CO_2 and, since CO_2 is a green house gas, higher atmospheric temperatures for the period. Data from the stomata count showed a rapid surge amounting to several times the normal atmospheric CO_2

concentration during the extinction interval, which could have caused high temperatures that dried and burned the plant leaves, thus depriving animals of their main food source. McElwain and four other scientists returned to Greenland in July 2002, to collect more fossils to establish the extent and timing of the event.

McElwain's group not only verified a Jurassic greenhouse catastrophe, they concluded that the cause was the same as for current greenhouse disaster fears; fossil fuels. Examination of stomata from fossil ginkgo leaves showed a spike in carbon-rich greenhouse gasses occurred in the latter half of the Jurassic. Furthermore, the kinds of carbon found in the fossils suggested that carbon dioxide from a huge plant-based source had caused the greenhouse effect; the ratio of carbon-12, the carbon isotope preferred by plants, to carbon-13 was substantially higher than normal. The source turned out to be coal, a fossil fuel born from ancient accumulations of plant life.

Great seismic activity had brought magma from deep in the Earth into contact with huge seams of coal. The location of the burned-out coal beds and carbon dating of the leaf fossils confirmed that the greenhouse event and the rise in carbon-12 in the fossil plants coincided. McElwain described the event as a "double whammy" on the atmosphere. First, a massive upwelling of magma from deep in the Earth dumped vast amounts of CO_2 into the atmosphere. Second, much of the magma did not make it to the surface but mixed with underground coal deposits baking them to produce even greater quantities of greenhouse gasses that eventually escaped into the atmosphere.

They also found evidence that a cycle of global cooling might have happened later in the Jurassic that sounded the death knoll for species evolved from the Tethys' tropical climates. The lower temperatures may have thickened the ice caps, causing major regression of the seas. This in turn replaced many of the friendly habitats of land animals, already in short supply due to the vastness and sameness of Pangaea, with marginal flatlands. However, the leading candidate for causing the Permian extinction may be something of a cop-out titled "a combination of all of the above." One could say the same about the extinction survivors. The ability of the more compact mammal-like reptiles to conserve body heat may have provided an advantage over spindly reptiles by helping them cope with cold. Some fish could withstand the more brackish water of the recessed oceans. What we can be certain of is that all the animals now inhabiting the Earth are beneficiaries of the extinction since it opened new paths for them to evolve.

A Ruling Class — Dinosaurs

The vast extinction of Permian species left a vacuum in the pecking order of the established food chain. Animals, once in short supply due to predators higher on the chain, could now redirect resources and energies previously employed in mere survival toward their advancement. Shortage of predators allowed their numbers to increase much faster thereby enhancing the probabilities of genetic mutations, some of which altered and improved their ability to survive. A new order of diversity and predator/prey relationships emerged. Predatorial survivors of the Permian became bigger, faster, stronger, and equipped with more efficient teeth and claws. These developments placed dinosaurs at the top of the food chain and facilitated their habitation of all Pangaea.

The street of evolutionary advancement is not one-way, however; there are built-in checks that prevent development of a hunter so efficient that he eliminates his own food supply. Enhanced predator skills place new evolutionary pressures on prey animals that must also improve or become extinct, as some in fact do. Many not fortunate enough to evolve adaptive mechanisms are hunted to extinction. However, even though the culling process reduces the numbers of some species by eliminating the slowest, weakest, or least cunning, others among them adapt and survive. The survivors mate and pass along the improved genetic information responsible for their survival, thereby strengthening and gradually stabilizing the species population.

The great extinction that closed the Permian era provided such a scenario. The more competitive environment of the Triassic meant natural selection mechanisms would reward genetic mutations that endowed animals with improved speed, agility, better eyesight, defense mechanisms, and camouflage characteristics—all to improve their likelihood of survival. The makeup of the animal world changed rapidly, causing both predator and prey animals, not blessed with the fortuitous mutations, to disappear from the scene. In addition, the breakup of Pangaea was well underway, providing more tropical coastal areas and climates conducive to growth of vegetation and habitation of animal life.

The first fossilized evidence identified with a dinosaur came from a group of huge chisel-shaped teeth measuring several centimeters in length, found by an amateur geologist named Gideon Mantell, in southern England in 1822. The teeth edges were fluted in a manner similar to those of Iguana type lizards, initially leading scientists to classify the owner as reptilian and assign him the name *iguanodon*. Over the following ten years, paleontologists assembled a complete skeleton of the animal from new fossil finds and from previously unidentified ones taken from amateur geologist collections. Using the collected bones as a starting point, and adding muscularity and skin consistent with his preconceived lizard notions, Mantell attempted to reconstruct iguanodon. The result was an enormous four-footed creature with a moderate length tail that walked low to the ground on all fours.

Many other bones belonging to Cretaceous period animals of similar size and appearance began to surface and by the early 1940s, scientists had identified nine different such giants of the Cretaceous/Jurassic periods. The term "dinosaur," meaning terrible lizard, was coined to identify them. In 1877, discovery of the first complete skeletons of Iguanodon in Belgium forced scientists to make drastic alterations in their perception of the dinosaur's appearance.

The tail contained many more bone segments than Mantell had assembled, and the rear legs were far larger than the forearms, implying the animal had bipedal ability; it could rear up or perhaps walk on its back legs.

The public image of dinosaurs went through a number of makeovers in the late nineteenth and early twentieth centuries, one of the most prominent being the appearance and function of the tail. A huge appendage that trailed along the ground replaced Mantell's shorter tail, and museums around the world followed suit. Later research into the mechanics of dinosaurs, using *Diplodocus*, the giant herbivore from Jurassic Park as the structural model, revealed that the tail actually served as a counterbalance for the rest of the body weight. He held it high where it functioned as a cantilever for balancing the long neck.

Pangaea began separating into isolated landmasses during the Lower Jurassic about 180 million years ago and over the next 30 million years became two super-continents, separated east and west by the Tethys and from north to south by what was at the time a very small Atlantic Ocean. The northern segment, *Laurasia*, included landmasses that would later become North America, Europe, Asia, Siberia, and Indonesia while the southern one, *Gondwanaland* included present day South America, Africa, Arabia, India, Australia, and Antarctica. Since Pangaea had not completely separated in the Mesozoic, dinosaurs populated nearly the entire globe.

The earliest known dinosaurs come from the early Triassic. Considerably smaller than the giants that produced most of the museum fossils seen today, they separated into two distinct types according to hip structure, the bird-hipped and the lizard-hipped. The lizard-hipped variety evolved into the giants of both the carnivorous and herbivorous dinosaurs, and strangely enough into birds as well. Meanwhile, the bird-hipped became the duck billed *Hardrosaurs* and horned dinosaurs such as *Ankylosauria*. Herbivores continued to increase in size throughout their entire 200 million year history on Earth, with one giant named *Seismosaurus* measuring 250 feet and weighing fifty tons. The humongous size has several explanations including natural selection advantages stemming from the ability to reach tree leaves out of reach of shorter species. Size also provides a more efficient use of energy and is relatively easy to attain in evolutionary terms. Therefore, as genetic mutations accumulate over the tens of millions of years, those involving larger size produce animals better equipped to survive and pass on their genes.

The hot versus cold-bloodedness of dinosaurs, a characteristic that separates mammals from reptiles, presents another long lasting and hotly debated question that remains unresolved today. Cold-blooded reptiles have variable body temperatures and normally need warmth from the Sun to function. They do not sweat and as a result, require very little water to survive. Hot-blooded mammals on the other hand, need to maintain body temperature and use sweating as a mechanism for cooling; hence, they require substantially more food and water. The sheer size and body weight of many dinosaurs provided the warmth needed for activity in cool weather even if they were cold-blooded. Some dinosaur bones have canals for the passage of blood, a feature of warm-blooded mammals, but dinosaur skin looks very much like that of cold-blooded reptiles, and shows no evidence of sweat glands. The strongest indication of dinosaur hot-bloodedness came from fossils of dinosaur eggs, some with babies inside with bone structures very similar to hot-blooded birds. However, it is possible they were hot-blooded and vulnerable while young but became more like cold-blooded animals as they grew.

If not cold-blooded reptiles, then what kind of animal are they? Most of the evidence points to birds being the dinosaurs of today. This may not be as preposterous as it sounds if you compare one of the smaller dinosaurs with a featherless chicken. The beady eyes and large rear legs are not too different from those of some Jurassic animals, and the chicken footprint closely resembles those left by dinosaurs. Convert the wings to small arms with claws, add predator teeth to the beak, then toughen and color the skin, and the comparison is not as far fetched. In 1995, paleontologists discovered an exceptionally detailed fossil of a small Cretaceous dinosaur named *Oviraptor*, found still squatting over its nest of eggs. As indicated by its name (meaning "egg hunter"), based on previous fossil evidence connecting Oviraptor with eggs, she had originally been classified as a regular reptile who happened to survive by robbing the nests of other mammals. A complete nest of eggs was not the only aspect of the new find connecting Oviraptor with birds however, as the fossil also showed her in a squatting position identical to that of the present day ostrich.

Even better evidence came when paleontologists compared the newly found fossil of a completely feathered *Archaeopteryx* with *Velociraptor*, the intelligent hunter of Jurassic Park. The Archaeopteryx feathers were the only significant difference between the anatomies of the two animals. Archaeopteryx had a full set of teeth, and the forelimbs that supported the wings had talons nearly identical to those of Velociraptor. Archaeopteryx was the missing link between dinosaurs and birds that led to the conclusion that not all dinosaurs were birds, but all birds were dinosaurs. Since all living birds are hot-blooded, we could easily infer that Archaeopteryx, and therefore Velociraptor and his closest dinosaur relatives, are as well. This is far from substantiated however, since evidence from other studies define Cretaceous birds as cold-blooded.

The Cretaceous was a period of unprecedented evolutionary change. Vocalization began; animals made honking, snorting, and singing noises. The ability to produce sound had a variety of purposes including defending territory, attracting mates, and scaring off enemies. Some believe the crests of crested dinosaurs served as resonating chambers designed by nature to produce a fearsome cacophonous roar. These sounds evolved from the same mutation processes as all other evolutionary change; the genetic alterations responsible for the effect could only survive if it provided benefit for the recipient.

The first animals to give birth to their young alive also lived in the Cretaceous. The Natural History Museum in London displays the fossil of a marine reptile called *Ichthyosaur* with skeletons of its unborn still inside. The Cretaceous also saw the evolvement of great varieties of shellfish including predator varieties. Hunting snails equipped with tooth-like radulae able to bore through shells appeared on the scene. So did lobsters and crabs with claws for crushing shells of smaller prey to expose the flesh within, and starfish that earn their dinner by prying

282

open the shells of clams. Human ancestors from the period were a group called *Xenarthrans*, mammals whose descendents also include anteaters, armadillos, and, sloths.

Flowering plants first appeared in the Cretaceous along with the insects necessary for their mutual survival and perpetuity through a process called *co-evolution*. Plants developed flowering mechanisms to attract particular insects that could facilitate their own pollination, while insects evolved characteristics that gave them access to plant nectar. Current examples of this include certain varieties of orchids that have developed flowers imitating particular species of wasps or bees. Other plants have flower tubes (corollas) of specific lengths, precisely fitted to accommodate a particular species of butterfly. Others have even evolved poisons to discourage unwanted pollinators, but guarantee admission to favored insects that have became tolerant to the poisons—and in some cases even internalized the poisons into their bodies in order to become poisonous to their predators. The specialization was so complete that neither the individual plant nor its particular insect could exist without the other.

The extinction that marked the end of the Cretaceous 65 million years ago was not as far reaching as earlier catastrophes in terms of the number of vanishing species, but it was definitely the most well known due to passing of the giant dinosaurs. The dinosaur has become as familiar and intriguing to us as lions, elephants, and many of the other noble animals of today. The demise of these great creatures strikes closer to home than the loss of trilobites and brachiopods because it leads us to ponder our own vulnerability. French scientist/philosopher Jean-Anthelme Brillat-Savarin wrote the following reflection on "The End of the World" in his *Physiologie du Gout* in 1825, without access to today's extensive fossil record.

> *Incontrovertible evidence tells us that our globe has already suffered several absolute changes, each one nothing less than an end of the world; and an indefinable instinct warns us that more revolutions are to come. Many times already it has been thought that such revolutions were close at hand, and there are still people alive*

whom the watery comet foretold by the worthy Jerome Lalande sent hurrying to confessional. Judging by what has been said on the subject men are prone to invest this catastrophe with vengeance, and destroying angels, trumpets and other no less terrifying accessories. Alas, there is need of no such fuss for our destruction; we are not worthy of so much pomp; and if the Lord so wills He can change the face of the globe without the help of ceremonial apparatus...

Let us suppose for example that one of those wandering stars whose course and mission are alike unknown and whose appearance has always been accompanied by a display of terror; let us suppose, I say, that some comet passes close enough to the Sun to be charged with superabundant heat, and comes close enough to the Earth to cause for the space of six months a temperature of 168 degrees... At the end of that fatal season, all living things both animal and vegetable will have perished; every sound will have died away; the Earth will revolve in silence, until new circumstances develop new germs.

In 1830, Charles Lyell developed his *Principle of Uniformitarianism*, based on the premise that all past geological events are comprehensible through scientific observation of current phenomena. Uniformitarianism grew out of a belief system from the early 1800s, called *Catastrophism*, which attributed the varied landscape of the Earth to multiple catastrophic occurrences. Biologists had already accepted that the Earth had been around for millions of years, and the presence of volcanoes led them to believe it had begun as a hot perfectly shaped sphere of molten rock that had cooled over time to give the crusted surface of today. Catastrophism held that mountains and canyons came into existence when the cooling of the planet triggered volcanic eruptions and sudden violent mountain-sized uplifts of crust. These were instantaneous and often cataclysmic events, frequently accompanied by the extinction of entire groups of animals, followed by their replacement with more new species. The continued visitation of such catastrophes over time, continued to alter the planet's surface leading to present day Earth.

However, by 1832 Lyell became convinced that religious goals motivated catastrophism, and its primary aim was to establish linkages between scientific observations and biblical events such as Noah's flood. Lyell wanted to divorce geology from religion, making it a true science of its own; one built on observations independent of the supernatural. *Uniformitarianism* held that catastrophic processes were not responsible for the varied features of the Earth's surface but the landscape had developed slowly over time through a variety of small but continuing geologic processes.   In the absence of knowledge concerning the movement of tectonic plates, Lyell assumed that all mountains result from molten volcanic rock that breaks through the surface; then over the ages, wind and rain erode and shape them to what we see today, while the eroded sediments form new layers of rock. The principles stated that such physical processes operated at the beginning of the world and continue today.   Rather than catastrophic events, they involved tiny changes that, given enough time, drastically altered the face of the planet, eventually making it suitable for humankind. One of the effects of the principles was to negate previous interpretations of marine fossils discovered on mountainous terrain as evidence for Noah's Flood. It concluded that it should be unnecessary to invoke extraordinary circumstances when readily observable phenomena provide simpler, more satisfactorily explanations.

Lyell's view of Uniformitarianism also applied to life itself and served to inspire Charles Darwin in developing his theory of evolution. Darwin's assertion that the study of existing information can provide evidence as to how new species formed in the past, and the leisurely time scale in which to accomplish it, both resided in Lyell's theory.   During his voyage aboard the Beagle, Darwin developed a history of the Canary Islands by applying Lyell's ideas to the volcanic rock he found there. Darwin may also have borrowed on Lyell's principles in extrapolating his observations of finches on the Galapagos Islands to evidence for species evolution.

The Principle of Uniformitarianism continued to hold sway throughout the nineteenth century and past the middle of the twentieth. However, in 1962, Catastrophism enjoyed a kind of rebirth when Professor Otto Schindewolf of the University of Tübingen published a paper titled *Neokatastrophismus*, suggesting that the great extinction of the Permian may have had an extraterrestrial cause. Although he presented an impressive array of evidence supporting his theory, the most compelling was that the damage wreaked by the event was too great to result from any known earthly source. Nevertheless, geologists ignored Professor Schindewolf's proposal for several years, but events of the next decade would confirm his suspicions and bring a new Catastrophism into the scientific arsenal.

Scientists have proposed dozens of explanations for the Permian extinction, including such intriguing candidates as death by indigestion caused by changes in plant life. There is actually very little evidence in the rocks of the Cretaceous/Tertiary boundary (65 million years ago, usually referred to as the K/T boundary where K comes from the Greek word kreta, an alternative to C for Cretaceous) that tells precisely what happened to cause the extinction. Most geological strata show the world before and the world after the event but shed little light on the critical moment.

The Scaglia Rossa Formation of Gubbio in central Italy, which contains marine fossils from the very late Cretaceous and the early Tertiary (65 – 40 million years ago), provides the best fossil evidence for the boundary period. The boundary between the two layers is amazingly thin, amounting to only one-centimeter thickness proving that the change happened very fast. Some abundant marine animals in the Cretaceous layer, including Ammonites had completely disappeared in the Tertiary. Marine animals in the rock just above the boundary were primarily microfossils, some carrying through from the Cretaceous along with new species that seem to have existed briefly and disappeared. Rocks several meters higher contained animals destined to become part of the Tertiary world; but the thin rock layer at the boundary held the clue that would establish the currently accepted cause of the extinction. Walter Alverez, while analyzing the boundary rock for the element iridium as a part of work unrelated to the mass extinction, found it had ten times more iridium than rocks just above and below the boundary layer. Subsequently, geologists began finding other K/T boundary layers with similar iridium levels in places all around the globe. Since iridium is a very rare element, and most of it comes to Earth aboard meteorites, a vast influx of meteoric material must have accompanied the extinction. Calculations revealed that a meteor with a diameter of ten kilometers, or about six miles, was necessary to deposit a worldwide layer of rock with the K/T boundary iridium concentration.

Finding the real "smoking gun" evidence of the extinction however required locating the hole left by such a massive object. In 1981, Geophysicists employed by Pemex, the national petroleum company of Mexico, noticed circular shaped gravitational and magnetic anomalies in a crater on the Yucatan Peninsula of Mexico. The crater is about 100 miles wide, the size expected from the impact of a 10 kilometer diameter object striking the Earth at a speed of 72 thousand miles per hour. The collision impact was equal to a force of 100 million megatons of TNT, an amount 10,000 times greater than the present nuclear arsenal of the world. Sixty-five million years of sedimentation and tropical forest vegetation had hidden the impact crater from detection. Boreholes of the crater revealed the meteor impact had vaporized sulfur-bearing deposits, which would have released enormous clouds of acid rain over the entire Earth. Glass balls, formed from molten silicates by the heat of impact, littered the crater perimeter.

Since Alverez's breakthrough discovery, geologists have accumulated mountains of supporting evidence for the meteor-caused extinction. These include pieces of shocked quartz, a form of silicon dioxide created at extremely high temperatures and pressures, found in the K/T rock sections. K/T boundary rocks from numerous sites have also shown a layer of graphite resulting from the residue of huge wildfires that burned away vegetation that had survived the loss of sunlight and acid rain attack. All elements of the apocalypse, fire and brimstone, sulfurous acid rain, and darkness over the Earth, defined the extinction.

The main cause of death came from large clouds of dust thrown up by impact of the giant meteor that blotted out the Sun and brought on an eternal night, destroying terrestrial and marine plant life that eliminated the food chains of many species. Without leafy food, the herbivores starved and—after one last grand feast—so did their predators. Greatly reduced temperatures caused the death of millions of species. Torrential rains of sulfuric acid, enormous wildfires, and great tidal waves added to the nightmarish event in different areas around the globe.

After about a decade following the meteor impact during which the Earth was a gray, nearly lifeless wasteland, surviving fern spores began to spread and germinate and, in the absence of competition, proliferate, making the Earth green again. Except for the bird-like flying reptiles, no dinosaurs survived the Cretaceous. However, many birds and mammals, and even some plants did survive. Numerous scientists have offered reasons to explain their success, but all have detractors. The hot-blooded nature of mammals likely helped them survive cold that would kill cold-blooded reptiles. Crocodiles were better suited to the conditions since they feed on a large variety of small mammals and go long periods without food. Peculiar seed designs of flowering plants allow them to withstand long cold periods. Insects on the other hand, have short life cycles and cannot shut down their systems for extended periods since they require living plants for food and shelter. To explain their survival, it is necessary to assume niches on Earth that sheltered plants from the worst of the disaster.

6   Humanity

Mammalian World

Mammal-like animals first appeared in the Cretaceous but didn't flourish in numbers, until after the dinosaurs had passed. These ancestors of current day mammals had been second-class Cretaceous citizens, probably relegated to the role of scavengers living off the leavings of a ruling-elite dinosaurian society; however, the misfortune of these now extinct adversaries was the gain of surviving mammals and birds of the Tertiary since they became heirs to the vacant ecological niches. Not surprisingly, surviving mammals were small creatures and few in number, resulting in a paucity of complete fossil skeletons. Most of the evidence of their early history is in the form of tiny teeth. Other animals, such as crocodiles, lizards, and turtles survived but the new cooler climate, along with limited food resources and shelter limited them to lush tropical regions. These factors allowed the smaller, warm-blooded, and fur-insulated mammals to escape their reptilian adversaries by occupying cooler regions.

Both the numbers of species and their sizes increased rapidly in the early Tertiary. Paleontologists' files include fossils of dog-size shrew relatives that lived only three million years into the Tertiary. Unprecedented variety in mammal life marked the period. Their progeny today includes animals ranging from swift gazelle to plodding rhinoceroses, from camels to tree sloths, kangaroos and rats. Anteaters that prefer tiny insects to a fresh slab of flesh, and pigs that root for dinner abounded. Land mammals, once content to walk on all fours, returned to the sea and became present day manatees, seals, porpoises, and whales. Due to the great size and relatively high numbers of whales and whale ancestors, the fossil history recording their evolution is more complete than that of other species. DNA of modern day whales and hippopotamuses show them very close relatives; in fact, hippos are closer to whales than to the pigs they so closely resemble. Whale and hippo genealogies trace to a group of land mammals of the early Tertiary that separated, with one group remaining on land while the other returned to water. In the nearly gravity-free environment of the sea, natural selection found little use for the hippo-like legs and favored mutations over the succeeding millennia that gradually converted them to useful fins.

Basilosaurus, a twenty-foot long ancestor of the hippo, who lived in the Eocene about 50 million years ago, represents a near mid-point hippo/whale transition relative. Basilosaurus still retained well defined, if not functional, rear limbs and arms complete with fingers. Rocks from slightly earlier in the period contained partial skeletons of *Ambulocetus* with legs arranged similar to those of a sea lion.

Only a partial skeleton, consisting of a skull and jawbone, is available from *Pakicetus*, probably the oldest whale relative, from still further back in the Eocene. However, his ear region shows him closer to his land bound ancestors since it had not yet evolved deep diving capability characterized by later whales blessed with specialized structures to withstand high water pressure. The Pakicetus fossil came from a discovery that included snails and other freshwater animals, indicating that whales may have lived and hunted in shallow streams and rivers early in the Tertiary and only later completed the transition to sea creatures.

Throughout the Permian period, Earth's land animal species population had been relatively homogenous. The climate and resources caused some variation, but there were no boundaries limiting the spreading of animals since the continents belonged to Pangaea. As the landmasses continued to separate during the Triassic and through the Cretaceous, animals on the various continents took separate evolutionary paths; but the K/T extinction leveled the diversity somewhat since survivors from the various continents all shared one thing: unique attributes that aided survival. The isolation afforded by separate environments had a major impact on new species that evolved after separation of the continents. Continents such as Australia and Antarctica became completely isolated with others such as North and South America, Europe, and Africa less so. Panama occasionally served as a connecting point for the Americas but was more often beneath the sea, while the Mediterranean frequently changed its shape, providing land bridges between Africa and Europe.

The prolonged isolation of Australia however, was responsible for its qualification as the ideal laboratory for studying the creative effect of separation on mammalian evolution. Evolution responded by producing a series of animals unmatched elsewhere for their uniqueness. *Marsupials*, animals that house and feed their young in a special pouch, were part of the Permian fauna worldwide when Australia separated from Pangaea. However, isolation caused Australian marsupials to remain marsupials while *placental*, or womb-bearing mammals, replaced them throughout the rest of the world.

In the latter part of the Oligocene Epoch (23-34 million years ago), the area around Riversleigh, a barren region of the Australian Outback, was a lush tropical forest landscape supporting a wide variety of even more mysterious marsupials. In fact, they appeared so strange, for years fossil collectors called them "*thingodots*", only later taking the scientific name, *Yalkaparidon*. Marsupial species continued to multiply throughout the Miocene, many of them strikingly similar to some of the great animals found on other continents today. *Thylacoleo*, a lion-like predatory marsupial roamed the outback in search of *Diprotodon,* a huge herbivore that looked very much like an elephant. Even more recently, marsupial mice that needed close inspection to distinguish them from the placental variety, and the *Tasmanian wolf,* a creature that would look comfortable in a dog show, roamed the plains of Australia.

By the early Miocene, continental drift was changing global weather patterns. Isolation of Antarctica affected the mixing patterns of ocean currents reducing the normal blending of warm tropical water with cold polar water. Thus segregated, the cold water gave rise to the buildup of Antarctica's ice cap to the south, while the warmer water heated wind currents and dried Australia's climate to the north.

These changes were reflected in a major fossil discovery in 1976 along the Gregory River at Riversleigh, revealing a very different history for Australian marsupial life than previously imagined. The find was unique in that the fossils represented animals belonging to more than one age. Moving from one side of the field to the other, a story emerged. One that detailed Australia's entire mammal history as the drying climate altered its animal population. Reduced rainfall caused the forests to recede, destroying the habitats of large herbivores like Diprotodon, thus reducing their populations. At the same time, it produced an overall increase in the variety of separate species. Stream areas for example, contained a greater variety of turtles than found any place else in the world, and swamps teemed with numerous species of crocodile. The trees housed a wide array of opossum and kangaroo species, many carnivorous. The picture that emerged from the Riversleigh find was one of a very diverse marsupial world. One filled with a wide range of species; however, most did not survive the ensuing desert conditions of the continent.

South America has been an isolated island much of the time since the beginning of the Tertiary period 65 million years ago, and like Australia, it evolved a series of unusual placental mammals. These include hundreds of placentals classified under two group names, *Notoungulates* and *Lipoterns,* frequently referred to as horse-like, deer-like, camel-like, rabbit-like and rhino-like, and generally more closely related to humans than marsupials. The isolated continent also welcomed a great line of rodents found nowhere else called *Dinomyids* (meaning terrible mice), one of which grew to the size of a bear, and *Glyptodon*, a tortoise-like animal with a domed shell measuring ten feet across. Another called *Ankylosaurus* was an armored mammal with an appearance best described as a cross between a dinosaur and a modern armadillo.

As in Australia, large, cumbersome creatures like the giant sloth added further evidence that animals evolved in isolation are frequently slow and not particularly intelligent, likely due to the absence of a variety of predators. This does not say genetic mutations that improve the speed of such isolated populations did not occur. They did. However, in the absence of predators, the mutations provided no advantage that would affect natural selection. In the presence of predators, a small advantage in quickness relative to other members of the species causes the mutation to fix in the population in a surprisingly small number of generations, since more of those with the "quick gene" will survive to pass it on. A mutation occurring in an individual member of a random mating, intermingled population, that improved its chances of survival to reproducing age by 5%, could become fixed (has reached 100% of the population) in only 20 generations. Of course, the same rule applies to predators. A newly mutated tiger gene that gives him extra speed to overtake wildebeests will provide an advantage that will spread throughout tigerdom in the same manner. Fast tigers will be less likely to starve when prey are scarce. In the absence of predator/prey relationships, the "quickness gene" will provide no natural selection advantage, so it will not spread and may even disappear.

At the end of the Miocene about 3 million years ago, the Isthmus of Panama emerged from the sea to permanently—or at least until now—link South America with North America, ending the isolation and drastically changing the ecology of both continents. Horses and deer, along with carnivores such as bears and dogs, even elephants, moved south passing armadillos and porcupine going in the opposite direction, many settling down in Central America on the way. Mixing of the formerly isolated populations had an enormous effect on the fauna ecology of the two continents. Many predators found new and easier prey, while others became prey themselves. Natural selection smiled on speed, strength, and efficient weaponry in the form of large sharp teeth and claws; their owners prospered while the slow and inept perished. The oceans on both sides of the Isthmus felt an opposite effect with marine inhabitants assuming different characteristics due to the impact of isolation on their evolutionary paths.

Animals of the early North American Tertiary included survivors from the Cretaceous such as *Diatryma*, a giant carnivorous bird that had given up its ability to fly to pursue a terrestrial hunting career. It had muscular ostrich-like legs for overpowering smaller herbivores along with a huge bill designed to shred them. Animals from the first half of the Tertiary were exotic creatures, most of which have no living relatives today. The second half, however, saw the emergence of species that we would clearly recognize as relatives of our current fellow animal neighbors on the planet. Some of these, including the giant sloth, became a contemporary of modern humans, surviving until the Pleistocene ice age (40,000 to 13,000 years ago).

The appearance of grasses was another Tertiary development of fundamental importance to modern mammals. Although not conspicuously so, grasses are wind-pollinated flowering plants, whose leaves grow from concealed bases that can withstand continuous cropping of its leaves without damaging its ability to replenish. The great expanses of current day grasslands like the savannah, prairie, and pampas established themselves about 20 million years ago during the Miocene, due in part to climatic cooling, coincident with the growth of Antarctic ice sheets. The development of grasslands greatly benefited ruminant animals, those that chew their cud such as cows, camels, and deer. These animals have several stomachs, allowing them to separate the activities of gathering and digesting their food. Pulp from grass hurriedly accumulated in the first stomach, can later be returned to the mouth for more leisurely chewing, thus providing an efficiency for converting grass to flesh not available to their single stomach contemporaries. The relationship between grazers and grasses paralleled that of predator and prey since incessant feeding by animals "weeded out" plants without rapid regenerative growth capability. Like slow prey animals, plants without rapid regenerative growth capability were "weeded out" due to incessant feeding, eventually giving way to plants that returned rapidly.

There were hints that intelligence played a larger role in the ecology of the Eocene than in previous epochs. Hot-blooded mammals need more food for generating body heat than the reptilians preceding them. This caused greater competitive pressures and increased the advantage afforded creatures capable of utilizing strategies born of instinct and experience to capture prey and avoid predators. The more intelligent animals developed social systems requiring mutual understanding that aided their ability to hunt and defend themselves in packs. This increased intelligence needed more neural pathways in the brain, which meant greater quantities of cortical tissue and a larger cranial cavity to house it, factors reflected in the fossils of the period.

Primates

In spite of many global pilgrimages, mammals destined to be the ancestors of modern humans were African. There, near the beginning of the Tertiary period (65 – 40 million years ago), a speciation event occurred that sent one group down an evolutionary path that produced a plethora of current day lemur species, while the other led to all primates, including humans, chimpanzees, gorillas, monkeys and gibbons. At about the end of the Tertiary, another such event caused a split in a species of African primates. One of the two groups somehow made it out of Africa to South America where they further branched into the hundred or so species of New World Monkeys, while the group that stayed home became the ancestors of humans and most of the other primates. No one knows precisely how monkeys accomplished this navigational feat; however, speculators believe they went by raft. No, they did not use slices of tree bark to lash logs together and set out to discover America. In fact they did

Columbus one better (or more accurately, one worse); not only ignorant of where they were going—they didn't know they were going anywhere. When a particularly violent storm swept a small group of them off the West African shore, they probably attached themselves to fragments of mangrove swamps and survived at sea for a period before being deposited on some small island. The fact that Africa and South America were considerably closer then and currents flowed in a westerly direction made the feat more doable. Furthermore, the lower sea levels of the period presented the possibility that a chain of small islands might have served as intermediate rest stops followed by further excursions brought on by future storms.

The original seafarers probably never saw, or even came close to South America. The first arrivals were their descendents; perhaps hundreds of generations removed. The occurrences of such events appear highly improbable. However, once again, we need to consider the number of opportunities presented over the millennia; put into the perspective of millions of storms and millions of monkeys, it is not too unlikely that it happened once. Adding to the probable side of the equation, seafarers have reported numerous sightings of animals well out at sea on such makeshift raft material. These explanations are admittedly speculative; however, DNA analysis of surviving New World Monkeys and African primates provides indisputable proof that migration by rafting or some similar method did happen. The similarities are far too great for them to have evolved separately.

Another speciation event happened within the African group in the Quaternary about 25 million years ago, that produced our last common ancestor with a hundred species of the stay at home Old World monkeys, a group consisting of macaques, baboons, langurs, guenons, and colobus monkeys. As shown in figure 52, our last common ancestor with 12 species of gibbons—probably our first ancestor to practice upright walking—lived in Asia 18 million years ago. Migration from Africa to Asia is not nearly the daunting task presented by the Africa/South America trick since land bridges connecting Africa and Arabia have come and gone numerous times over the last 30 million years.

Figure 52 Phylogenetic Tree of Recent Human Ancestors

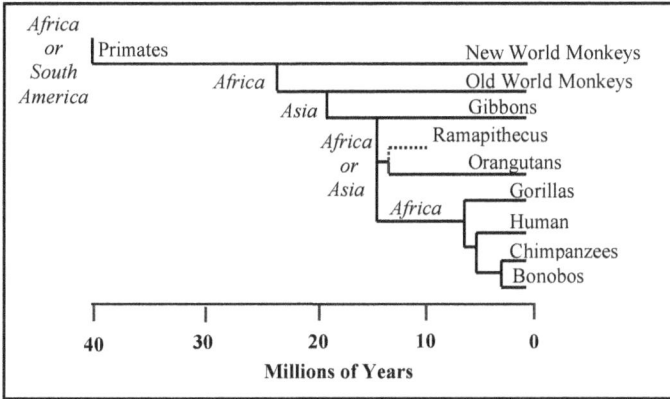

It would be wrong to picture the emigration as a continuous line of Ape-like creatures strung out to the horizon with their leader urging them on to the Promised Land. Like the ancestors of Old World monkeys, theirs was a rambling journey with no destination other than better feeding or hunting grounds. They probably found many such places on their way and settled there for several generations, but scarcities brought on by changing climates, increased population, or competitors forced sub-groups to break off and continue their eastward movement. Once separated, the groups began to diverge genetically and in time produced a variety of new species, most extinct but some still with us. Evidence for the migration theory comes from genetic comparisons that show greater similarities in human and current day Asian gibbon DNA than that of humans and African Old World monkeys. A speciation event within the Gibbon family fourteen million years ago sent a group that became modern orangutans on their own evolutionary path.

Migration from Africa to Asia was not a one-time event; in fact, it probably happened several times during the latter half of the Quaternary. Furthermore, one-way signs did not mark the road; reverse migrations happened as well. About eight million years ago, a

297

member of such a returning group became the common ancestor of two Gorilla species (Eastern and Western gorillas) and another group that included humans and two species of modern chimpanzees (Common chimpanzees and Bonobo chimpanzees). All of the remaining evolutionary change along the lineage that would lead to the emergence of modern humans about 200,000 years ago happened in Africa.

Speciation

The popular view of human evolution shows a man in a business suit leading a parade followed by a cave man, a few bipedal apes, and finally a chimpanzee bringing up the rear. This view implies an ever-improving straight-line path with each species evolving from the one behind, ever-advancing toward creation's goal, the perfect creature, you and me. Evolution in the lineage of a single species progresses through a continuing series of helpful mutations within the population responding to the impact of natural selection. In the absence of reproductive hindrances, the process of evolving into a new species progresses slowly with all helpful mutations spreading, thereby maintaining "sameness" in all members of the population. Eventually, members of the species will have changed enough that—if magically given the opportunity—would not be able to mate with their earlier ancestors. In this type of evolution, a new species is born at the expense of a previous one; with no proliferation of new species involved. More often, new species result from *speciation events.*

Speciation events can happen in two ways, the more common of which is through the *allopatric* model. Allopatric speciation occurs through a stepwise process, the first step happening when some members of a group become isolated from the main population, often via geographical boundaries. Separation can result from barriers brought about by geological changes, such as when a river carves a canyon, or a mountain range forms due to tectonic activity, isolating members of a species from a parent group. *Colonization* can also isolate a new population when a small group leaves the mainland to colonize a distant island. Physical separation prevents individuals from the two populations from mating, halting all gene flow between the two groups. In the next step of colonization, genetic change continues within each population; however, the fact that mutation is a chance event, changes that accumulate in one population will be different from those that accumulate in the other, and so the populations begin to diverge genetically.

The speed at which genetic divergence takes place depends greatly on the environments of the two separated groups. Differences in the environments accelerate divergence since natural selection will exact different rewards and penalties on subsequent mutations. Eventually, the populations may reunite, providing the potential for renewed gene flow. If genetic divergence has not advanced enough, as was the case with the chimpanzee ancestors of humans, individuals from the two populations may be able to breed. If this happens, speciation has not occurred and, in the absence of other hindrances to reproduction such as appearance differences that discourage mating, the two populations will eventually merge, and then accumulated genetic divergence will disappear as interbreeding continues. On the other hand, if genetic divergence has progressed to the point that mating produces no offspring or only sterile hybrids, then speciation has happened, and the two populations have evolved into separate species.

Speciation events have been the leading mechanism in human evolution, relegating gradual change within continuous lineages to a minor role. The first Homo species did not result from a gradual transformation from *Australopiths*. Instead, after separation from the Australopith population, genetic change within the group destined to be humans caused them to be better adapted to their environment, eventually allowing them to replace the Australopiths by being more competitive. The combination of reduced competition and the newly acquired advantage possessed by all its members allowed the new group to flourish. The process then repeats as further mutations produce new groups and finally new species.

Biologists had long thought evolution in humans and other

animals required prolonged isolation in order to affect speciation; that they needed to separate long enough for evolution to proceed to the point where they could no longer interbreed. This is not always the case however, due to another phenomenon called *sympatric speciation*, or reproductive isolating mechanisms—things that prevent or discourage mating outside the group even though biologically feasible. Sympatric speciation happens when groups that are still capable of mating become separated for an extended period, and as a result the two groups develop different mating rituals and other behavioral patterns. After passage of a sufficient number of generations, behavioral differences such as mixed signals during the partner selection process and inappropriate responses to them hamper mating attempts among members of the separated groups. While sympatric speciation has had only minor impact on modern human evolution (although among humans, religious taboos that forbid interbreeding have resulted in detectable gene pool differences between otherwise homogenous groups), it has proved quite prominent among other animals, and as we will see later, may have impacted human evolution through our primate ancestors.

When groups become different enough, reproductive isolation can happen at the molecular level even if mating does occur. Compounds on the egg surface can recognize sperm belonging to a different species and prevent entry through the cell wall. Furthermore, even if the sperm does successfully penetrate the cell wall, chromosome incompatibility can prevent embryo formation; finally even when there are offspring, the hybrid animal may be sterile (e.g. mules born from horse and donkey mating). The fact that these different reproductive isolating mechanisms exist tells us that evolution of new species is not a rare phenomenon. The continual evolvement of new species is nature's way of experimenting with life; new, better-adapted species flourish while deficient ones disappear.

So how do such mutations pose a barrier to mating and gene flow between population subgroups that once belonged to the same species? Chromosomal rearrangement is one answer. Chromosomal rearrangements happen somewhat frequently when large portions of DNA within the chromosome become inverted or repositioned. The most recent common ancestor to humans and mice lived about 60 million years ago, and since then about 300 rearrangements have occurred in the genomes of the two species, or about one rearrangement every 200,000 years. Human and chimpanzee DNA, has shown substantially less divergence, amounting to only 10 rearrangements over the 6 million years since our common ancestor lived, or 1 every 600,000 years. The time was sufficiently long to allow significant rearrangement in the chromosomes but not long enough to affect all of them.

Hominids

Until as recently as 2006, paleoanthropologists believed the fork in the road that sent chimpanzees and humans down separate evolutionary paths happened 6 million years ago. The timing had been statistically estimated, via a technique called a *molecular clock*, as the time required to account for the differences between the two species' DNA; a difference that amounts to only about 2%—another way of saying we are 98% identical to chimpanzees. Timing the points where a genetic divergence occurs is possible even though individual gene mutation is a random event. Each gene has its own mutation probability, so within the larger population, mutations occur in a predictable fashion. Knowing the number and type of differences in the DNA of two related species allows one to back-calculate to the point when their DNA was identical—the time when they were a single species. The 2% DNA difference between chimpanzees and human DNA makes chimps more closely related to us than orangutans. You might not expect this from an appearance point of view; however, our greater similarity to chimpanzees does make evolutionary sense since the most recent common ancestor of chimpanzees and orangutans lived 8 million years earlier than that of chimpanzees and humans.

In April 2006, a group of researchers from Harvard and MIT were able to establish the timing of the events involved in the human-chimpanzee separation through a massive study comparing DNA differences among humans, gorillas, and chimpanzees. The study involved collecting more than 20 million nucleotides—800 times more data than available to previous studies. The new data concluded that a group of primates who were common ancestor to humans and chimpanzees separated about 10 million years ago; however, the separation was not permanent. After taking separate evolutionary paths for the next 4 million years, members of the two groups reunited. The prolonged separation produced substantial genetic divergence between the two groups—but not enough to prevent interbreeding. Subsequently, the hybrid offspring of the reunion, sharing traits from both lines, became human ancestors 6 million years ago.

The type of DNA differences between humans and chimpanzees was consistent with the interbreeding capability of the two chimpanzee groups. The greatest degree of similarity between the genomes occurred in the X chromosome, the location of the many genes governing fertility; this is the expected outcome for any two species capable of mating with each other. Our ancestral interbreeding is also not too surprising from the sympatric speciation point of view since modern humans, particularly the male of the species, are known to exhibit considerable flexibility in their choice of mating partners. Alfred Kinsey's report on *Human Sexual Behavior*, published in the 1940s found nearly 50% of American men raised on farms, and 8% of men overall, reported having had sex with animals.

Since we normally trace our relationship to other species from our own ethnocentric point of view, we begin with modern humans and trace the steps backward, identifying points where our genealogy diverged from another species or group of species. Interestingly, the starting point picked for such a backward journey is irrelevant. If we begin with any one of the two hundred or more species of monkeys, and if we follow its lineage backward to its most recent common ancestor with humans, we would arrive at the same point of divergence that we get by following the human lineage. In fact, this holds true for squirrels, butterflies, amoebas, and oak trees; using any living thing as a starting point, we can trace its lineage back to the same minute organism that was the ancestor of all earthly life.

A recent discovery in the desert region of Chad in 2001 may require adjusting the molecular-clock-estimated evolutionary split between humans and chimpanzees to a point at least a million years earlier. The skull belonging to a new hominid species (hominids are defined as all species on the human side of our last common ancestor with all living apes) named *Sahelanthropus tchadensis* turned out to be the oldest hominid skull on record at 7 million years. Paleoanthropologists had long considered *Australopithecines*, who lived between 1.7 and 4.4 million years ago, in the human ancestral lineage; however, the new fossil find showed Sahelanthropus had several more humanlike features than Australopithecus in spite of being more than 2 million years older. Factors like the point at which neck muscles attach to the back of the skull suggest bipedal capability, while smaller canine teeth and thicker tooth enamel are traits more human than ape-like, all pointing to human lineage. However, accepting tchadensis as a relative is problematic, since his fossils predate the reunion, a fact that contradicts the chimpanzee/human separation/reunion theory. Some have proposed solving the contradiction through a compromise; why not place tchadensis among the inter-breeders who produced the population of ancestral hybrids?

Orrorin tugenensis, who lived between 5.8 and 6.1 million years ago, is probably the second oldest ancestor of modern humans. His leg structure suggests he had advanced to walking upright at least part of the time, while full molar teeth give evidence to a human-like diet of mostly fruit and vegetables with an occasional dinner of meat. The jury is still out on a species named *Ardipithecus kadabba* who lived between 5.8 and 5.2 million years ago who may or may not be a human ancestor. Kadabba shares several traits with the African great apes, leading some to rank them in a grouping with chimpanzees rather than humans. Most however point to a similarity in teeth with a later hominid of 4.4 million years ago, and already accepted in the human ancestral lineage, named *Ardipithecus ramidus*. Both Ardipithecus species were about the size of current chimpanzees but the toe structure of ramidus showed more features associated with walking upright.

Australopithecine

By 4 million years ago, one of the species in the group destined to count modern humans as descendants began to use upright walking as the normal mode of locomotion. The series of genetic mutations that led early hominids to an upright stance was a major evolutionary event; it freed the owner's hands for other activities and allowed his field of vision to include creatures and objects above the underbrush when hunting and being hunted. Paleoanthropologists considered upright walking such an important factor in human evolution that they assigned the species a separate genus called Australopithecus. Fossils of several species identified collectively as Australopithecines date as early as 4.4 and as late as 1.7 million years ago. They include Australopithecus *(Au) anamensis*, *Au afarensis*, *Au africanus*, *Au aethiopicus*; and their descendents *Paranthropus (Pa) robustus*, *Pa aethiopicus*, and *Pa boisei*.

The Genus Homo

The Quaternary period, the one in which we live, began 2 million years ago with another climatic event, the Pleistocene ice age; a period characterized by repeated glaciations that covered nearly one-third of the Earth's surface. Ending only 13,000 years ago, much more information is available about the Pleistocene than previous ice ages. Pollen, preserved in sediments supplied data that allowed accurate estimates of temperatures throughout the period. These showed a Pleistocene that alternated between extended cool periods followed by brief warming trends lasting 1,000 years or so, causing ebb and flow of the ice sheets, moving toward the equator then receding again toward the poles. Plankton moved back and forth across the latitudes in response to the temperature cycles, providing paleontologists another source of data for measuring climatic variation. Core samples of sea floor plankton deposits taken at various latitudes provide more accurate data on their movement than deposits of land-based fauna from higher in the ice flow structure, since subsequent ice flows may have moved the latter southward. More recently, techniques involving decay rates of oxygen isotopes have provided the most accurate measurements of all. The new information shows, that at its greatest extent, the northern ice sheet of the Pleistocene reached slightly beyond the Great Lakes in North America and covered much of England, Germany, and Russia in Europe.

Massive migrations characterized the Pleistocene. Vast herds of warmth-loving land animals gradually moved away from the encroaching ice masses during the cooling phases, only to have their progeny return 50 or more generations later. Some species perished in the cold, while others, like the mammoth and cave bear, prospered. The icy climate forced elephants, hippos, and other warmth-seeking animals to occupy the most equatorial regions, only advancing marginally toward the poles during warming phases. New species, better equipped to cope with cold, evolved during the Pleistocene. These included giants like polar bears and Irish elk, whose increased

305

size provided an advantage for conserving heat.

Figure 53 Phylogenetic Tree of Homo Species

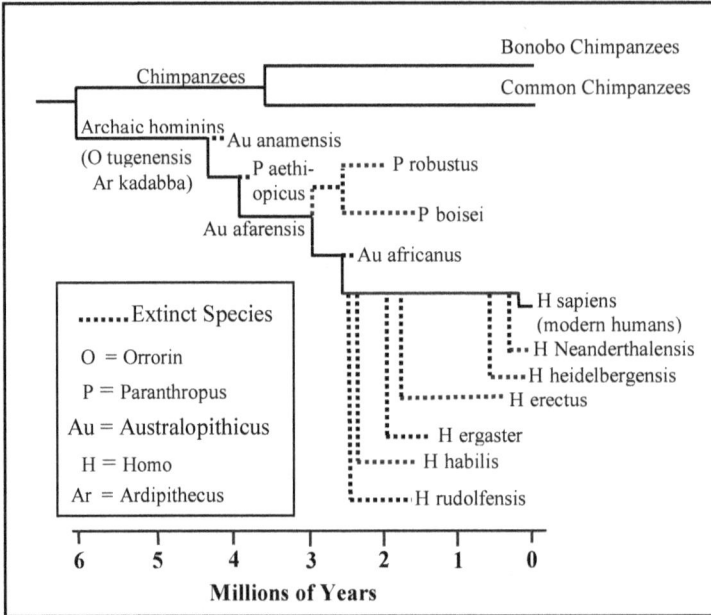

Intelligence also played a roll in coping with these variations in habitat. The beginning of the Pleistocene marked the emergence of a particularly brainy biped called *Homo*, who developed the ability to alter natural objects such as stone and leftover animal parts to make tools for lifestyle upgrading. Homo found he could knock certain rocks together until they split, forming sharp-edged implements for killing and butchering other animals. He used still more precisely honed rocks to separate skin from flesh to provide a warm coat and shoes that allowed him to venture beyond the equatorial regions and survive the ice age cold. They converted bones of their larger prey into hammers and anvils to use for breaking other bones.

It is not unusual for several variations of a species to coexist for an extended period. In fact, evolution in most animals proceeds in this manner. The evolution of genus *Homo* (frequently abbreviated as simply "H"), which spun off new varieties during the early Pleistocene, bears this out. Speciation events spawned several human species that existed at the same time, and some may have known of each other's presence. Figure 53 shows that two different populations of Homo, *Paranthropus robustus* and *Paranthropus boisei*, shared an area in eastern Africa about 1.8 million years ago with *Homo ergaster*, a species destined to be a modern human ancestor. More recently within the last 200,000 years, *Homo sapiens* (modern humans), *Homo neanderthalensis* (better known as Neanderthal), and *Homo heidelbergensis*, coexisted in Africa. H heidelbergensis overlapped with his more ancient cousin, *Homo erectus* who lived between 400,000 and 1.8 million years ago, who in turn was contemporary to their relatives, Homo ergaster and *Homo habilis* from 1.4 to 1.9 and 1.8 to 2.5 million years ago respectively.

Long ape-like arms and short stature gave Homo habilis an overall body structure that least resembled modern humans of all Homo genus species. However, when we include the head and facial characteristics, it becomes obvious he was a major step in the evolution from primate to modern man. The habilis face had lost much of the protruding quality of earlier species making him more "human-like", but the reduced sloping angle of his forehead brought about by his 650 cc brain size truly set him apart. While still well short of the 1400 cc modern human brain, it is substantially larger than the 450cc brain of the earlier Australopithecines. Artifacts found with H habilis suggest his larger brain and higher intelligence gave him an improved level of social organization, manifesting itself via a more sophisticated usage of tools. Most of the tools took the form of stone flakes used to carve meat from leftover carcasses of other predators' victims. Obviously, not the hunter his descendents would be, Habilis was more often the hunted falling prey to *Dinofelis*, a large, now extinct leopard-like cat.

Of course, animal brain size is not a direct measure of intelligence. If so, whales and elephants would rule the world. A great portion of all animal brain capacity goes to managing basic body functions like breathing, digesting, and controlling a million or so muscle groups. Scientists account for these factors by estimating the *Encephalisation Quotient* or EQ (also called the *braininess index*) of humans as well as other animals. EQ compares the actual size of a species' brain compared to what it 'should be' based on body size and a number of other factors unrelated to intelligence that impact brain size. The calculation does not use a simple ratio of brain to body size since the geometry of animals change as they progress from the very small to the very large; e.g. if a spider grew to elephant size, the weight of his body would crush the spindly supporting legs. Since we are not linear creatures but grow in three dimensions simultaneously, our weight is a cubic function; therefore, if the height of an animal increases by a factor of 10 with all of its parts increasing proportionally, its weight would increase by a factor of 10^3 or 1000. The animal's leg strength however is a function of the cross-sectional area of its supporting leg muscles, making it a squared rather than a cubic function. So, with the miraculous growth spurt, the creature's leg strength has increased only by a factor of 10^2 or 100—leaving him with 10 times more weight than his poor legs can handle. Therefore, legs of huge land animals such as elephants and rhinoceroses must be substantially thicker in proportion to the rest of their bodies to accommodate the added load. The brain capacity needed to service all of the bodily functions, on the other hand, relates more to the brain surface area than mass, and this increases by a factor closer to the square than to the cube of single-dimensional growth. Therefore, while larger animals require larger brains, volumetric brain size is somewhat smaller than one would estimate by increasing body size proportionally.

When comparing Encephalisation Quotients for different species, we must also account for portions of the brain that deal only with the animal's body functions For example, cold-blooded reptiles do not require as much brain surface to support their lower metabolism as warm-blooded mammals. These factors become especially important when comparing animals from broad categories like all vertebrates. Ranking the EQs of human species uses a baseline that sets the combined average EQ of all mammals at 1 and examines how individual species rank numerically compared to the average. Some species of Old World Monkeys rank near the base EQ of 1, indicating their intelligence is comparable to the average of today's mammal population. Modern humans score a whopping 6 on the EQ scale meaning our brains are six times larger than would be necessary to perform the functions required by the average mammalian. Brain size was definitely increasing during the time of *Homo afarensis*, who measured just above 2, and later still, Homo *(H) boisei* and *H robustus* skull measurements gave them a 3. However *H habilis*, a contemporary of boisei, and a species more likely to be on the direct ancestral line of modern humans, was closer to 4.5.

Homo erectus, blessed with an EQ similar to habilis, became the first migratory humans when they left Africa for what is now Indonesia. A direct descendent of H ergaster, erectus was probably the first species that most of us would consider far more human than ape, both in terms of appearance and lifestyle. About the size of modern humans, H erectus had an upright stance and a larger cranium, which further lessened the steepness of his sloping forehead. Throw in smaller teeth and shorter arms, and he appeared very human-like indeed. Erectus was also the first human to invent hunting tools and to take control of his own destiny by developing an organized approach to hunting; factors that for the first time allowed him to function more as hunter than prey. Evidence found near their remains (hand axes sharpened on both sides by chipping away bits of stone from flint-like materials) suggest sophisticated stone tool making ability. Some evidence indicates erectus may have even taken to the sea on self-assembled rafts and possibly mastered the controlled use of fire.

About a million years ago, a population of Homo neanderthalensis inhabited Europe and continued to do so until about 24,000 years ago. Until very recently, paleontologists considered H neanderthalnesis the latest surviving relative of modern humans. This may very well have changed in 2003 when they unearthed a nearly complete 18,000-year-old fossilized skeleton of a species named *Homo floresiensis* from a cave on the Indonesian island of Flores. Found with parts of other similar sized individuals dating from 94,000 to 13,000 years ago, the fossil measured about 1 meter in length. The site also contained tiny but sophisticated stone implements appropriate to smaller humans. However, many researchers are not convinced that floresiensis represents a new species but may simply be a dwarf version of H erectus. Regardless of whether the discovery turns out to be erectus or a newly discovered species, the date when our last surviving relative lived moves forward by another ten millennia.

Modern Humans

Discovery of the oldest modern human fossil could well be attributed to the former Ethiopian Emperor Haile Selassie, who in an off-hand conversation on the subject of fossils, questioned why none had been found in Ethiopia. The emperor's casual inquiry led paleoanthropologist Louis Leakey to form an expedition headed by his son to excavate on the Omo River. The mission, launched in 1967, succeeded beyond anyone's wildest dreams. The group, headed by Louis's son, Richard Leakey, discovered a single skull that remains today the oldest known modern human fossil. The face had smaller brow ridges with a much steeper forehead that gave it a flatter appearing face than skulls from all previous finds, leaving little doubt that it belonged to a species common to modern humans.

Selassie's fossil turned out to be 130,000 years old; however, this does not mean the speciation event producing modern humans happened at that time, as any date based on a fossil find will necessarily be low. The initially small population of any new species makes discovery of a fossil representing one of its very early members unlikely. Therefore, when possible, other techniques are used. One of the most reliable is a genetic technique based on a section of DNA called *mitochondria*. It turns out that the remarkable similarities in mitochondrial DNA in humans from all over the world make it possible to measure the timing of speciation.

A group of molecular biologists working out of the University of California, Berkley in the 1980s, developed a technique that used variation in mitochondrial DNA to establish the timing of evolutionary events. Mitochondrial DNA, which contains 16,569 nucleotides arranged in a circular loop, is unique in animal genomes in that we inherit all of it from our mother by transmittal through the egg cell. Cells of all living organisms use mitochondria to break down complex organic compounds into a form they can use for energy to run the cell's many chemical reactions. The California group wanted to determine if they could detect trends in human evolution by comparing DNA sequences in the mitochondria of different groups of people around the world. Although there are trillions of mitochondria within an individual, each with the same DNA sequencing, there are differences between individuals that give clues to ancestral commonality. Since we inherit all mitochondria from our mother through the egg cell, we can state with absolute certainty that all 6 billion people on Earth descended from a common female ancestor.

Steve Olson outlined the logic of this surprising statement in his 2002 book *Mapping Human History*. The logic goes something like this: Of the 6 billion people, 3 billion are women and 3 billion men; however, men do not factor into the calculation since they only receive mitochondria but do not transmit it. Now consider the generation of women one generation removed, which would include the mothers of all females living today. Not all of this generation had daughters; some had only sons and others had no children at all. Therefore, the mitochondrial DNA of today's 3 billion women must derive from a subset of the mitochondrial lineages from the previous generation. Of course, there are more females alive today than in previous generations (true only because some women had more than one daughter). The grandmothers of all the women alive today are a similarly reduced subset of their mothers, so the mitochondrial lineages from each preceding generation must be smaller than the number of such lineages from the succeeding one. This generational reduction in contributors to mitochondrial DNA continues, reducing the contributors from the billions to the millions and further into the thousands, hundreds, tens and finally the single digits. The number of extant mitochondrial lineages eventually shrinks to two—two women from whom all 6 billion of us descend. The two female lineages could have extended back in time for a considerable period where each had a separate mother, but eventually the two women would have to be sisters, and their mother would be mitochondrial Eve, the female ancestor who supplied all of the mitochondrial DNA on Earth.

As Olson points out, naming this woman Eve is in a sense misleading; it implies that she was the only modern human female alive at the time, whereas there were actually many. All of them, however, received their mitochondrial DNA from another female; perhaps a non-modern human who lived further in the past, but it's important to note that none of their mitochondria has survived. This process, in which a DNA sequence in many people traces back to a sequence belonging to a single person from the past, is called *coalescence*. Mitochondria is but one of many nucleotide sequences in our chromosomes and coalescence can be applied to the other sequences as well, although few provide such good examples for illustration since most are complicated by being received from both mother and father.

Genealogists have analyzed the sex determining Y chromosome in a similar manner. Just as mothers pass mitochondria only to their daughters, fathers pass the Y chromosome only to sons—unaffected by a mother's contribution. If a man has only daughters or no children, his Y chromosome dies with him, and the same winnowing process traces backward and coalesces with the Y chromosomal Adam. Eve and Adam never had offspring together however, and probably did not even live at the same time since his Y path was completely independent of the one taken by Eve's mitochondrial DNA. Coalescence is more complicated for DNA in other chromosomes due to the shuffling of genetic contributions that takes place in the recombination process (where random sections of matching DNA are exchanged between chromosomes). Still it applies to particular segments of the chromosomes. For example, all 6 billion of us inherited the ten-nucleotide string at the end of chromosome 6 from a single person from the past. Geneticists have estimated that the entire DNA of all humans alive today came from about 86,000 ancestors, two of which were mitochondrial Eve and Y chromosomal Adam.

Even though we can trace our entire DNA to individual ancestors, it is not identical to theirs since DNA copying is not a perfect process and mistakes occur. In fact, these mistakes serve as the engines of evolution, without which bacterial blobs alone would populate the Earth. Although fortunately rare, DNA copying mistakes, or mutations, are statistically constant, occurring only once in billions of copies. This property of DNA allows scientists to estimate the number of generations and hence the length of time since coalescence, by comparing the number of changes in DNA from different periods or populations.

An analysis of the worldwide variation in mitochondrial DNA showed that mitochondrial Eve lived about 150,000 years ago. Strangely, Y chromosomal Adam lived a similar but slightly shorter time ago. Geneticists are not sure why, but much of the DNA coalescence occurs in this time frame, about 150,000 years ago; however, they do know that it happens more rapidly in small populations. In these populations, a few people become the ancestors of large numbers of descendents, which causes their genes to pass through a genetic bottleneck. This is consistent with what one would expect when the first anatomically modern humans arrived on the scene. A small tribe became isolated for a long time, allowing the group to acquire certain characteristics that set them apart from other humans. DNA evidence supports this view and shows a genetic bottleneck occurred between 100,000 and 200,000 years ago; human population numbered about 20,000, and from this group, located somewhere in Africa, we all descend. Considering such nuclear and mitochondrial DNA evidence, molecular biologists place the date that modern humans first appeared in Africa at slightly less than 200,000 years ago.

While humans have become somewhat smaller, lighter, and

compact due to fewer physical demands on the body, brought about by the advent of agriculture—minor physical changes in evolutionary terms—no significant evolution has occurred within the human species in the last 100,000 years. Humans from all races and locations on Earth share an equally distant kinship with these early Africans; the same number of generations separates all of us from them. Expansions away from eastern Africa, however, have resulted in human groups with minor differences in genetic histories that hold the clues to understanding our settlement of the planet.

The Bushmen of southern Africa provide the best evidence of the constancy of human evolutionary history over the last thousand centuries. The Bushmen were one of the first groups to move away from eastern Africa to the south, where they have remained to become ancestors to all of the region's original people. This isolation from other human groups for much of the last 100,000 years has made them genetically closer to the first modern humans than any other group. However, comparisons of their genetic profiles with those of other groups have not shown differences that would indicate divergent evolutionary paths. Furthermore, other than characteristics born out of their living circumstances, Bushmen are fundamentally the same as people from all other parts of the Earth, including their intelligence, sense of humor, affection, and love of music and art. Therefore, if isolation has caused the Bushmen to be closer to early modern humans, then we can conclude that either modern humans have not changed in the last 100,000 years or that separate groups have all changed in step—an unlikely scenario.

Kinship

The fact that all humans received their mitochondrial DNA from a single woman, all of their Y chromosomal DNA from one man, and in fact our entire genetic makeup from only 86,000 people from our past, begs the question: Why are we so different? Why such variation in skin color, nasal width, eye shaping, and hair color? The answer lies in the process of cell division and is fundamental to the heart of biological evolution in all living things. The copying of DNA during cell division is an amazingly accurate process, however; replication errors do occur. The copying process can continue error free through millions of nucleotides before a difference occurs. When a woman's mitochondria are copied to make an egg cell, the 16,500-nucleotide sequence is normally the same as in all of her other cells. Occasionally though, a mistake, or mutation, occurs; perhaps a G shows up where a T should be or an additional sequence is either added or missing from another spot. Mutations can also result from exterior disturbances such as altering the DNA sequence by radiation or presence of foreign chemicals. Most of these changes occur in non-functional parts of DNA, but occasionally, changes are important and result in a new effective mutation.

Comparisons of these mutations among groups are a key factor in the process that allows us to reconstruct our genetic history. Steve Olson gave an interesting illustration of this by beginning with an assumption that mitochondrial Eve had two daughters, the younger having a mutation in her mitochondrial DNA. Then all of the people alive today who descended from that daughter would share her mutation, while those descended from the older daughter would not. Therefore, Mitochondrial Eve would have produced two separate mitochondrial lineages. Lineages produced in this manner are called *haplotypes* and a group of related haplotypes, a *haplogroup*. Geneticists use both designations to establish genetic relationships between human groups. Now, suppose the younger daughter with the mutation moved to southern Africa and became the source of all the

315

Bushmen's mitochondria, while the other daughter remained in eastern Africa, making her the mitochondrial ancestor of the rest of the world (since this group was the source of all of the subsequent global expansions). All Bushmen would therefore have the same mutation as Eve's younger daughter, but the rest of the world would not. In fact, geneticists used haplotype analysis to establish that Bushmen were one of the earliest distinct populations of modern humans. They own the oldest mitochondrial DNA haplotype found in the world.

Mutations in mitochondria are comparatively rare due to its relatively small size compared to other organelle sequences; however, DNA sequences of chromosomes are 400,000 times larger than mitichondria, making them much more likely mutation candidates. In fact, our cells contain so much DNA that every act of human procreation produces at least one unique mutation. The average union will result in a child with nucleotide sequences that differ from the parents chromosomal DNA in about one hundred locations. Even identical twins have a few unique mutations that occur after division of the fertilized egg. When the children mature, the inherited mutations will remain in their egg and sperm cells and pass on to their children along with new mutations that will define their genetic uniqueness.

Unknown billions of mutations have combined to form the human race as well as all other species; however, every one of them resulted from just one individual. Mutations cannot appear simultaneously throughout a population; they occur in a single cell within a single individual and then pass on, and in some cases spread, through succeeding generations. While true that genetic mutations are the prime movers, evolution of species involve only a small fraction of them. Only 2% of human DNA has any known purpose, making mutations within the remaining 98% irrelevant; furthermore, most of the mutations in the functional regions have no effect on our bodies due to built-in redundancies in the genetic system.

Harmful mutations are the second most common kind; but we do not pass most of these on to succeeding generations since their recipient either does not survive or is incapable of producing offspring. Beneficial mutations, the rarest of all, survive since they help the organism in its battle for survival. A mutation that aids reproduction also is passed on since the recipient will have more offspring allowing the beneficial mutation to spread faster, giving it an advantage in species survival.

Considering the complexity of genetic structure and the myriad possible consequences that can result from alterations, it makes sense that beneficial mutations occur far less frequently than either neutral or harmful ones. Nucleotides within a gene can combine a nearly infinite number of ways but only a tiny portion of these produce a viable living entity (or as Richard Dawkins put it in *The Ancestor's Tale*: "...there are many more ways of being dead than of being alive"). The genomes of every species have undergone countless mutations, each either helpful or neutral to the surviving members. This is true since recipients of harmful mutations perish immediately, or natural selection eliminates the harmful trait. Even mutations that might otherwise be beneficial will be harmful if their effect is too extreme; for example, longer legs are frequently advantageous, however a four-foot increase in the length of a human thigh would create not only a peculiar looking individual, but would exact evolutionary penalties as well.

Helpful genetic mutations played an important role in adapting human groups to the varying environments on different parts of the Earth. Olson provides an excellent example of such a mutation that deals with skin pigmentation. Dark skin is a major advantage for people living in equatorial regions where they face greater risk of skin cancer from exposure to harmful ultraviolet radiation from the Sun. West African blacks who inherit a mutation that suppresses the production of melanin, the chemical responsible for dark skin pigmentation, usually develop skin cancer at a very early age. If they die before reproducing, they do not pass the mutation on and it disappears from the population until it happens again in another individual.

On the other hand, dark skin can be a liability in locations where the Sun is less intense. The body uses ultraviolet light absorbed through the skin to synthesize vitamin D needed for proper bone growth and for avoidance of rickets. Dark skin can absorb the high intensity ultraviolet light near the equator and produce vitamin D, but it blocks out the dimmer light of more polar latitudes. Today, we can solve this problem in arctic-like climates by adding vitamin D to milk, but dark skinned children from higher latitudes are still subject to rickets if their diets do not include vitamin D.

Rickets, a disease that disrupts the normal mineralization process in some children, can result in disfigurement by softening and distorting bone structure. Women suffering from rickets are also subject to a side effect, which narrows the pelvis and may cause premature death of the mother in childbirth. After early Africans had migrated north into Europe, this genetic susceptibility to rickets probably resulted in increased death rates among the female population. If not for genetic mutations, it is likely that absent another source of vitamin D, rickets would have presented such a hindrance to procreation that our ancestors would not have survived at northern latitudes, limiting human population to tropical regions. While there is no supporting evidence, it may be that a mutation occurred in one of the women that lightened her skin, allowing her to produce more vitamin D and greatly decrease her susceptibility to rickets. She would therefore have had a better chance to survive childbirth, bear more children, and pass along her genetic mutation. Such a genetic advantage would spread throughout a group until the entire population possessed the trait. On the other hand, a negative genetic disposition to reproductive success of only two percent can cause the responsible gene to disappear from a population in a mere twenty generations.

Typically, when people discuss their family tree, they place greater emphasis on the male lineage since, as the search extends further back in time; information about the father's surname is easier to obtain than from the ever-changing female side. However, since groups of humans result from the mixing of many other groups preceding them, any given person descends from a great host of ancestors. We have two ancestors—our mother and father—from one generation ago while our grandparents give us four from two generations ago. This number goes on to double with every preceding generation giving us eight from three generations ago, sixteen from four and so on. If your family happened to be making history over the last few centuries, your search should identify 1,024 great-grandparents living only 10 generations ago or—since we measure generations as 20 years—200 years ago.

The number of generations increases exponentially, resulting in more than a million ancestors 20 generations ago since each of our 1,024 ancestors of ten generations ago also had 1,024 ancestors, or 1,024 x 1,024. To determine our number of ancestors thirty generations ago, we multiply the number again by 1,024, and find we had more than a billion ancestors only 600 years ago in the year 1400. However, this is obviously not the case since the Earth's population at the time was only about 375 million people. We do not because we have been counting some of our ancestors twice. For example, a female ancestor from several generations ago may have had two daughters, each of whom married and bore children. If a child from each of these families, even though cousins, married, they could only have six instead of eight grandparents since one set is common to both sides. Today, marriages between cousins are relatively rare in most countries, but in many, second cousin marriages are common. Children from these marriages have fourteen instead of sixteen great-great grandparents.

The vast complexity of our ancestral heritage makes it difficult to appreciate how closely we are related, even to the most distant populations around the world. But any attempt to amass a complete genealogy of one person to verify this would be doomed to failure since obtaining complete records of an individual's lineage, even for the most notable, becomes unmanageable after going back only a few generations.

However, by using a computer programmed with a set of reasonable assumptions, statisticians estimated the number of ancestors for a "statistical any person" (SAP). They found that SAP had several million living ancestors about forty generations ago. In fact, Yale statistician Joseph Yang calculated that 80% of the people who lived in a given part of the world at that time are direct ancestors of everyone living there today. Lineages for the remaining 20% have since become extinct, making them ancestors of no one. For generations beyond 800 years ago, computer simulations show the individual has more ancestors than the existing population of the time. This represents a critical threshold since the world human population in 1200 totaled 350 million, scattered mostly throughout Asia, Africa and Europe. The facts also suggest that most of the adult population in the region of the world where our particular ancestors lived are our relatives.

Therefore, if we know the homes our ancestors, we have the right to claim kinship to any of the historical figures living there more than 800 years ago who had children. Just as there is a very high likelihood that your next breath will contain at least one oxygen atom from Julius Caesar's dying breath, if you are of Middle Eastern descent, there is an equally high probability that Caesar is an ancestor. This does not say you have any of Caesar's DNA, only that if your ancestry was for example, thirty percent South European, then approximately that percentage of your DNA came from people of that region alive during Caesar's time; and, for the reasons stated above, makes it highly probable that he is an ancestor. Steve Olson likened our DNA to a "patchwork quilt" "...stitched together from the DNA of our ancestors." Mitochondrial Eve supplied our mitochondrial DNA, but the other DNA in our chromosomes came from thousands of different individuals living at various times in history.

About 100,000 years ago, a small group of modern humans left Africa and settled in what is now the Middle East. The mitochondria of this group, who were to be ancestors to most of the human populations of the world, had many of the mutations of the larger remaining African group, but since DNA of no two people can be exactly the same, they left many of the mutations behind. In 1932, archaeologists working in Skhul and Tabun caves—two limestone formations located about 200 yards apart near Tel Aviv, Israel—uncovered a series of fossilized skeletons. Skeletons from the Skhul cave had long limbs and skulls with concave cheekbones and a high rounded forehead. Their brow ridges still resembled those of archaic people; however, their overall appearance suggested they were modern humans. In contrast, skeletons from the nearby Tabun cave had short thick limbs—an adaptation to cold climates—and skulls with sharply slanted foreheads and heavy brow ridges that set back from the lower part of the face, indicating they belonged to Neanderthals. The name comes from the Neander Valley near Düsseldorf Germany, where quarrymen working in a limestone cave first discovered Neanderthal fossils in 1876. The workers thought they had uncovered the bones of a cave bear, but fortunately, they called in a high school teacher who also happened to be an amateur naturalist. The teacher knew immediately that the bones, while human, were not like any human alive today.

The relationship between occupants of the two caves remained a mystery for the next fifty years. Some believed that Neanderthals occupied the Middle East first and evolved into modern humans, which would explain why paleontologists had not found any Neanderthal fossils younger than 30,000 years old. Another possibility was that modern humans arrived first when advancing glaciers of the last Ice Age forced them to a warmer climate. The answer had to wait until 1980 and the development of two new dating techniques, thermoluminescence and electron spin resonance. Application of the new techniques to the Skhul and Tabun fossils produced completely unexpected results showing Neanderthals there before and after the modern humans. Neanderthals had lived in the Tabun cave as long ago as 200,000 years and as recently as 45,000. All of the fossils from the Skhul cave were nearly 100,000 years old, making them the oldest modern human fossils ever found outside of Africa. The fossil record of modern humans in the Middle East breaks off about 80,000 years ago, indicating that group did not survive or perhaps migrated back to Africa. About 45,000 years ago, however a group of modern humans returned, and this time the Neanderthals disappeared forever.

Anthropologists have not adequately answered the question of how modern humans interacted with the Neanderthals; perhaps they had a symbiotic relationship, exchanging stone tools and food or even mates. However, the more popular scenario sees modern humans as the bitter enemies of Neanderthals that eventually cause their extinction through use of higher developed tools and weaponry to eradicate them in war over hunting lands.

Geneticist Svante Pääbo led a team of scientists who shed light on these questions through laboratory experiments performed on Neanderthal mitochondrial DNA at the Max Planck Institute for Evolutionary Anthropology in Leipzig, Germany. Testing was limited to mitochondrial DNA since different cells contain from 100 to 10,000 separate copies compared to only a single copy of nuclear DNA (DNA located in the cell nucleus), making it much easier to extract testing size quantities of mitochondrial DNA. Biochemists had discovered earlier that significant quantities of DNA could remain unchanged in bones for thousands of years if kept in a cool dry place; but it was usually not found in quantities needed to run the many laboratory tests required by geneticists. By 1996, biochemists had solved the problem with new techniques that provided the large samples needed for testing by making multiple copies from small amounts of DNA. The German Museum of Natural History, which housed the Neanderthal bones the quarrymen had found in the Neander Valley, gave Pääbo and his workers a small portion cut from an arm bone to provide material for testing.

The technique, which is extremely elaborate and requires precise laboratory savvy, involves grinding about one gram of the bone into a powder then mixing it with a liquid solution that extracts the DNA material. The liquid portion containing the DNA is then separated by centrifuge and combined with the four DNA bases—adenine (A), thymine (T), cytosine (C), and guanine (T). As in the transcription process for protein manufacture described in chapter 5, the bases attach to their complimentary pairs (As to Ts and Cs to Gs and vice versa) producing a complimentary copy. A second copying procedure similar to the translation process of chapter 5, then converts it to the original configuration, thereby doubling the quantity of DNA. A centrifuge then separates the resulting mixture and the process repeats numerous times with the amount of test material doubling with each cycle, until there is enough DNA to conduct the tests. Pääbo's initial analysis revealed a thymine at position 16,223 where modern humans have a cytosine, and an adenine at 16,254 instead of the guanine in human DNA. As Pääbo's team continued testing, they uncovered more and more discrepancies, leading them to conclude there were far too many DNA differences for the Neanderthal to have descended from the same mitochondrial Eve as modern humans. In fact, they gathered enough information to calculate that the lineages leading to Neanderthals and modern humans diverged about 315,000 years ago.

Mitochondrial DNA ranks as one the most important parts of all life processes although its function is quite limited. Its sole purpose is to provide energy for the cell by generating ATP (adenosine triphosphate) which transports chemical energy for metabolism. While mitochondrial DNA differences show that humans and Neanderthals did not interbreed, they do not tell how much the two species have diverged in other areas. To answer these questions, scientists had to develop techniques to copy nuclear DNA, the portion housed in the cell nucleus that carries the information for constructing living organisms. Pääbo and his team used a novel sequencing method that employs a small machine, the heart of which is a credit-card-size chip containing 1.6 million tiny wells that function by sequencing nuclear DNA fragments. Each well contains an equally tiny bead, and each bead accommodates 10 million copies of a small DNA segment. Sequencing takes place by washing the four nucleotides of DNA across the chip one base at a time. When a base finds its complimentary nucleotide in a DNA fragment, it immediately pairs with it—A to T, C to G as well as T to A and G to C—and emits a flash of light, detected by fiber optics programmed on the chip. The DNA source material came from male Neanderthal bones found in the Vindija Cave located outside Zagreb, Croatia, that had been hidden by glaciers for most of the last 45,000 years. Over time, Pääbo and his colleagues sequenced a whopping million base pairs or about 0.03% of the Neanderthal's entire genome.

A comparison of the nuclear DNA from the 45,000-year-old Neanderthal bones with samples from 35,000 and 100,000-year-old modern humans showed DNA of current humans much closer to either of the early modern humans than Neanderthals. As work progresses with the new cloning techniques, geneticists should identify many biological differences between Neanderthal and humans.

Many scientists believed language was the barrier that separated the two species; Neanderthals were thought to be genetically incapable of speaking. However, in October, 2007 Paleogeneticist Johannes Krause of the Max Planck Institute in Leipzig, Germany, showed Neanderthals may have been at least biologically capable of complex languages. A study reported in the journal *Current Biology*, compared FOXP2 genes—which as we will show in the next chapter is intimately involved with language ability in humans—of Neanderthals and chimpanzees with that of humans. While chimpanzee FOXP2 DNA was different, the Neanderthal and human DNA were identical.

Whether or not language impeded their interaction, the preponderance of data suggests no significant interbreeding took place between modern humans and Neanderthals even though they coexisted in numerous locations for several millennia. This is somewhat surprising since modern humans, particularly the male of the species, are known to demonstrate considerable flexibility in their choice of mating partners. Alfred Kinsey's report on Human Sexual Behavior, published in the 1940s found that nearly half of American men raised on farms, and 8% of all men, said they had sex with animals. Possibly, the long separation between Neanderthals and modern humans made their DNA incompatible for reproduction. There could have been alignment problems with the egg and sperm chromosomes of a modern human Neanderthal couple, or perhaps chemical signals on the egg surface blocked sperm from penetrating the cell walls. It is also possible that mating did produce children but both groups rejected them due to their strange appearance, so the offspring never survived to pass on their genes. On the other hand, they may simply have found sheep more attractive.

A number of theories can explain the disappearance of Neanderthals and their DNA, the most obvious being that modern humans wiped them out in territorial wars as they expanded their occupation of the globe. However, the only fossil evidence of Neanderthal injuries that imply war-like activity comes from periods when modern humans were not present, and therefore appears to be associated with internal conflicts.

The most likely scenario for their demise is that they suffered from inability to compete with certain technological advantages acquired by modern humans. These advantages probably gave humans an edge in food gathering or bearing and rearing children. Computer models show that a modern human advantage in child mortality of only two percentages points would be adequate to drive Neanderthals to extinction within a thousand years. The competitive advantage that has received the most agreement among scientists is language. This would seem to imply that the superior intelligence of the modern human endowed him with abilities denied to the brute Neanderthal. However, the Neanderthal brain size was actually larger than that of modern humans, so there must have been numerous factors.

7   Civilization

Language

Language presented humans with an important new tool for solving problems previously beyond their reach by allowing them to conceptualize things into words. The ability to speak and think using individual words greatly increased the efficiency of all efforts involved in group survival. Words that identified places, oneself, one's group, and one's enemies, as well as animals and edible plants, along with other words that represented activities from sleeping to eating to cooking to killing provided an enormous advantage over other humans who had not developed language. Language allowed the task of group hunting to be better coordinated, providing more food to support even larger human groups. Language was necessary to develop logic that allowed modern humans to understand cause/effect relationships and to foresee and prepare for problems in their future like the approaching winter, gathering storms, and drying of water sources. This capability with symbols and language had a tremendous impact on human interaction with natural selection, allowing rapid advancement in all dealings with other species. With language, rather than 'waiting for' favorable genetic change to improve ones place in the food chain, interactions among humans, made possible by their newfound ability to communicate, improved human capabilities in all competitive areas.

Regardless of how modern humans developed language, it had an enormous effect on human history. Until recently, scientists thought the first usage of language took place between 65,000 and 100,000 years ago. However, a study by W. Enard published in the August 22, 2002 edition of Nature indicated modern humans had the ability to communicate verbally closer to 200,000 years ago, well after the split between Neanderthals and the ancestors of modern humans. Enard showed there was a genetic connection to language, and that Neanderthals did not possess sophisticated speech skills. The study involved a family (named family KE in the literature) in which about 50% of its members, spanning three generations, suffered from a unique hereditary language defect that negated their ability to articulate by causing problems with movements of the lips and tongue as well as in selection of the correct word tense. Geneticists traced the problem to a single gene located on chromosome 1 called *FOXP2* that was normal in unaffected family members, but had mutated in those who suffered the defect. This very close association of the FOXP2 gene makeup with the family language capability strongly suggested the gene was the key to language in all humans, and the factor that separated us from other animals in this regard.

A comparison of the human FOXP2 gene with those of mice and chimpanzees—animals with which we share the great majority of our DNA—showed a difference of only one amino acid displacement between chimps and mice, but a difference of three amino acids between humans and mice, and two between humans and chimpanzees. Having established the FOXP2 gene intimately involved in the language differences between humans and our nearest ancestors, scientists used a technique called *triangulation* to estimate the time when the human FOXP2 evolved to its present form.

328

Geneticists normally use triangulation to establish the point in time when existing species shared a common ancestor by comparing the differences in molecular sequences in DNA from the two species. Different mutations in genes occur at different rates. However, the rate for any given gene is predictable. Therefore, knowing the number and kind of differences in the DNA of two species, one can determine the age of their common ancestor by calculating the time needed for that number of the particular mutations to occur. The relatively low number of DNA differences between close relatives such as chimpanzees and humans (about 2%) translates to a recent common ancestor, while the 50% difference between bacteria and us means a very distant one.

Use of molecular clock evidence for triangulating backward to an evolutionary event works quite well for estimating the divergence point of two living species, but determining the age of a beneficial genetic mutation needs a more indirect method. Molecular clock triangulation requires genetic evidence from groups representing a wide range of variation in a particular gene. However, except for very rare groups, such as family KE, there is no variation in the FOXP2 exons (active sections of DNA containing genetic instructions) to triangulate from. Fortunately, passage of time has resulted in plenty of variation in the FOXP2 introns (or silent code letters), and since introns are not subject to natural selection pressures, their mutations remain in the gene pool. Therefore, geneticists examine the patterns of intron variations to estimate when the mutation not shared by the two species happened.

While the estimate that humans acquired language capability 200,000 year ago coincides very well with the date of modern human emergence, it does not mean the individual with the first mutation began verbally communicating immediately, only that she had the capability. This presents a puzzling and contentious aspect of the FOXP2 theory. As we have seen before, in order for a mutation to spread throughout a population, it must be immediately beneficial in some respect; and since the first recipient of the 'talking gene' had no one to communicate with, no obvious benefits come to mind.

Agriculture

By 45,000 years ago, as humans were emerging from Africa, they began making major changes in the type and design of tools. Tools became smaller, more sophisticated, and task specific. They varied in their construction from one region to another and utilized a much wider variety of materials, including rocks, bones, tusks, and antlers. Humans of the era created and wore jewelry made from shells, bones, and animal teeth; and religious ceremony had become part of their existence.

But in spite of their newfound abilities, the basic methods of human existence did not change greatly for the next 35,000 years. Humans of the Middle East, like the Neanderthals, remained essentially hunter-gatherers relying on a variety of animals, including deer and gazelle, while gathering indigenous fruit and seeds like pistachios, almonds and chickpeas. This dependence meant continually moving to follow animal migrations and the seasonal availability of edible vegetation.

The ability to communicate by language proved an essential factor in the transition from this lifestyle to one of the most important developments in human history, the emergence of agriculture. The oldest evidence of humans supplying their food through agriculture dates to about 10,000 years ago in China where the inhabitants harvested rice and millet in the Yellow River region along with lentils, chickpeas, and grasses in the Fertile Crescent. The transition of modern humans from hunter-gatherers to planters of seeds and tenders of animals required new ways of interacting with each other. Hunter-gatherers probably organized their activities at very basic levels, much as the African Bushmen of today, in bands consisting of ten to fifty people belonging to a few closely related families that may or may not have a leader.

Agriculture required different relationships with more interaction and communication among individuals; one person, or group, could trade the excess production of a crop for meat that another group did not need. Bands could unite to form alliances for protecting their crops and territory from other more nomadic humans who had not yet learned to respect property rights. Larger groups also meant increased opportunity for trading tools, and for exchanging ideas on how to improve them. However, in terms of human and societal evolution, the most important aspect of alliances was the increased opportunity for exchanging mates. Prolonged separation of groups produces genetically dissimilar populations, and if continued long enough, results in distinctive physical differences that make interbreeding more difficult or impossible. In fact, most of the physical differences among current humans originated during the tribal phase of our history.

The roots of plants tended by these Pleistocene ancestors also became the roots of western civilization, and they were sown in the balmy climate of the Mediterranean. The year round moderate Mediterranean climate warms the surface, causing high evaporation rates, but inward flows from rivers, seas, and rainfall replace only about 30% of the loss. This situation, combined with a huge subsurface outflow, makes the Mediterranean surface nearly a foot lower than the Atlantic. This in turn causes five mile-per-hour inward flows through the straits of Gibraltar, creating a current that extends two hundred and fifty feet deep. The shallow underwater strait at the Atlantic entrance causes very low Med tides of less than two feet, which guarantees that only the warmer surface water of the Atlantic enters. These factors combine to produce the beautiful blue-green Mediterranean waters that average about eighteen degrees F warmer than the neighboring Atlantic.

The resulting sunny climate and balmy temperatures of Mediterranean coastal regions, combined with an abundance of fertile soils enriched by volcanic residue, provided an ideal environment for humans to prosper. Probably more importantly, it gave man the spare time to take advantage of his greatest edge over competitive species, his brain. Acquiring the ability to think by assigning words to represent things combined with a benevolent environment that eased the daily task of survival, marked the beginning of human civilization.

The first evidence that humans had begun to abandon the hunter/gatherer lifestyle to settle in a given area dates back about 12,000 years to Jericho, the biblical town on the Mediterranean, just north of Jerusalem. It is not surprising that the Middle East would be the source of the first evidence of agriculture since no other section of the world has as many naturally occurring large seeded plants and grasses suitable for agricultural purposes. Animal husbandry was certainly not an overnight invention. It probably began with human hunters migrating with animal herds and guarding them against raiders, both human and non-human. Next, they began herding the animals and later on, storing grasses to feed them in winter, and later still, preventing their escape by constructing crude fences and shelters.

The next serendipitous advancement was domestication of the wild beasts. Confined to corrals and protected from predators, speed and other survival skills no longer influenced the animals' gene pools, causing successive generations to become fatter and tamer. This turned out to be a much easier job than one might expect based on our perceptions of wild animal appearance and behavior. D. K. Belyave highlighted this fact through an amazing experiment designed to see if he could domesticate silver foxes through selective breeding. By breeding only the tamest animals with each other through successive generations, in just twenty years he produced a breed of fox that behaved exactly like border collies, complete with wagging tails and a craving for human companionship. They also took on the appearance and characteristics of border collies including black and white coats and floppy ears, and changed from seasonal to year-round breeding habits. Of course, these animals were still foxes not dogs, which descend from wolves; but original domestication of wolves probably proceeded in a similar albeit un-orchestrated manner.

Most anthropologists believe the advent of agriculture and animal husbandry had a similar taming effect on humans. We can reasonably assume that prior to these advancements, when human survival required reliance on many of the same physical skills as other animals; man too was a far "wilder" or feral creature. The more sedentary life involving daily communication with the animals produced a more docile, but more lovable human creature as well.

Lactose intolerance is the textbook example of the impact of animal husbandry on human evolution. Lactose from milk normally has a negative effect on adult humans. Prior to agriculture and animal husbandry, only infants drank milk and even today, people from many places around the world suffer from lactose intolerance. The only reason lactose intolerant people can digest cheese is that bacteria from the production process has already digested the lactose. A study showed that the common thread among those who can tolerate milk is they all have pastoral histories. Lactose intolerance stems from a gene that produces an enzyme called *lactase*, which the digestive system needs in order to digest the lactose in milk. Humans and all other mammals are born with the lactase gene switched on but a control gene normally turns it off sometime after infancy when, in response to its evolutionary history, it is no longer required. All genes can mutate however, and the control gene in some of those early humans must have carried the mutation. During periods of draught, people desperate for food probably resorted to eating the same food as their babies, and those with the helpful mutation could drink milk giving them a survival advantage over the lactose intolerant individuals. The result of course, was that the mutation spread through the population.

Agriculture was to gathering, what herding was to hunting. At some point, people became aware that when they had left seeds of a given plant in warm ground, new plants of the same type eventually emerged at that spot. Later, they noticed that the presence of plenty of water helped them grow while competition from other plants that we call weeds was harmful. In times of plenty, they probably dined selectively on the tastiest and plumpest of each plant's harvest and used their seeds for planting next seasons crops, bringing natural selection pressures to bear that improved the plant species.

The Jericho site showed evidence of wild ancestors of various modern domestic animals including goats, sheep, cows, and pigs.

Archaeologists digging by a spring at a site called *Al Natuf* near Jericho uncovered stone mortars for grinding grain along with cups and bowls for storing goods. People built permanent homes here by digging shallow rounded holes in the Earth and covering them with brush. They fashioned sickles for harvesting the wild grasses by imbedding small, sharpened stones into bone or wooden shafts. When the growing season ended, they travelled to hunting areas and brought their kill back to the base camp. These changes had a profound effect on human habits. They now had more discretionary time, and since they no longer had to carry all of their possessions with them from one camp to another, they could make and accumulate assets. The archaeological remains at Al Natuf show the original occupants abandoned the site after a few hundred years, but another group resettled it again about 10,000 years ago. This time however, the settlement amounted to much more than a camp; it was more like a town—probably home to several hundred people who subsisted on their domesticated livestock along with the wheat, barley, chickpeas, and lentils they had planted near water sources. They were agriculturists.

Jericho has been called, and lays claim to being, the oldest town in the world, although several similar towns flourished along the Mediterranean coast at the same time. However, Jericho was one of the largest and the wealthiest of these towns. About 10,000 years ago, Jericho's occupants decided to build a wall about 5 feet thick and 12 feet high to encircle the 2,000-foot circumference of the town, including a 30-foot wide and 10-foot deep moat. They built a 30-foot tower in one section of the wall with a stone stairwell spiraling through its center. One proposal suggests the townspeople erected the wall to protect grain supplies from marauding nomads; but evidence shows no other kind of military fortifications being built for another 3,000 years. The wall may also have served to protect against flood-induced mudflows brought on by over-harvesting of brush and cutting too many trees from the surrounding areas. Whatever the intended purpose, such a project shows a surprising level of social organization, and serves as the earliest evidence of the human potential to accomplish great feats through the combined efforts of large numbers of people.

The greater abundance of food gleaned from agriculture allowed human population to increase rapidly over the next 6,000 years, with agricultural societies flourishing all around the world from southeastern Asia to South America. A new evolving climate became man's friendliest ally in what turned out to be the agricultural revolution. Glaciers from the last ice age were rapidly receding, bringing in a planet-wide warming trend that greatly enhanced agricultural endeavors. Hunter-gatherer women living before agriculture had to limit their family size to the number of children they could care for while wandering from place to place. With agriculture, large family size became an asset; the availability of domesticated goat or cow milk meant children could be weaned from breast-feeding and become assets by helping with planting and harvesting chores. When the population of an area became too large, many of the grown children would move away to start their own farms. Vastly superior food supplies meant better fed, larger, and stronger individuals. Hunter-gatherer societies, previously free to wander in search of better hunting grounds, had to withdraw when they encountered the stronger, better-armed agriculturalists. Many of those who did not convert to an agricultural economy perished because of inability to store food supplies for and during periods of draught. The end-result of this most important revolution was the first population explosion. A species that had grown to 6 million through the first 190,000 years of its existence would increase to 250 million people in only 9,000 more years.

Seeds of Modern Religion

Very early on, humans must have been aware of their immense vulnerability to events beyond their control and strived to understand and mollify them. The shortage of rain could dry up plant life they depended on, or an excess could mean destruction by flood. Hurricanes, earthquakes, and volcanic eruptions, along with plagues and forest fires, were even more destructive to humans then than now. They probably could not conceive that such events, being so crucial to their survival were due to mere chance. Mysterious forces beyond their understanding had to be causing such chaos and catastrophe. In the absence of direct knowledge of the source of such happenings, the most logical explanation would seem to involve great powers that manipulate events to punish or reward their behavior. Since these gods affected their existence in so many fundamental ways, it made sense to get on their good side by building edifices or offering sacrifices of food or even members of their own group.

Anthropologists have found evidence of ritualistic behavior suggesting that archaic humans believed in the existence of such powers more than 500,000 year ago. Human bones from caves near present day Beijing, China, appear to be those of defeated combatants. Victors had eaten the contents of their foe's skulls, likely to internalize the strength of the original owner. Early humans living in caves near Charente and Les Elyzies, France shaped skulls of their opponents into drinking cups for use in sacramental ceremonies. While such findings indicate awareness that non-physical elements influenced their existence, the small brain size of these pre-modern humans limited their ability to conceptualize and personify them as deities. It was not until modern man emerged that evidence of behavior we could consider in theistic terms began to appear.

Sophisticated cave art from caves found in Spain and France, dating to the Upper Paleolithic Age of 45,000 to 12,000 years ago, reveal early man's concerns for forces that impact both his earthly existence and his afterlife. Locations of some paintings revealed the severe hardship they were willing to endure in order to influence these forces. Artists often positioned paintings high on cave walls or in nooks and crannies; places they could only access by standing on the shoulders of an assistant. In order to provide light for painting in dim caves, they fashioned marrow or animal fat into candles with dried moss serving as a wick. Such extremes make it inconceivable that a people so hard pressed for survival were merely decorating their cave, or simply loved art.

Like the birth of the species itself, human religious belief was an evolutionary process. While its starting point must have sprung from a belief in a sacred and transcendent power, the capacity of the human mind to conceptualize the higher attributes of gods and spirits could not have been in place until language had developed to a comparable level of sophistication. In the absence of language, pre-modern human religion was probably little more than unstructured beliefs based on fears and anxieties, which eventually took the form of *animism*, the belief that individual spirits dwelt in all living things; every tree, every flower, every stream had its own individual spiritual identity. Animism evolved into *polytheism* as these many spirits that inhabited every element of nature eventually merged into "departmental gods". Each god represented an entire natural grouping such as *Silvanus* and *Aeolus*, gods of the woods and the wind. Eventually, a god represented every element of nature with which man interacted, from the different seasons to sex, fertility, birth, death, and his dependence on animals for food.

Nearer the end of the Paleolithic, female representations of deities began to dominate, resulting in a proliferation of figurines called 'Venuses' that placed strong emphasis on the maternal aspects of the female. Anthropologists found one such painting, apparently used for fertility rituals, in a rock shelter in Catalonia of the Spanish Pyrenees, which showed nine women with no facial features, each wearing knee length skirts, dancing around a small naked male.

Black skinned mother and child figures are a common element of Egyptian mythology. In early representations, the female was the primary Deity while the male appeared as a child dependent on the female for existence and support. In time however, as men were relieved of hunter-gatherer chores and free to assume a larger role in human affairs, the male aspect of the Deity began to reach parity with the female. By the end of the second Iron Age, Jordanians from the town of Petra worshiped the mother-goddess and her son as dual gods in the same sanctuary.

By 3,000 years ago, the male aspect of the Deity had absorbed much of the female's powers, relegating her to a supporting but nevertheless necessary role. However, the male dependence on the female as the creative agent was such an obvious factor that she remained part of the Deity concept. However, the embodiment of the male within the major Deity spread to become the seed of modern monotheism.

The very nature of polytheism limited its staying power since it inevitably gave rise to many uncomfortable questions: Which deity is most powerful? Which came first? And. Who created the others? One must be supreme. The interaction of natural elements such as ebb and flow of tides, the change of seasons, and movements of Moon and stars may also have helped shift belief toward monotheism. This meant there must be a single cause behind the creation and management of Earth. The 1997 discovery of the 3,600 year-old Migdol Temple in the Jordan River Valley provided a textbook history of the transition from polytheism to belief in a single god, all within the confines of the 95 x 72 foot structure. Excavation of the Temple required digging through layers of material to a depth of nearly 100 feet that contained hundreds of religious artifacts and other remnants of civilizations that spanned both the Late Bronze Age and Iron Age I. Evidence covering an 800 year period beginning in 1650 BCE, shows five distinct phases of building and rebuilding the Temple that reflect the gradual change from Bronze Age polytheistic beliefs into Iron Age henotheistic beliefs (belief in a single God while accepting the existence of other gods).

While developments of religions and agriculture were essential elements, the creation of civilizations involve a great number of factors. These include trade, craft specialization, record keeping, religion, public ideologies, factionalism, warfare, wealth differentiation, and distribution of natural resources. The most fundamental factor, however, is one made possible by agriculture; namely population density. Evolvement of diversified social and economic systems requires large numbers of people occupying a limited area. Therefore, it comes as no surprise that the first civilizations of the Middle East began in the same regions as the first agricultural activity. Farmers cooperated with each other to expand arable land by diverting flows of the Tigris and Euphrates rivers to create fertile floodplains, allowing them to produce more crops than they could consume. This marked the beginning of specialization. Individuals with exceptional tool-making abilities became craftsmen, making plows, hoes, pottery, and weapons they could trade for excess crops and meat. Economic and political systems emerged, bringing bureaucrats and scribes to keep track of and distribute the surplus agricultural production. Religious leaders formalized group activities and appointed priests to establish and interpret dogma and oversee rites and services.

By 3,500 BC, Middle East civilization had grown to the point where people of the city-state of Uruk—now northern Iraq—built a massive temple to serve as a religious center and a food distribution hub. Within another hundred years, scribes had invented a crude form of writing to keep track of community activities. These initially took the form of symbols representing relevant common objects such as animals and pots of grain inscribed on clay tablets. Over the next millennia, they were gradually simplified and abstracted, and by 2,000 BC had evolved into the wedge shaped symbols known as cuneiform. Having a reasonably efficient writing technique, scribes advanced from functioning as mere record keepers to more demanding tasks, like writing the details of war victories, poetry, and hymns to the gods.

Other regions of the Middle East, as well as China, Greece, and the Incans of Mesoamerica were also in the early stages of civilization development. Egypt had its first pharaoh, while regions of the Indus Valley engaged in trade enterprises that would eventually lead to the Harappan, Babylonian, Assyrian, Canaanite, and Israelite civilizations. The most advanced of these were the Sumerians of the region now known as Southern Iraq, who by 3,000 BCE had developed a sufficiently complex lifestyle to identify them as a legitimate civilization even by current standards.

Freedom from the daily necessity of finding adequate food for immediate survival had a profound effect on all subsequent human endeavors. Whereas before, only babies were free from the single task of food gathering, now many individuals could devote their time and efforts to learning and improving specialized skills. Interactions among people had been limited to members of their own family or tribe. Now they could interact with many people with whom they shared common interests like religion, occupation, or the town they lived in. With the advent of civilization, freedom from every-day survival concerns meant more time, which they could use to interact with one another and learn craft skills suitable to their abilities.

New interactions also meant new allegiances. Obligations and allegiances previously limited to a person's particular family or tribe, now belonged to the town, state, or king as well. Not surprisingly, many of these obligations proved contradictory and led to friction between the long established family affiliations and the demands of the state for furthering the common good. In order to create the kind of cohesion needed to defend their country from invading armies, leaders had to replace kinship-based loyalties with allegiance to the state. The Romans accomplished this by first developing a national history that portrayed a single person, Romulus, as the ancestor of all the inhabitants of Rome, thereby creating a sense that the state itself was one great family or tribe.

Of course, populations of Bronze Age states were not constant. Those with healthy economies needed more workers to sustain their growth, a need satisfied by the influx of newcomers who obviously did not fit into the common ancestor scenario. To counter this multitude of ethnicities and conflicting allegiances, they needed a system in which loyalty to the state took precedence over, or was independent of kinship. The Romans, Greeks and many others successfully developed cohesive societies in this manner. But bonds of kinship that have sustained groups throughout their known history were not easily broken. Many groups continued to maintain closer affiliations with their own kind than with the state; those with whom they shared common cultural attributes, be it ancestry, language, or homeland. Today these communities, which we refer to as ethnic groups, are very large and diverse, and intermarriage and conversions within them has greatly diffused actual kinship. However, they continue to claim a common biological origin. Some states successfully merged this real or perceived group kinship with nationalism in order to obtain loyalty. The Jews of the early Middle East provide an excellent example of this.

The earliest settlements in the territory now occupied by Israel happened about 5,500 years ago at the beginning of the Early Bronze Age near the desert fringes of the eastern highland regions. The settlers, Nomadic desert tribesmen, who had previously survived by following flocks of sheep and goats, found the fertile soil and availability of indigenous edible plants of the region allowed them for the first time to remain in one place for extended periods, only seasonally following herds. Archaeological evidence shows settlements in the same areas in three separate periods over the next 2,400 years. The evidence indicated that each settlement initially experienced growth but eventually suffered substantial population loss before being abandoned. However, succeeding resettlements brought even larger populations with more complex villages. The earliest indications that the people of the region had unified as a country come from the beginning of the Late Bronze Age (c. 1550-1150 BCE) when the highland tribes of both regions were under frequent attack by invaders from the western lowland and coastal Canaan, now referred to as Philistines. To protect themselves, the tribes formed an alliance that succeeded in forcing the invaders to retreat to a narrow section of coastal area. The success of this action became the catalyst that over the next five hundred years led to the emergence of the State of Israel.

By the end of the late Bronze Age, more or less permanent occupation had taken place with a combined population of about 45,000 people. Supporting this many people was not possible in a simple system where individuals provided for themselves by herding and tending small farms. They needed to utilize the economic potential of the entire region by trading with people of the western

Canaan lowland and coastal regions that had already developed larger, more economically sophisticated towns. The highland people could raise larger herds than needed for their own use and trade the excess meat and milk for grains needed to supplement their animal product diets. This barter system worked fine as long as production was plentiful for both sides; however during periods of famine or political unrest, the lowland villagers needed all of their grain supplies for their own use. This was the situation in the 12th Century BCE, when the political system of Egypt collapsed, destroying Canaan's economic networks and their ability to produce excess grain supplies. Since lowlanders could survive without meat better than highlanders could without grain, highlanders had to either produce their own grain or return to nomadic life. Many of them left the region to return to pastoral existences, signaling the end of a settlement wave. However, by the 9th Century BC, agriculture dominated, periods of migration had shortened, and the settlements had stabilized.

In each of the periods, settlements occurred in two separate regions of Canaan, one in the northern and the other in the southern highlands, roughly occupying the areas of Israel and Judah respectively. The town of Jerusalem emerged as the major urban center in Judah while Shechem became the center of social and political activity in Israel. Archaeological surveys revealed that during each period, the two regions had developed distinctly different societies. Blessed with a better agricultural climate and more fertile farmland, the northern region became more densely populated and developed a more sophisticated economy. In spite of their differences, the people of both regions spoke similar dialects of Hebrew, wrote in the same script, and shared many heroes, legends and folktales suggesting a common, ancient nomadic past.

Until recently, the absence of archaeological and historical information meant that details of the social and political structure of Israel and Judah, from the Bronze Ages through the Babylonian Period, came almost entirely from biblical references. Beginning in the mid 19th Century AD and, more extensively, in the mid 20th Century, following the establishment of the present state of Israel, scholars began amassing volumes of information from archaeological

342

digs in the Holy Land and other Middle Eastern countries that interacted with Israel. Much of the earlier work tended to confirm the existence of the biblical places and events. However, more recent studies that employed more advanced techniques, revealed too many contradictions between archaeological finds and biblical narratives to accept that the Bible could be considered a factual description of actual occurrences.

The historical and archaeological records of this period, encompassing Iron Ages I and II (ca.1150-586 BCE), show that the pharaoh Shishak of Egypt invaded Canaan, completely destroying more than a hundred towns and villages. While the invasion included the towns of Jerusalem and Shechem, the primary target was the richer Canaanite cities like Rehov, Beth-shean, Taabach, and Bethel. The invasion aftermath presented a great opportunity for people of the northern highlands that they took full advantage of. They expanded into the lowlands by the end of the 10th Century BCE and ushered in a two hundred year period with Israel a full-fledged kingdom while Judah continued a pastoral existence.

By the mid 7th Century BC, the kingdom of Israel had fallen with its cities destroyed and many of its residents deported to places throughout the Assyrian empire. By this time however, Judah was prospering and coveted a united Israel with its center in the Temple of Jerusalem. To provide an ideological validation for unifying the two kingdoms, with Judah at its center, a group consisting of officials from the Judahite court in Jerusalem, along with various scribes, priests, and prophets assembled to write a scripture of their shared dogma that would suggest a common ancestry. Central to the document was the task of allying the religion of the two peoples with the state while placing Judah at the center of it all. To accomplish this, it proclaimed the Temple in Jerusalem the only place within the two lands where they could worship. Further, it included a prophecy stating that all of Israel's pagan priests and shrines would be destroyed by a man of God named Josiah.

The result of the effort, and subsequent expansions and embellishments by scholars over the centuries, was the creation of the *Pentateuch*, the first five books of the *Bible*—the *Old Testament*. It consisted of a collection of historical writings, poetry, memories, legends, folktales, anecdotes, and prophecy; all combined to illustrate how the people of Judah and Israel all descended from the same ancestor, Abraham. Most important however, at one time, they had all belonged to one great kingdom ruled from Jerusalem.

The Biblical Pentateuch's version of the ancient history of the two regions described a combined kingdom that lasted from ca.1025 to 931 BCE with King Saul as its first leader, followed by King David, and succeeded by his son, Solomon. Solomon built the temple in Jerusalem to serve as the religious and political center of the combined kingdom. However, following Solomon's death, the northerners, resenting their subjugation to the kings in Jerusalem, unilaterally seceded from the united monarchy, thus creating two rival kingdoms: the southern kingdom of Judah and the northern kingdom of Israel. The biblical narrative describes the following two hundred years as a period of decline for Israel with God sending Aramean invaders to destroy it in 720 BCE, and to exile many of its inhabitants as punishment for their sins of worshiping foreign deities.

The biblical explanation that the northern kingdom had been a sister state of Judah in the mythical united monarchy provided King Josiah the justification to take over the territories and to require that the Israelites worship only in the temple of Jerusalem. With Josiah already established by Prophesies as a direct descendant of King David, the stage had been set for Josiah to rule a united kingdom from Judah. During his reign from 639-609 BCE, Josiah conquered and controlled much of the northern territory; but he never achieved his goal of uniting Judah and Israel. The army of Egyptian King Necho II, marching northward across Judean territory on its way to fight the Babylonians, killed Josiah. The kingdom of Judah survived for another twenty-three years, but suffered a fate similar to that of the northern empire in 586 BCE when the Babylonians destroyed the Temple of Jerusalem. However, the great book, created to accommodate Josiah's ambitions, would not only survive; it would prove to be the most important document ever created by civilization and marked the beginning of modern monotheism.

8   Our Future

Whatever Nature has in store for mankind, unpleasant, as it may be, men must accept, for ignorance is never better than knowledge.

Enrico Fermi

The history of our 13.7 billion year old Universe is replete with preternatural events, without which our solar system would have never emerged from the chaos of circumstances that characterized the Big Bang. Another equally unlikely sequence was necessary for even the most rudimentary life to emerge from the resulting accumulations of matter, and yet another for the evolution of intelligent life. A successful outcome for just one of these events meant overcoming overwhelming odds; and that all should have happened is nearly beyond comprehension. Nevertheless, in keeping with the anthropic principle, we are here; therefore, these events did happen. Furthermore, the right events occurring as they did can be explained by the fact that regardless of how remote the chances, there have been an equally mind-boggling number of opportunities. If only one Universe had ever been created, it would seem incredibly foolish to believe that in this single attempt, all the forces of nature would be so precisely balanced that even a single star would have formed, much less stars circled by planets supporting intelligent life. Therefore, the fact that we know of one such universe, it seems highly unlikely that it would be the only one. Perhaps it has happened before, not once, but many times, producing unknown trillions of duds or sterile universes for each one like ours where all of the pieces of the complex puzzle fell magically into place. This too seems a preposterous proposition when we consider the spatial extent of such a multi-universe. However, we know from relativity that space and time are vaporous concepts; shrinking to insignificance depending on the status of the observer. Billions of billions of stars reside within each of these successful universes, each with some incredibly small opportunity to produce rudimentary life, and of those successful, some overcome the equally remote odds and evolve beings with the awareness and intelligence to question, "Why do I exist?"

Man has always been a thinker. This ability is the single factor that has set him apart from other animals and afforded him a unique place on the planet. Certainly, an upright stance, prehensile thumb, and other characteristics provided advantages over other creatures, but rather than being responsible for his advancement to the top of the food chain, these factors only hastened it. Man would have arrived eventually because of the ability to think, to reason, and to question the workings of the world in which he found himself. Since the first humanoid with a genetic mutation that set it apart from its parents, siblings, and other tribal members and made it more like us, all of his or her descendents have had this ability to ponder, question, reason, and decipher.

However, there were always a few exceptionally gifted in this capacity, even before historians existed to record their achievements. They were the Newtons, Einsteins, Bohrs, and Darwins of their day. Instead of fathering physics, relativity, quantum mechanics, and evolution, they were the first to use fire instead of run from it, to sow and cultivate seeds rather than rely on nature's whims, and build shelters rather than depend on the availability of caves when winter arrived earlier than expected during their nomadic searches for food.

346

If their accomplishments seem trivial when compared to the mental feats of more modern geniuses, it is only because they did not have the shoulders of giants to stand on. Had the means to record their feats been available, the benefit of those endeavors to humankind would have placed them in history alongside Curie and Galileo. Today, the fruits of such creative thinkers have brought us from a time two hundred thousand years ago, when our naked forbearers competed for sustenance on a one to one basis with the rest of the animal kingdom, to become masters of the globe with the capability to create the wonders of civilization, or to destroy the world and ourselves with it.

While man has always had an insatiable thirst for knowing his history—as distant and as complete as possible—he has also pondered his destiny, not only his individual future but also that of humankind. We look past our own mortality with plans for the welfare of our children and grandchildren of course, but deep down, we also feel a responsibility for the continuation of human existence and of life itself. We take comfort from the assumption that humans will continue to exist indefinitely. Life has proven remarkably adaptable to surviving extreme attacks on its environment. It is at the same time fragile and durable—fragile due to its complexity at the individual species level, yet durable in its totality due to the enormous diversity inherent within that complexity. During the Permian era about 250 million years ago, 96% of all species became extinct due to environmental changes. Yet life has emerged from ice ages that froze 90 percent of the planet and greenhouse gas disasters that seared the vegetation, robbing animal life of its sustenance. It has survived debris-filled clouds wrought by apocalyptic meteor bombardments that darkened the Earth for decades, playing havoc with the food chain of all species. Life has even found accommodation in boiling hot sulfurous waters and the icy confines beneath polar caps.

While life itself has exhibited a quality of near indestructibility, individual species have not shown the same resilience and have proven extremely vulnerable to change, some becoming extinct from relatively trivial variations in their habitat. While ice ages, greenhouse disasters, volcanic eruptions, and meteor catastrophes have had

347

relatively minor effects on microbial and smaller animal life, larger, more complex creatures have not fared so well, some dying out from the loss of a single plant on which they placed total dependence for sustenance. Scientists estimated the giant meteor that struck the Yucatan Peninsula of Mexico 65 million years ago caused the extinction of nearly half of all species existing at that time; however, destruction of mammalian species was nearly total.

Modern humans came dangerously close to extinction as recent as 71,000 years ago when a volcano erupted on the West Indonesian island of Sumatra, spreading nearly 3,000 times the quantity of volcanic ash into the atmosphere as emitted from Mount Saint Helens in 1980, blocking out the Sun across much of the globe. The ensuing six-year volcanic winter completely destroyed 75 % of Northern Hemisphere plant life and played havoc with the food chain over much of the human occupied world, reducing the total population by more than 90% to a mere 5,000—a number that would today have placed us on the list of animals near extinction. Interestingly, only about 500 of the survivors were females. Some believe the intense rivalry for food made cannibalism a common mode of survival, the major victims being the physically weaker female sex. This particular disaster is responsible for the incredibly small amount of variation in the DNA of humans throughout the world. There is less variation in the DNA of humans from all races and regions of the globe than within a single tribe of chimpanzees. The most likely explanation is that a small region inhabited by closely related humans escaped the most devastating damage allowing its occupants to survive; and the entire feminine portion of our DNA came from this tiny group of very similar females.

Meteorites probably present an even greater threat to the long-term future of humankind. For the first billion years or so of earth's existence, giant meteorites with diameters of 300 kilometers or greater visited on a regular basis; objects that, if they struck the planet today would not only eliminate intelligent life, they would essentially obliterate all life forms on the planet. Such an impact would completely evaporate all the oceans and reduce much of the solid surface to a molten lava-like state. While the frequency of meteorite

348

impacts with Earth have lessened dramatically from this early period, and no such giants have visited for the last 3.5 billion years, the outlook is not as rosy for smaller, though still deadly, objects. Within the asteroid belt that crosses the solar system between Earth and Jupiter, there are more than a thousand objects greater than 1 kilometer in size with the potential to strike Earth. Statistically, one such object should hit our planet every three hundred thousand years, and every thirty million years an object of 10 K or greater, similar to the Cretaceous giant should arrive.

While there is no evidence of visits from these behemoths in the last 65 million years, there have been narrow escapes. NASA astronomers actually photographed the most recent of these, a 4 K asteroid named *Toutatis*, which came within 2.2 million miles of Earth on January 3, 1993. Four years earlier, a 1 K object came much closer missing by only 700,000 miles (about 10 hours away if travelling toward us). In 1992, Harvard-Smithsonian Astronomer Brian Marsden predicted a 1 in 10,000 chance that a giant called the Smith-Tuttle comet would wipe out all life when it collided with Earth precisely on August 14, 2126. The prediction, which had gained wide recognition in the press being dubbed the doomsday rock, later, had to be recanted because Marsden had not properly accounted for trajectory changes that would result from pockets of ice on the comet's surface vaporizing explosively as it came near the Sun.

Paleontologists David Rupp and John Sepkoski prepared a chart plotting the number of known species on Earth versus time that revealed a pattern in mass extinction events; every 26 million years there has been a precipitous drop in the number of species. The Cretaceous period event was part of the cycle, as well as one that occurred during the Eocene 35 million years ago, that wiped out many of the land mammal species and greatly reduced the population of others.

A theory explaining this phenomenon suggests our Sun is part of a double-star system, partnered with another star called *Nemisis* (meaning Death Star). In the course of Nemisis' circle around the Sun every 26 million years, it passes through a cloud of comets believed to exist beyond the orbit of Pluto called the *Oort cloud*, disturbing their trajectories and causing the Earth to be showered with comets. Some strike the Earth with enough force to generate quantities of debris sufficient to block out the Sun. The evidence does not verify extinctions at each 26 million year interval, but this is likely due to intervention by the laws of chance sparing mass extinction during some cycles with narrow escapes instead of direct hits by large objects from the Oort cloud.

The ability of modern humans to survive collisions with these intermediate size meteorites is not completely predictable since it involves many other variables. However, even one K meteorites certainly have the potential to end human existence. Body size has proved a major liability in previous environmental disasters, and human beings rank in the top one percent among all species in body mass measurement. The major extinctions wiped out nearly all of the large species, though many of the smaller creatures survived. While these potential disasters present a horrific outlook for our distant future, we should keep in mind that we have not tested the creativity of the human mind under such duress and should hesitate to sell it short. We may well find a way.

Yet, just as the laws of science tell us our Universe had a beginning 13.7 billion years ago, they show just as definitively that it will some day have an end. It seems incongruous that our understanding of the universal physics is inadequate to tell us next week's weather with more than 50% confidence, but we can predict the destruction of the Earth 4 billion years from now with certainty. Perhaps the prognosticators of such phenomena do not feel the same pressure as our weatherman since they have less concern about having their feet held to the fire if they turn out to be wrong. Yet as Albert Einstein said, "…the most incomprehensible thing about the Universe is that it is comprehensible." While complexity reigns in the ambient microenvironment of Earth, there is simplicity in the order of nature at the macro-scale of the Universe that accommodates prediction.

In terms of its life expectancy of some hundred thousand trillion years, the Universe is still an infant; however, like cells in our bodies, various parts of it have relatively brief existences of only a few million years. The expected lifetime of the Sun is about 10 billion years. It is now 5 billion years old and therefore middle aged. This does not insure us safety for the next 5 billion years since changes in the Sun during its senior years will be very inhospitable to earthly life. Just as the Sun now shines about 30% brighter than it did 3.8 billion years ago, it will be another 40% brighter 2 billion years hence. This brightness level will provide Earth with a heat source similar to that now experienced by the planet Venus, where surface temperatures exceed 400° Celsius (750° F). However, the presence of vast quantities of water on Earth will cause an enormous greenhouse effect, even greater than that of Venus, which will send Earth temperatures to 1,200° Celsius, evaporating all the oceans and creating a hellish environment by melting great portions of the crust.

By about 4 billion years from now, the Sun will have used up most of the hydrogen nuclear fuel in its core by converting it to helium. Since the Sun is not sufficiently massive to generate enough core heat for helium fusion to carbon and other heavier elements, there will be no nuclear reactions to support the star against the inward crushing pressure of gravity. The Sun will then begin to collapse, generating tremendous core pressures and temperatures, causing heat to flow to the outer regions that still contain hydrogen. Temperatures will become high enough in these areas to resume hydrogen burning, causing the Sun to expand in size. The resulting force of the nuclear explosions will eject enormous quantities of matter through its surface and into galactic space.

By this time, the Sun will have expanded enough to consume the inner planets and encompass Earth's present orbit; however, the Earth will no longer be located there. Expulsion of solar mass will have reduced the Sun's gravitational attraction, causing the Earth to move outward to a more distant orbit. Even at this greater distance, the expanded lower density Sun will fill nearly half of the earthly daytime sky. The expansion will cool the Sun surface from its normal temperature of 7,000 to 4,000 degrees C giving it the reddish glow of a Red Giant star.

The Sun will continue burning hydrogen and ejecting matter into space for another billion years before its nuclear fuel resources will have become too low to maintain its size; then it will begin to shrink and to dim. Explosions will have depleted the Sun's mass to such an extent that it will no longer have sufficient gravitational attraction to hold planets; and after 11 billion years in orbit about the Sun, those not previously absorbed by its earlier expansion will spin off into galactic space. Our solar system will not be alone in this fate. By this time, other stars similar in size to the Sun and larger will have suffered the same misfortune, and any surviving life within the Milky Way will view vastly darker nighttime skies than now. Stars larger than the Sun will have died billions of years earlier and will exist as neutron stars or black holes.

At about this time, the Milky Way and the Andromeda galaxies, which have been drifting toward each another since their formation 12 billion years earlier, will finally collide. Due to the great distance between galactic objects, most stars and planets will avoid direct collision and pass by one another without impact. However, gravitational effects brought about by their close proximities will alter their paths, slinging them in one direction or another. If the Earth's trajectory takes an inward path, gravity will pull it into orbit around a giant black hole at the galaxy's center, where it will ultimately be absorbed in a few million years. If it takes a course directed outside of the galaxy's influence, its final fate will be delayed.

By 10^{18} years from now (1,000,000,000,000,000,000), all 200 billion galaxies of our visible Universe will have suffered a similar fate. The Universe will consist of billions of monstrous black holes wandering through space, sucking in all matter and energy, including our tiny Earth along with anything that comes within their ever-growing reach. The Universe will be absolutely black.

At the risk of making the mother of all understatements, these facts present a gloomy long-term outlook for humankind. As individuals, we are aware of the temporary nature of our existence but take comfort from the fact that we do not know the time or circumstances of our demise and have within our power the ability to intervene in ways that may postpone it. In a sense, we feel the same way about our species as a whole, and natural cataclysmic scenarios present a horrific final curtain for the human drama. However, an even more depressing and immediate concern lies in the potential that man might be the instrument of his own destruction. We find the concept of destruction by forces beyond our control somehow more palatable than committing species suicide. The discovery of the energy potential stored in atomic nuclei, along with the development and proliferation of weapons to release that energy, and the massive quantities of deadly radiation that accompany it, represent a far more depressing and immediate threat to man's existence.

This does not say we will give in to the apparent inevitability of cataclysmic events. Humankind may well be in a race against time to develop technology for surviving them; and a case can well be made that we will win. The ability to think and create may well carry us through these seemingly insurmountable challenges to our existence for billions, and perhaps even trillions of years.

We may not even be aware of—or need to be aware of—our entry into such a race; the natural human instinct to acquire knowledge may be the only declaration of intention to compete necessary. The phenomenal rate at which scientific knowledge has grown should make us reluctant to believe anything within our ability to imagine is impossible. In the last century, we have acquired more knowledge than in all our previous 200,000-year modern human history. As illustrated in figure 54 below, scientific knowledge increases at an exponential rate, roughly doubling every 15 years.

Expansion of knowledge follows this exponential path since it essentially grows out of itself. As new ideas develop, other thinkers assimilate and improve on them; or in some cases scientists and laymen alike combine these ideas with their own thoughts, thereby creating even more knowledge. As a result, within a single century, a mere blinking of the eye in human evolutionary terms, our knowledge base has multiplied by a factor of more than 100. This rapid advancement of technology may well prove to be the source of man's ability to overcome many of the apocalyptic events on our calendar. Some of these matters concern us immediately, while others are in our distant future. Nuclear self-destruction is a possibility at any time, while other self-imposed environmental holocausts are at least centuries away, and the deaths of our solar system and galaxy will not

353

happen for billions of years.

Figure 54 Worldwide Scientific Publications

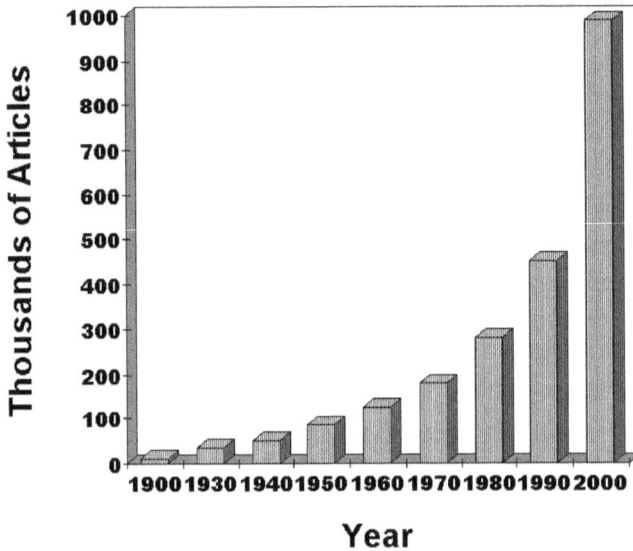

While we do not know where future technological advancements will take us, we do know it will involve the ability to control greater and greater quantities of energy. Strongly correlated with man's technological advancement is the energy at his disposal. Prior to 100,000 years ago, his energy resources were limited to the power he could muster through the strength of his own two hands. Under these conditions, it required about eight individuals to generate the equivalent of one horsepower. One of the first steps toward energy efficiency came from man's invention of simple hand tools, which effectively lengthened his arms, doubling his horsepower rating.

A continuous pattern of Ice Ages over the following 90,000 years presented a major impediment to man's progress; scarce food supplies hampered population growth and man spent much of his creative energies in the fight for mere survival. With the receding of polar ice caps 13,000 years ago, the Earth warmed, increasing the availability of plant life providing more food. A wide expansion of all animal species followed, and human populations began to grow at the fastest rate in its 2 million year history. This greater population, combined with some relief from day-to-day problems of survival, meant more thinkers with more time to think. With discretionary time at his disposal, man invented agriculture and learned to domesticate animals, allowing him to cease his wandering hunter-gatherer ways and greatly ease the burden we call survival.

Man also found his animals useful for purposes other than food. He could increase the energy resources under his control to one full horsepower by harnessing them for transportation and to ease the numerous tasks associated with agricultural life. He could grow more food to support even more humans with more thinking time; the sparks of the human knowledge explosion had become a smoldering fire.

The advent of agrarian living also presented inventive and industrious individuals another energy resource—the exploitation of other humans to increase production. The more effective agriculturist could have several horsepower under his control by convincing less entrepreneurial neighbors that working together under his competent direction would produce far more than each could do separately. This activity led to division of labor, separating workers according to their talents and preferences, allowing them to hone their individual skills.

The cooperative exploitation of agriculture led to people gathering in villages where they could build shelters nearer their work and be in a better position to protect their assets from wandering nomads. Trading became common between villages but so did war, and out of war emerged slavery. Conquered villagers often became slaves, thus endowing hundreds of horsepower to the command of a victorious leader. Further conquests by the most skilled of the warring leaders gave them the power of kings with control over thousands of equivalent horsepower, which they used to turn villages into towns.

Using only the energy generated by human and animal labor and the ingenious use of tools, humans continued making enormous progress. This continued for about 98% of the following 10 millennia in civilizations around the Earth, erecting major cities, highway and aqueduct systems, along with beautiful palaces and massive monuments. However, the rate of change began to change dramatically about two hundred years ago.

Although knowledge expands exponentially, occasionally certain events can alter its basic rate. New concepts come together in a fortuitous manner, creating a kind of critical mass of ideas; then, instead of available knowledge doubling every 1,000 years, it begins doubling every century and eventually every decade. Such a watershed event led to a quantum leap in man's control over energy resources in the early 19th Century, sparked by the scientific advancements of Isaac Newton a century earlier. Newton's development of equations representing laws of motion touched off a flurry of inventiveness that culminated in the industrial revolution. The invention of the coal-powered steam engine gave a single worker the ability to control hundreds of horsepower through machines like steamships and railroads. James Clerk Maxwell's electromagnetic force discoveries, which led to the invention of dynamos powerful enough to light-up entire cities, further magnified the available horsepower. The popularization of automobiles powered by hydrocarbon burning internal combustion engines gave nearly every adult human of the industrialized nations control over a hundred or more horsepower. Some segments of science expand at even greater rates. For example, in 1965, Gordon Moore accurately predicted in Moore's Law that computer chip technology would progress at a rate that would double the calculating speed of computers every 2 years.

For centuries, the study of science has for the most part been limited to gentlemen of leisure who merely included science with other pursuits of philosophy and the arts. However, the great inventions spawned by the industrial revolution heaped economic rewards on their creators and inspired people from all walks of life to seek their fortunes as inventors and entrepreneurs. Universities began adding large departments dedicated to a wide array of scientific specialties. Successful inventors and companies that built their inventions dedicated entire departments to research and development of new ideas that might lead to more and better products.

In the last century, humans arrived at the current state-of-the-art status in terms of control of energy resources with the discovery of the force that holds nuclear particles together, and the subsequent harnessing of nuclear energy. The magnitude of the enormous increase in energy control represented by nuclear technology versus chemical combustion of hydrocarbons becomes obvious when comparing the yardsticks used in their measurement. We measure the energy gained or released in chemical reactions in electron volts, whereas quantification of nuclear fusion and nuclear fission reactions requires millions of electron volts.

While we have come a long way in energy development, our highest energy achievement, nuclear fusion is nowhere near adequate to the task of surviving these foreseen ominous threats to human existence. Nikolai Kardashev, a Russian physicist, arrived at some very interesting prospects for humankind's future by extrapolating science and technology advances to describe future civilizations built around the energy sources they had developed. He defined the future civilizations according to their technological development, and hence the energy resources they will be able to control, as Types I, II, and III. The technology of our current civilization was considered too primitive to rate a numeral and simply called Type 0.

A Type I civilization controls all the energy resources on its planet and has perfected technology for mining the heat and materials beneath its planet's crust. A Type II civilization will have exhausted the energy resources of its own planet and found ways to use a far greater portion of the energy output of its host star. A Type III civilization controls energy of an entire galaxy and has mastered Einstein's equations to the point of being able to manipulate spacetime.

Energy available to each civilization Type differs from its predecessor by a factor of 10 billion, or 10^{10} watts. Energy supply for a planet based Type I civilization is 10 million billion, or 10^{16} watts; therefore type II, with access to direct solar resources uses 10^{26} watts, and a Type III with an entire galaxy at its disposal has 10^{36} watts. While our current level of 10^{13} watts of energy usage leaves us far short of a Type I civilization, we certainly do not deserve a zero-rating either. Carl Sagan provided a more accurate assessment of our cosmic ranking by modifying Kardashev's scale to provide finer gradations between civilization types. Sagan's scale made room for differentiating between civilizations by more modest differences, using factors of 1 billion as opposed to Kardashev's 10 billion, with ten increments between Type I and Type II (e.g., I.1, I.2,). On the Sagan scale, our present civilization, which uses 10^{13} watts, ranks as a 0.7 instead of a zero? However, we still fall short of being in charge of our planetary resources by a factor of 10^3, or 1000. At our rate of knowledge and energy demand growth, earthlings should become a Type I civilization in a few centuries, and a Type II in another 1,000 to 5,000 years. Control of galactic resources is a daunting task indeed, and even with the exponential rate of technology development, a Type III civilization may need a million years.

Advancing to a legitimate Type I civilization over the next few hundred years—without having to round to the nearest whole number—by harnessing energy from the inner earth, would free man from many of the natural perils we face on a regular basis such as draught, floods, earthquakes, and volcanic disasters. Multiplying our energy resources by the required factor of 1,000 would provide the ability to explore our entire solar system, build cities on, and cultivate the ocean, and provide energy necessary for moving toward a Type II civilization. However, we will not likely achieve Type I capability with the status quo level of world organization. Major problems such as population growth, availability of food resources, pollution, global warming, disease, religious extremism, and threat of nuclear war are global in nature and require global solutions. Although difficult to see due to the parade of chaotic world events reported daily, whether we like it or not, we may already be well on our way to such a world order. The recent explosion of advances in technology, global communication, and information transmission has helped bring about a situation that many believe will have an impact on civilization equal to that of the industrial revolution.

The result has been a "flattening" of the world, erasing national boundaries and forcing all nations to compete in commerce on a flat playing field. Like many revolutions, chaos has accompanied its youth, and we are now experiencing its severe growing pains. However, countries involved in such free commerce eventually become less war-like. China is a prime example of this; having invested so much in their goal of achieving economic parity with the West through free commerce, they may be very reluctant to sacrifice their gains to war. Many other countries have taken advantage of this flattening; and when it has spread throughout the world in a few generations, the need for vast expenditures on defense will be less necessary, leaving more resources to solve real problems. This does not mean nations will not exist. They probably will exist, but organized around commerce much as the European Union, and to some degree NAFTA, today. Economic integration of the various commercial entities should greatly diminish or eliminate warfare since interdependency of economies will make the cost of war too great. On a Universal basis, however, Type I status civilizations will have found solutions to these political problems through necessity, long known as the mother of invention, but also the mother of trade and commerce, and perhaps one day, the mother of peace.

An "out of the box" thinking senior physicist named Freeman Dyson from the Institute for Advanced Studies at Princeton University proposed a series of devices for capturing solar radiation that could provide Type II energy. The devices called *Dyson spheres* involve erecting a shell that surrounds the sun at a radius of 90 million miles, constructed from material obtained by disassembling the planet Jupiter. The sphere would reflect solar radiation, which appropriate devices located on Earth would collect. Without arguing the seemingly impossible engineering task involved, critics originally shot the idea down by emphasizing the instability of such a creation; even minor impacts by meteorites or comets on its surface would cause it to drift toward and eventually collide with the Sun. However, such minor details could not dissuade Hollywood and science fiction writers from using the design in numerous creations including episodes of Star Trek, Andromeda, and Crest of Stars. Dyson however pointed out that he had not envisioned a solid sphere but rather a loose collection, or as Dyson called it, a "swarm" of up to

100,000 individual objects called collectors, each measuring a modest 6 million miles in diameter and placed in its separate orbit around the Sun.

There is considerable flexibility inherent in the construction of a Dyson swarm since it allows production of a wide variety of individual panel sizes, compositions, and designs. Furthermore, one can add panels incrementally over a long period, as energy needs demand. The initial installation could begin with a single collector and be expanded to include perhaps a thousand panels laid end-to-end, forming a ring that orbits around the Sun. Later still, Type II engineers could add additional rings, arranged to resemble the longitudinal lines of a globe.

Scientists are so convinced that building a Dyson sphere is not only feasible but will be the preferred source of future energy, they believe it is inevitable that any alien Type II civilization will have already done so. In fact, the federally funded SETI Institute (Search for Extra-Terrestrial Intelligence) has so much faith in Dyson spheres that they have made them the basis for their search to contact alien civilizations. SETI is counting on the fact that the collectors would absorb both short and long wave length visible light from its host star. However, in accordance with the second law of thermodynamics, a portion of it would be re-emitted into space, all as long wave length infrared radiation. If while scanning the heavens, SETI telescopes pick up abnormally high ratios of infrared radiation in a star's spectrum, they could conclude that one of its planets houses an advanced civilization.

The exponential growth in knowledge combined with the enormous energy available to Type II civilizations would permit travel at near light-speed in starships powered by antimatter/matter fuel allowing us to colonize other planets of the solar system. By the time we have advanced well up the Type II ladder (say from II.6 to II.9), we may well have developed advanced knowledge of the gravitational force and be able to use it to warp spacetime, providing the ability to explore nearby stars of the Milky Way galaxy. Use of the 10-billion-fold increase in energy would eliminate fear of annihilation from impacts by giant comets and meteorites, and man would be able to manage the atmosphere commanding complete control of both short and long-term weather, including the ability to forestall ice ages and global warming.

Perhaps a million years from now, when the energy controlled by our civilization increases by a second factor of 10 billion, and we become a member of the Type III fraternity, we will be a galactic civilization, using the energy of complete star systems along with power harnessed from supernovae and the millions of black holes within our galaxy. With the rapid movement of the technological clock, only a few hundred thousand more years will pass before warp drive starships will use this energy to zip back and forth across the Milky Way on missions of commerce with colonies established by our species. As such, our civilization could endure for billions of years after our solar system dies, and through intelligent selection of the location of our new planet and star, survive the collision of the Milky Way and Andromeda galaxies.

But you might well ask, what about the inevitable final curtain waiting for humankind some billion billion years from now, when all the stars and galaxies have died and our Universe is in its final lifeless deep freeze stage? From our vantage point in time and the state of our knowledge, it may be impossible to predict how a civilization of that future period will solve a problem of such magnitude, but we may know enough to predict that man will have options to passing quietly into the night. The answer lies in the concept of multiple universes; and the fact that science fiction has dominated this concept does not mean true scientists are not interested; they are. However, all serious considerations of multiple universes involve the existence of extra dimensions, or dimensions beyond the four we know; namely, the three spatial dimensions of height, width, and depth, along with the fourth dimension, time.

The concept of multiple universes sounds more like a cheap movie than a serious scientific proposal, although not that improbable when considered in light of concepts inherent in relativity. The vast reaches of our single Universe alone can easily overwhelm our senses, so to consider a large or infinite number of them existing together may be incomprehensible. Part of the difficulty stems from our perception of space and the dimensions they comprise, namely length, width, and height. Most of us—including the scientists promoting them—cannot visualize spatial dimensions beyond these three; however, science tells us they do exist. The first credible scientific evidence for extra dimensions came in 1919 through brilliant work by a German mathematician and physicist Theodor Kaluza who showed that Einstein's theory of gravity and Maxwell's theory of light could be unified by introducing a fourth spatial dimension. At the heart of Kaluza's proposal was the assumption that light is merely a disturbance caused by the rippling of this extra dimension. When offering his controversial proposal to Einstein, Kaluza first presented a solution he had found to Einstein's relativity equations for gravity, calculated in five dimensions instead of the usual four. He then went on to show that the five dimension equations could be broken down to produce not only Einstein's four-dimensional theory of gravity; amazingly, it could include Maxwell's theory of electromagnetism as well. Kaluza's achievement ranks as one of the major scientific feats of the 20th Century since it effectively unified two of history's greatest theories by connecting them with a fifth dimension.

In 1926, Oscar Klein, a Swedish theoretical physicist, made several improvements on Kaluza's theory based on the newly emerging quantum theory. Klein concluded that while extra dimensions may be physically real, they were curled up to the extremely small size of 10^{-33} cm (the Planck length, or one millionth of a billionth of a billionth of a centimeter), much too small to be detected by any existing or foreseen equipment. The energy probe necessary to detect this tiny distance is 10^{19} billion electron volts, the Planck energy; a quantity a hundred thousand billion times greater than even the Large Hadron Collider (LHC), and a thousand times more than the energy available to a Type I civilization.

Furthermore, string theory, the most widely accepted scientific candidate for the highly coveted title, "theory of everything," depends totally on the existence of as many as eleven dimensions. In the standard model of physics, all of nature exists at its most fundamental level as zero-dimensional point particles. However, in string theory all components including matter, energy, the natural forces, and gravity are tiny 10^{-33} cm, infinitely thin, vibrating filaments that differentiate themselves from one another by the frequencies of their vibrations. In fact, it is possible to convert any particle into another by "plucking" the string to change its vibration frequency, thereby converting an electron into a quark or vice versa.

String theory has come to be more of a generic term used when referring to a series of related theories. During its evolvement, string theorists developed five different versions of the theory, which as it turned out were related to each other through dualities. Effectively, the dualities allowed physicists to relate the description of an object in one theory to that of a different object in another. After several years of debating which of the string theories was on the right track, physicists concluded that each theory is but a different aspect of one underlying theory called "*M-theory*". The primary goal of string theory is to develop the "theory of everything", the long sought Holy Grail of world physicists, which involves unifying the known natural forces including the strong, weak, electromagnetic, and gravitational forces through a single set of equations. One of the major impediments to that goal has been inclusion of one of the major elements of the Universe, namely gravity. The standard model of particle physics, developed between 1970 and 1973 to unite the theory of relativity with quantum mechanics, has successfully described three of the four fundamental interactions between the elementary particles of matter. However, due to the standard model assumption that all particles are point-like, all attempts to include the fourth interaction, gravity, have failed. But when the infinitely thin filaments of string theory replace point-like particles, a sensible quantum theory of gravity emerges naturally. Such developments have heightened optimism throughout the scientific world that a theory of everything is in reach; however, one major concern still remains. One of the largest problems with string theory has been the inability to establish its validity through experimental evidence. High-energy probes of atomic particles in the recently completed Large Hadron Collider in CERN (European Organization for Nuclear Research: a consortium of twenty European member states operating the worlds largest particle physics laboratory) near Geneva, Switzerland could produce confirmatory results.

Although the extra dimensions might well exist, we are not aware of them since, being smaller than the Planck length, they are too small to see or even to detect experimentally. String theory describes the strings as being "curled up" in various configurations called *Calabi-Yau* spaces. Visualizing extra dimensional configurations is beyond even the abilities of Mr. Calabi and Mr. Yau, and they exist within the theory as mathematical equations. Having evolved in a three-dimensional world, our senses cannot fathom what the addition of another dimension to our Universe would entail. If one of these extra dimensions suddenly expanded, providing us with a fourth dimension, in which direction would we look to observe it? How would it influence our surroundings and our lives? We cannot answer these questions but we can get an idea of what a different world it would be by considering a Universe consisting of not more but fewer dimensions.

In his book *The Elegant Universe*, prominent string theory physicist and bestselling author Brian Greene explained the concept by viewing a garden hose stretched across a canyon from a distance of a quarter of a mile. From this vantage point, the hose is clearly visible but its lack of thickness makes it appear to be a single dimensional object. If we had information that an ant lived on the hose, we would conclude the ant's traveling options were one-dimensional; its movement choices would be on a line going from the left to the right. To specify the ant's location on the hose only one piece of information would be needed, its distance from a particular end of the hose. Therefore, if all of our life experiences with garden hoses involved those stretched out across canyons a quarter mile away; we would conclude they were one-dimensional objects. But since we know garden hoses have thickness as well as length, we could zoom our binoculars in on the ant and see that it has another movement option—it can also move around the girth of the hose. There is another dimension to the ant's world—it is two-dimensional. Therefore, in order for us to identify the ant's precise location, in addition to its distance from an end of the hose, we must specify its location around the circular girth (the hose is of course three-dimensional since the ant could burrow downward toward the center, but the ant doesn't know that).

While we can see that the ant's garden hose world is two-dimensional, there is a distinct difference between the two dimensions. The length dimension is extended and readily visible while the width dimension is very short or "curled up," requiring more precise identification measurements; in this case, binoculars. Since we can see the magnitude of the differences in scale of the two garden hose dimensions with minimal magnification, it is obviously not nearly as great as the difference between the extended and extra dimensions of our Universe. Nevertheless, the analogy is directly applicable; the garden hose has one extended and one curled up dimension, whereas our Universe has three extended dimensions and one or more curled up ones. The difference lies in the fact that the extra dimension of the garden hose Universe can be detected with a simple telescope or binoculars, whereas, there is as yet no instrument sensitive enough to detect the extra dimensions of ours. Like the ant, at the moment, we haven't the tools, but unlike the ant, we believe our dimension is there.

Figure 55 Extra Dimensions

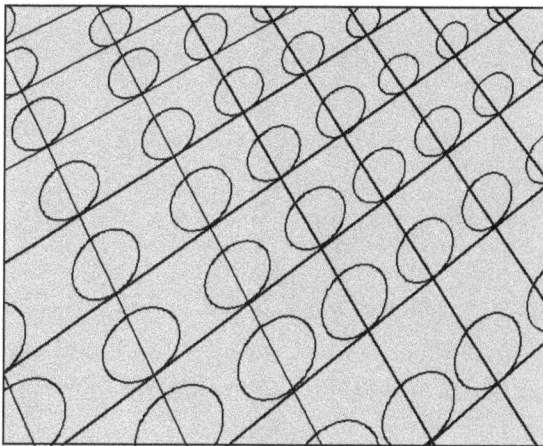

Drawing a representation of a new dimension added to three-dimensional space is a formidable task, so an illustration wherein one small "curled up" dimension is added to two-dimensional space will need to suffice. The grid-lines of figure 55 denote distance scales such as meters or miles and the circular loops represent the "curled up" extra dimension. For illustration, there are loops shown at regular points throughout the two-dimensional surface; however, we must be aware that an extra circular dimension exists at every point within the extended dimensions, and we would need an infinite number of loops to complete the diagram. Just as in our three spatial dimension world, up and down exists at all locations of left, right, forward and back.

Figure 56 Extended Dimensions

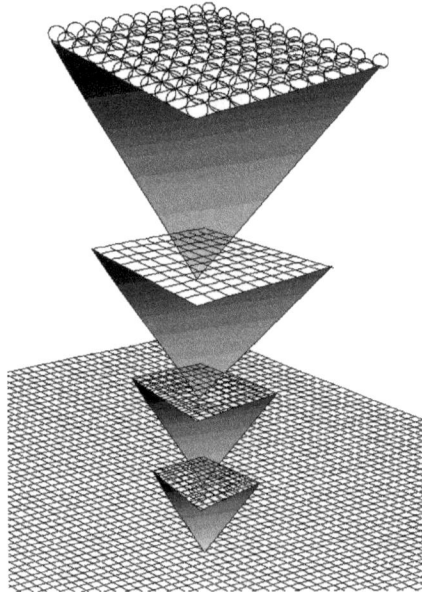

Figure 56 suggests the level of the magnification we would need to make visual contact with a new "curled up" dimension. The bottom plane is a two-dimensional representation of our Universe's three-dimensional space, with grid-lines denoting distance scales. Above the bottom level are a series of images that represent zooming in on a tiny portion of the grid through ever increasing magnifications, each by a power of about 10^8, or 100 million to 1. As indicated, there is no observable change in the basic structure of space as we advance through the first three levels of magnification. However, at the fourth level the new "curled up" extra circular dimension becomes visible looking much like the circular loops of thread that make up the pile of a carpet. Again, for illustration, the loops are shown at regular points throughout the two-dimensional surface; but the extra circular dimension exists at every point within the extended dimensions. A fourth spatial dimension of our Universe would also exist at every point within the visible three-dimensional space. With all four dimensions extended, specifying the location of an object would require information as to how far east or west, how far north or south, how far up or down, and how far into the fourth dimension.

In order to illustrate the impact of an extra dimension, Brian Greene took inspiration from Edwin Abbot's *Flatland*, a whimsical 1884 novel about a race of mythical beings who inhabit the world of Flatland; a place where the entire Universe is bounded by the two-dimensional surface of a tabletop. Before entering Flatland however, Greene briefly visited an even more bizarre, if simpler place, the single-dimension universe of Lineland. The Lineland universe is a humble place indeed, consisting of a single extremely thin line; in fact, it would be infinitely thin since Lineland has no height or width, only length. Lineland beings' anatomy consists of infinitely thin, dot-like eyes located at either end of an equally thin filament, which houses the remaining thin organs. A Lineland citizen needing an operation would have to depend on the medical expertise of either the individual to his front or rear since no one else would have access. Furthermore, the only way to reach the internal organs would be through the dot eye using a surgical instrument held by the eye of the operating neighbor. Remember, Lineland is not a string extended in space, the tiny line is space—it is the universe. Any life existing in Lineland would have to be the epitome of boredom with only a rear view of the person or object in front for entertainment throughout what you hope will be a mercifully short eternity.

Supposing however, the ruler of Lineland figured a way to add a second dimension to his universe, an accomplishment certain to garner a Nobel Prize and maybe even the ultimate recognition of human achievement, a super bowl ring. He found that Lineland actually contained another space dimension, existing as a tiny circular direction like that illustrated in figure 55, that no one had noticed due to its small size. He then discovered the secret to expanding the new dimension by bringing on a period of exponential inflation of the second dimension. The world had just grown from a 100-billion light year long filament to a plane nearly as wide as it was long. Freedom of movement had just increased dramatically and the citizens were ecstatic. For the first time in history, a Linelander could travel to new places. If the person in front moved too slowly, you would simply go around. You could even turn around and see what you have been missing from the rear, or get adventuresome and explore the vast frontier now visible for the first time to your left and right. Dancing might even be in your future since you could finally relax your straight-line posture and wiggle to your heart's content.

What a wonderful world Lineland had become; so wonderful in fact that the citizenry voted to give it a new name more befitting its glory; Lineland was renamed Flatland. In time, new creatures evolved in Flatland that took advantage of the new dimension available to them. An infinitely thin skin would grow to cover their internal organs previously protected by eyes at either end of their one-dimensional bodies. Meanwhile, limbs would evolve to aid travel about their new universe.

Of course, taking one more dimensional step would place Flatlanders into another universe, equally beyond their evolved senses, but one with which Earthlings are completely familiar—a three-dimensional universe. String theory physics tells us that extra dimensions do exist, but our ability to visualize them is just as limited as a two-dimensional universe to Linelanders or a three-dimensional one to Flatlanders.

So how do the extra dimensions of string theory help us escape a dying universe?

The answer may be in dimensional gateways, better known as wormholes. Physicist Michio Kaku, cofounder of a branch of superstring theory called string field theory, has examined the options available to an advanced Type III civilization in his book *Parallel*

368

Worlds. While the public's familiarity with other universes stems more from the likes of H. G. Wells and Gene Rodenberry than from scientists, they nevertheless have a scientific basis. When Einstein first released his theory of relativity, he was not aware that buried within its descriptive equations were solutions predicting the presence of a host of bizarre phenomena such as black holes and white holes, and Einstein himself fought vigorously against other scientists that were promoting them. One of these solutions predicts the existence of wormholes or dimensional gateways through which Type III civilizations might access other universes.

Although Einstein did not believe in wormholes, he nevertheless unwittingly provided the first insights that led to a possible solution to our future grand $x10^{12}$ children's problem of finding a new home when our Universe dies. In 1935, Einstein had not yet accepted the singularity prediction inherent in general relativity, that gravitational forces can cause matter to have an infinite density and zero volume. While working on ways to remove this perceived discrepancy, Einstein and a student at Princeton named Nathan Rosen were using black hole solutions from general relativity to model elementary particles. Since at the time, physicists viewed electrons as tiny point particles, Einstein and Rosen used them to represent the black hole's singularity. The starting point was a mathematical standard black hole solution, which in a physical sense, we can picture as a long vase with a sharply tapering stem. The throat of the vase is cut near the bottom where it merges with the tip of another vase shaped in the same manner but inverted to allow the two vases to join at their tips. Einstein hoped that the mathematical solution to this configuration, later called an *Einstein-Rosen bridge*, would be free of the singularity, and would behave as an electron, thereby serving to unify general relativity and quantum theory.

While the Einstein-Rosen bridge didn't work as intended, the concept is being looked at today as a description of a dimensional gateway between two universes where objects can be sucked in one end of a black hole and emerge on the other side. In 1962, a New Zealand mathematician, Roy Kerr, discovered a solution to Einstein's equations that predicted the existence of rotating black holes.

Rotating black holes form when a massive spinning star has spent its nuclear fuel and begins its collapse toward becoming a black hole. However, a rotating star never quite makes it all the way. Just as a figure skater spins faster after she brings her arms in, due to the conservation of angular momentum, the spin speed of a rotating star increases as it collapses inward. At some point during the collapse, the outward centrifugal force of the spinning star balances the inward gravitational force thereby preventing it from collapsing to a singularity. The result is a stable ring of spinning neutrons. Rather than being crushed and spaghettified, in a properly designed Kerr black hole, you could pass safely through the narrow neck of the Einstein-Rosen Bridge and out into a new universe.

Whether we believe in extra-dimensional space or not, science seems to demand its recognition since it appears to be the domain in which all of the natural forces are compatible. M-theory is the only theory that has the capability to solve the problems of interfacing nature at the very large and very small scales; and extra dimensions are basic to M-theory. The problem is that within the present state of our technology, M-theory is not provable. In "Parallel Worlds," Michio Kaku identified steps that a future civilization would need to take before attempting an exit of the Universe, and they begin by finding such proof and either validating or replacing M-theory.

The first step is to "…create and test a theory of everything" capable of calculating the quantum corrections to Einstein's equations. Kaku feels we might accomplish this within a few decades if M-theory turns out to be the long sought theory. Next, we must find ways to verify the consequences of the theory by building large atom smashers to create super particles, and detectors for proving the existence of gravity waves. The existence of gravity waves (predicted by general relativity to have formed during the first moments of the big bang and should be echoing through space) would confirm the inflationary period of the Universe, an essential component of M-theory. Physicists Russell Hulse and Joseph Taylor, Jr. have already proven gravity waves indirectly through measurements taken of circling rotating stars. As the orbits of rotating stars decay, they move closer to each other, which should cause them to emit gravitational waves.

By using equations from M-theory, Hulse and Taylor calculated the rate at which the stars moved closer to each other, and they found it matched the values predicted by relativity.

Our future civilization will need to verify the stability of dimensional gateways since any entering object would create a disturbance that might cause the Kerr black hole to collapse before it could pass through the Einstein-Rosen Bridge. Furthermore, radiation buildup encountered at the event horizon could be disastrous. Probably the greatest problem will be finding a source for the energy needed to open and stabilize the dimensional gateway. Negative energy, known to exist in small quantities, is the ideal candidate to solve both of these problems if the four digit IQ scientists of our future can find or create sufficient quantities of negative energy.

Physicist Lawrence Ford of Tufts University may have touched upon the technique that a Type II civilization would use to collect negative energy. His discovery begins with the fact that when a powerful laser pulse passes through a special optical material in a vacuum, it creates pairs of photons that alternately enhance and suppress quantum fluctuations, causing the emission of positive and negative energy pulses. Ford devised a scheme to separate the two pulses by shining the laser beam into a box designed with a shutter that alternately opens and closes as positive and negative pulses enter. By precisely timing the apparatus, such that the shutter opens immediately preceding a negative pulse, then closing again until the next positive pulse passes and a new negative pulse is ready, Ford might be able to collect negative energy. Theoretically, our descendants could create vast amounts of negative energy in this fashion; however, it too will be no easy task. It happens that the very act of closing the shutter creates another positive energy pulse inside the box, which could cancel out the negative one. The scheme is theoretically possible, but our descendents would need to find a way to prevent the secondary positive pulse formation for this method to be effective; a task well beyond a Type 0.7 civilization, yet possibly solvable with Type II technology.

Next, they would need to locate naturally occurring dimensional gateways and white holes in space. While to date there is no evidence for their existence, general relativity predicts white holes and hopefully, the next generation of space-based detectors will find them. Time reverses in a white hole, which presents the possibility that an object sucked into a black hole could eject from a white hole at the other end in the same manner it had entered. White holes will be a critical factor in escaping the Universe and microscopic versions may have formed naturally from the enormous energy released during the big bang, and then expanded to macroscopic size during the inflation period of the early Universe.

Next, they will need to send probes into black holes to obtain critical information; this will include measuring the amount of radiation, and determining if the entering probe makes the dimensional gateway unstable causing it to collapse to a singularity. They will need a supercomputer to analyze data collected from numerous probes in order to determine the distribution of masses in the various universes and calculate quantum corrections to Einstein's equations near a dimensional gateway. Furthermore, the possibility exists that a dimensional gateway located at the center of a *Kerr ring* (the central, "singularity" point of the Kerr ring, which has zero thickness but a non-zero radius) could take us to a new universe or to different points in time within our universe; and, since we cannot retrieve information from a black hole without actually entering, it is impossible to tell which way it will lead.

After gathering this information, the next step would be creating a slow motion black hole. Future geniuses can hope to accomplish this magical feat by artificially injecting enough energy and matter into an already spinning system to force the inner masses to gradually fall within the *Schwarzschild radius* (the radius any given mass would have if compressed to the point where it will collapse to a singularity). This trick would require our future Type III civilization to apply their control of galactic energy to corral a host of neutron stars into a swirling formation (keep in mind they have the energy of an entire galaxy at their disposal). They would control the accumulation of stars in such a manner that gravity would bring them closer but not all the way to the Schwarzschild radius. Scientist would then begin to adjust the mass of the mixture by injecting new stars until it eventually collapses to within the Schwarzschild radius and becomes a Kerr black hole. This controls the rotation speed and radius of neutron stars in the mix to make the Kerr black hole open at any speed the scientists would like.

Another problem results from deadly radiation caused by light that increases its frequency by gaining energy as it passes through the event horizon. Probes would need to determine the amount of radiation and protective shielding required. In addition, the white hole at the center of the Kerr ring may not be stable enough to permit objects to fall completely through. The necessary calculations and experiments needed to answer this question will have to wait the development of a proven quantum theory of gravity.

After completing their homework, our Type III civilization will move to the real work of harnessing the tremendous energy necessary to move to a new universe. This is a formidable undertaking and the comparatively pitiful state of 0.7 civilization technology is probably not up to the task of predicting how our descendents will handle it. However, scientists have identified options consistent with our state of the art science that future civilizations might consider; they just have not addressed all of the engineering aspects. The big problem is that of controlling an energy source equal to the Planck energy. At 10^{28} volts, the Planck energy is 10^{14} times more energy than our civilization can muster at CERN's LHC. However, by definition, when we achieve Type III status, we will be able to do so. One way they can do this is to build a stellar-sized collider. Colliders—also referred to as particle accelerators or just plain atom smashers— consist of long circular tubes into which particle physicists inject two streams of subatomic particles traveling in opposite directions. They then accelerate the particles to very high energies and smash them into one another when they achieve proper velocity. Large magnets placed at various points around the tube continually accelerate and bend the particles into a circular path. The radius of the circle through which the particles pass limits the energy achievable with a collider. The tube of the LHC at CERN has a radius of 3.5 kilometers giving it the capability to generate 10^{14} volts. Achieving Planck energy levels requires a collider with a radius greater than our solar system.

Our Type III scientists may have one major advantage in building their super collider in that construction of a tube may not be necessary since the vacuum of space may be all of the protection from contact with stray particles the traveling particles need. Magnets could be located on strategically selected planets, moons, and asteroids throughout space to energize and deflect the beam of particles guiding them through their circular orbit. Once completed, they could inject two beams of subatomic particles, one traveling clockwise and the other counterclockwise, into the collider and accelerate them to near light speed. They would then direct them on a collision course to create energy approaching the Planck level.

The above ideas are consistent with the current state of physics; however, that does not necessarily mean they will work. The concepts have been neither proven nor disproven. Stable dimensional gateways may well exist at the microscopic level but cannot be enlarged enough to allow humans to pass through. Furthermore, even if we could enlarge such gateways, there may be no level of protection that could shield the passengers from the tremendous stresses they would encounter.

If this should be the case several billions of years from now, Kaku has identified one more option our descendants would still have available to perpetuate our species and civilization. Rather than sending human beings through an enlarged gateway, we would send enough information through a small gateway to recreate our civilization on the other side. The information would include advanced DNA engineering, nanotechnology, and robotics, plus sophisticated software that would merge the thinking processes and personalities of individuals directly into clones created in the new universe. Hans Moravec of Carnegie Melon University claims that individual silicon transistors may eventually be able to replace the functions of neurons in our brains. They would first use the information to build a nanobot, a microscopic robot measuring about one-millionth of a centimeter, programmed to construct a spaceship on the other side. The nanobot would then search the new universe for a planet suitable for habitation. The nanobot's tiny size and lack of need for creature comforts would make it possible to build a proportionally small spaceship that would easily travel at near light speed using electric fields for fuel. Once established in our future home, the nanobot would then use the planet's raw materials to build a factory, which would create additional nanobots that would build a cloning laboratory to regenerate our civilization. They would use the laboratory to inject individual DNA sequences into cells to create whole organisms that would grow into adult beings with memory and personality of original humans.

Far-fetched—bizarre? Perhaps, but if we consider human progress in the last century alone, it may not be all that extraordinary. Genetic manipulation of IQ's and interfacing of the brain with computer memory and calculating power could produce individuals that will be our ultimate evolutionary survival. While the technology advances necessary for these accomplishments may seem more like science fiction—as in fact some of them are—than science, they are nevertheless products of the world's greatest physicists, and the theories are based on scientifically sound concepts that cannot be disproved with current knowledge. Granted, implementation is another thing. The role of the physicist is to establish whether the laws of physics allow the possibility. It falls to the engineer to arrange the resources to make possibilities come to pass. From our vantage point, controlling the universe may seem impossible. However, considering our advancement from the horse and buggy of two hundred years ago to sending unmanned spacecraft on exploratory missions to planets of our solar system, it doesn't seem impossible that one million years further down the technology road, the details will have been worked out. Furthermore, we should never rule out the indomitable will of all living things to adapt and survive.

Appendix

The Oregonator Reaction

The Oregonator is a series of chemical reactions designed to increase our understanding of the chemical processes that may have been involved in the synthesis of the first earthly life form. As discussed in chapter 5, there is considerable evidence that these processes proceeded through built in feedback mechanisms that assured repetitive behavior, and that through millions of years of trial and error reactions, culminated in the original life form. The process uses autocatalytic mixtures of chemicals designed to study systems that exhibit oscillating behavior. University of Oregon researchers found they could set up a mixture of six kinds of chemicals interacting in five steps, including autocatalysis that would begin in an unchanging mono-color state, and then by adding reactants, begin changing color in an oscillating manner.

The complexity of the model (nicknamed the Oregonator in

honor of the patron institution) prohibits its use here; however, the following simplified version below, consisting of 4 starting chemicals that undergo 7 reactions with 6 intermediate species, while less dramatic, encompasses all of the model's key aspects. The reactions separate into 3 processes identified according to the chemical function involved.

The reaction, shown below, begins in process A with a mixture of $HBrO_3$ (bromic acid), HBr (hydrogen bromide), $CH_2(COOH)_2$ (malonic acid), and a source of Ce^{+3} ions in a beaker. Process A involves 3 oxidation/reduction—or redox—reactions. Redox reactions occur when chemical entities move between electronic states in which electrons transfer between atoms or molecular groups, thus causing one to gain electrons (be reduced) and the other to lose electrons (be oxidized), hence redox. For example in step I of the Oregonator, the oxidation state of the bromine atom in $HBrO_3$ is +5 but –1 in HBr. The –1 charge for Br in HBr is readily seen since hydrogen always has an oxidation state of +1, so to make HBr electrically neutral bromine must be –1. Oxygen on the other hand carries a –2 charge; therefore Br in $HBrO_3$ (bromic acid) must have a +5 charge to balance the –6 charge contributed by the 3 oxygen atoms and the +1 from hydrogen. The bromine atom in HBr is said to have been oxidized in the reaction since it lost—in this case—2 electrons making its valence more positive, going from –1 in HBr to +1 in HOBr. The bromine atom in $HBrO_3$ gained 2 electrons, going from +5 to +3 while being converted to $HBrO_2$ (bromous acid), and therefore it was reduced. Notice that the electrical charges (highlighted for illustrative purposes) balance between the left to the right side of the equation; the +5 and –1 bromine atoms on the left balance the +3 and +1 on the right, both adding to +4.

When the reaction takes place in a stirred beaker, the entire contents regularly change color from red to clear and back again. The concentrations of the two oxidation states of cerium dictate the color at any given time, with Ce^{+3} imparting the red color and Ce^{+4} the clear. The starting solution will be red since the cerium ions in the initial reactants carry a +3 charge (usually in the form of $Ce(NH_4)_2(NO_3)_5$ where NO_3 has a –1 charge and NH_4 +1).

Process A – Elimination of HBr and bromination of organic compounds

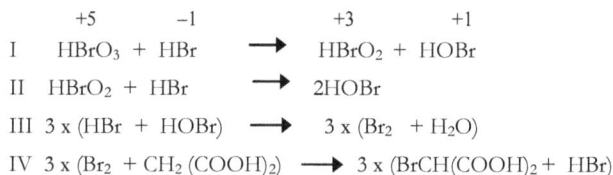

$$
\begin{array}{llll}
& +5 & -1 & +3 & +1 \\
\text{I} & HBrO_3 + HBr & \longrightarrow & HBrO_2 + HOBr
\end{array}
$$

I $\quad HBrO_3 + HBr \longrightarrow HBrO_2 + HOBr$

II $\quad HBrO_2 + HBr \longrightarrow 2HOBr$

III $\quad 3 \times (HBr + HOBr) \longrightarrow 3 \times (Br_2 + H_2O)$

IV $\quad 3 \times (Br_2 + CH_2(COOH)_2) \longrightarrow 3 \times (BrCH(COOH)_2 + HBr)$

Net Process A:

$HBrO_3 + 3CH_2(COOH)_2 + 2 HBr \longrightarrow 3 \times (BrCH(COOH)_2 + 3H_2O$

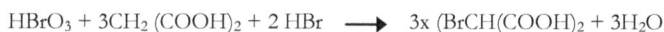

Process B – Radical generation with oxidation of Ce^{+3} and disproportionation of HBrO$_2$

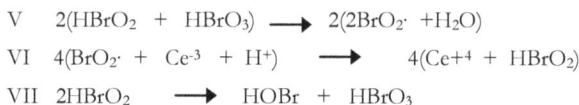

V $\quad 2(HBrO_2 + HBrO_3) \longrightarrow 2(2BrO_2 + H_2O)$

VI $\quad 4(BrO_2 + Ce^{-3} + H^+) \longrightarrow 4(Ce^{+4} + HBrO_2)$

VII $\quad 2HBrO_2 \longrightarrow HOBr + HBrO_3$

Net Process B:

$HBrO_3 + 4Ce^{+3} + 4H^+ \longrightarrow HOBr + 2H_2O + 4 Ce^{+4}$

Process C – Feedback through regeneration of Ce^{+3} and Br$^+$

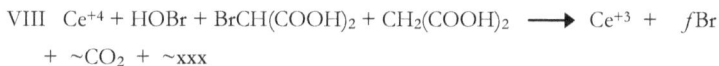

VIII $\quad Ce^{+4} + HOBr + BrCH(COOH)_2 + CH_2(COOH)_2 \longrightarrow Ce^{+3} + \int Br$
$+ \sim CO_2 + \sim xxx$

Net Rx I, II, III:

$HBrO_3 + 5HBr \longrightarrow 3 Br_2 + 3H_2O$

Beginning with a freshly prepared mixture of chemicals, the reactions of process A will dominate, lowering the concentration of Br^{-1} anions (from HBr) through reactions with HBrO$_3$ in the +5 state)

and $HBrO_2$ (+3) in reactions I and II. In reaction I, the electron rich Br^{-1} donates a pair of electrons to its high oxidation state cousin Br^{+5} changing them to Br^{+1} and Br^{+3} respectively. HBr continues to be consumed in reaction II with Br^{-1} being oxidized, and Br^{+3} reduced, both ending up as Br^{+1} in HOBr. The oxidation state of the Ce^{+3} in the starting mixture does not change since Br^{-1} serves as an inhibitor to the Ce^{+3} to Ce^{+4} oxidation reaction.

However, as reactions I and II progress, the concentration of HBr falls below a critical value, causing Process B to kick in, where reaction V converts $HBrO_2$ and $HBrO_3$ into $BrO_2\cdot$ radicals (an element or group with one or more unpaired electrons—represented by the dot). The unstable nature of the $BrO_2\cdot$ radical—brought about by the need of a companion for its unpaired electron—drives the rapid oxidation of Ce^{+3} in reaction VI by relieving it of one electron, converting $BrO_2\cdot$ to $HBrO_2$. The lost electron changes Ce^{+3} to the Ce^{+4} state, causing the solution to become clear.

Reaction VIII provides the negative feedback necessary for the cyclic behavior of the system. An oxidizable organic compound (in this case malonic, or propanedioc acid) oxidizes to CO_2 providing electrons to reduce Br^+ to Br^{-1} and Ce^{+4} to Ce^{+3}. Negative feedback effectively makes Process B self-inhibiting since it replenishes the supply of HBr (Br^{-1} + H^+), which acts as an inhibitor for the Ce^{+3} to Ce^{+4} oxidation. The HBr serves to switch control of the reaction sequence back to Process A while the Ce^{+3} is regenerated for Process B in the next phase of the cycle, Process C. The factor f in reaction VIII is a stoichiometric coefficient that denotes the ratio of the amount of Br^{-1} ions per Ce^{+4} ion generated. The symbols ~ represents the quantity of CO_2 to be produced that will be dependent on the stoichiometry (the relative amounts of the reactants and products; e.g. stoichiometric quantities of the atoms of H_2O would be 2 atoms of hydrogen and one of oxygen) while ~xxx represents organic byproducts of malonic acid—also dependent on stoichiometry. In order for the system to oscillate properly between red and clear, the value of f must always be greater than 1 (e.g. more Br^{-1} than Ce^{+4} ions) but less than some maximum value to prevent either Process A or B from gaining permanent control and preventing the oscillating behavior.

The truly fascinating aspects of the Oregonator however, occur in a quiet environment. Unstirred, the interaction of different diffusion rates of the reactants, combined with the autocatalytic reaction of $HBrO_2$, generates traveling waves of reaction and therefore color. As HBr from process A becomes spent in small regions, Process B kicks in with reactions V and VI producing BrO_2· radicals that diffuse rapidly through the liquid, oxidizing Ce^{+3} ions to Ce^{+4}. This creates a clear wave front in the red liquid and places the region immediately behind the front under the control of Process C. The vessel takes on the appearance of a pond into which stones are dropped, but instead of generating water waves, waves of alternating color propagate outward from the point of entry. Even more dazzling color displays come from boiling away volatile portions of the mixture and placing the concentrated reactants between two glass plates. This causes beautiful patterns of concentric or spiral waves of color to pass through the liquid.

The oregonator represented a significant step in understanding chemical processes that may have been involved in the synthesis of the first life on Earth. Adding reactant chemicals that cause complex but stable mixtures to oscillate, changing the system from a one period to a two period state, verifies that bifurcation has taken place and demonstrates the chaotic nature of the reaction. By gradually increasing the rate of reactant addition, while at the same time removing non-critical reaction products, the system reaches critical thresholds where further bifurcations bring on a four period state, eight period, and so on into chaos. These accomplishments led to developing a mechanism that connected the mathematics and chemistry of the BZ reaction and proved that negative feedback in autocatalytic processes can cause homogeneous chemical oscillations, and that complex patterns of self-organization can arise from a few simple interactions.

SOURCES AND ACKNOWLEDGEMENTS

Bloom, M., Mouritsen, O. G. *The Evolution of Membranes*. Handbook of Biological Physics, Volume 1, 1995

Bodanis, David. E=Mc²: *A Biography of the World's Most Famous Equation*. New York: Walker and Company, 2000

Cram, Donald J, Hammond, George S. *Organic Chemistry, Second Edition*. New York San Francisco Toronto London: McGraw Hill Book Company, 1964

Cotton, Albert F. *Advanced Inorganic Chemistry: A Comprehensive Text*. United States of America: Interscience Publishers, 1962

Dawkins, Richard. *The Ancestors Tale: A Pilgrimage to the Dawn of Evolution*. Boston-New York: Houghton Mifflin company, 2004

Dumiak, Michael. *Neanderthal Code*: Archaeology: A Publication of the Archaeological Institute of America,Volume 59 Number 6, November/December 2006

Ellis, Richard. *Aquagenisis: The Origin and Evolution of Life in the Sea*. The Penguin Group, 2001

Enard, W. et al. *Molecular evolution of FOXP2*. Nature, 418(6900):869-7 Aug 22, 2002

Field, R. J. *A Reaction Periodic in Time and Space*. J. Chem. Ed. 49, 308, 1972

Field, R. J., Noyes, R.M. *Oregonator Model of Oscillating Chemical Reactions*. J. Chem. Phys. 60, 1877, 1974

Fields, Helen, *Superfast DNA Sequencing*. U.S.News & World Report, Posted 8/4/05

Finkelstein, Israel and Silberman, Neil Asher. *The Bible Unearthed: Archaeology's New Vision of Ancient Israel and the Origin of Its Sacred Texts*. The Free Press, 2001

Fortey, Richard. *Life: A Natural History of the First Four Billion Years of Life on Earth*. New York: Vintage Books, 1997

Frazer, James George. *The Golden Bough: A Study in Magic and Religion*. Oxford World's Classics, 1890

Friedman, Thomas L. *The World is Flat: A Brief History of the Twenty-first Century*. New York: Farrar, Straus, and Giroux, 1960

Gamble, Eliza Burt. *Alternative Religion*

Gleick, James. Chaos - *Making a New Science, Harmondsworth*, Middlesex: Penguin Books Ltd 1987.

Goldberg, Elkhonon *The Wisdom Paradox*. New York: Gotham Books, 2005

Gould, Edwin S. *Mechanism and Structure in Organic Chemistry*. New York: Holt Dryden, 1960

Gould, Stephen J. *Full House: The Spread of Excellence From Plato to Darwin*. New York: Harmony Books, 1996

Gould, Stephen J. *The Evolution of Life on Earth*. Scientific American, October, 1994.

Greene, Brian. *The Elegant Universe: Superstrings, Hidden Dimensions, and the Quest for the Ultimate Theory*. New York: Vintage Books – Random House, 1999

Gribbin, John. *Deep Simplicity: Bringing Order out of Chaos and Complexity*. New York: Random House, 2004

Grube, G.M.A. (Translated by) *Plato's Republic*. Indianapolis: Hackett Publishing Company, 1984

Hawking, Stephen W. *A Brief History of Time*. New York Toronto London Sydney Auckland: Bantum Books, 1996

Hawking, Stephen W. *The Universe in A Nutshell*. New York Toronto London Sydney Auckland: Bantum Books, 2001

Hawking, Stephen W.; Thorne, Kip S.; Novikov, Igor; Ferris, Timothy; Lightman, Alan: Price, Richard; *The Future of Spacetime*. New York, London: W. W. Norton & Company, 2002

Itzaki, Jane. *The FOXP2 Story*. Human Genome Website, April 24, 2003

Jeffery, Simon. *Beyond Maturity: Stellar Evolution Beyond the Main Sequence*. Now Magazine, June 1998

Kauffman, Stewart. *At Home in the Universe: The Search for Laws of Self-Organization and Complexity*.

Krauss, Lawrence M. *Atom: An Odyssey from the Big Bang to Life on earth…and*

Beyond. Boston New York London: Little, Brown, and Company, 2001

Kaku, Michio. *Hyperspace: A Scientific Odyssey Through Parallel Universes, Time Warps, and the 10th Dimension.* New York: Anchor Books – Doubleday, 1994

Kaku, Michio. *Parallel Worlds: A Journey Through Creation, Higher Dimensions, and the Future of the Cosmos.* New York: Doubleday, 2005

Latimer, Wendell M. *Oxidation Potentials*: Second Edition. Prentice Hall Inc. Englewood Cliffs, N. J., 1964

Leavitt, David. *The Man Who Knew Too Much: Alan Turing and the Invention of the Computer.* New York: W. W. Norton & Company, 2006

Magueijo, Joao. *Faster Than The Speed of Light: The Story of Scientific Speculation.* Cambridge: Perseus Publishing, 2003

Maxwell, J. B. *Data Book on Hydro*carbons. Robert F. Krieger Publishing Company, Huntington, New York. 1968

Neanderthal Yields Nuclear DNA. BBC News Website, Tuesday, 16 May 2006

O'Neill, Ian *Large Hadron Collider Could Create Wormholes: A Gateway for Time Travelers?* Universe Today, February 7th, 2008

Olson, Steve. *Mapping Human History: Discovering the Past Through Our Genes.* Boston – New York: Houghton Mifflin Company, 2002

Orgel, Leslie E., Miller, Stanley L. *The Origins of Life on Earth.* Prentice- Hall Inc. New Jersey, 1974

Rees, Martin. *Just Six Numbers: The Deep Forces That Shape the Universe.* New York: Basic Books, 2000

Rees, Martin. *Our Cosmic Habitat.* Oxford and Princeton: Princeton University Press, 2001

Ridley, Matt. *Genome: The Autobiography of a Species in 23 Chapters.* Harper Collins, 1999

Saldama, Stephanie. *Temple Reveals Secrets of the One God.* Daily Star, Lebanon, March 5 2002

Singh, Simon. *Big Bang: The Origin of the Universe.* New York: Fourth Estate, 2004

Uthman, Ed. *Elemental Composition of the Human Body.* Uthman Website

Waddell, P. J., Hirohisi, K., Risso, O. *A Phylogenitic Foundation for Comparitive Mammalian Genomics.* Genomics Informatics 12: 141-154, 2001

Günter Wächtershäuser "*Origin of Life: Life as We Don't Know It*", Science, 289

(5483) August 25, 2000 pp. 1307-1308.

Wolfson, Richard. *Einstein's Relativity and the Quantum Revolution.* The Great Courses

Yang, Lingfa Epstein, R. *Oscillatory Turing Patterns in Reaction-Diffusion Systems with Two Coupled Layers.* Physical Review Letters, Volume 90, Number 17, 2 May 2003

INDEX

Lactose intolerance. See
   mutation, genetic, See
   mutation, genetic
Lagomorphia, 264
Lake Baikal, 272
lambda. See cosmological
   constant
lancelet. See vertebrates
language, 5, 339
   earliest usage of, 327
   family KE, 327
   FOXP2 gene, 327
Large Hadron Collider, 78, 361,
   363
Laurasia, 279
lava
   magnetic allignment, 253
Leakey, Richard, 309
Leavitt, Henrietta, 100, 98–101, 103
Lemaître, George, 15, 93, 95, 106
leucine, 225
Lewis, G.N., 157
light
   constructive and destructive
      interference, 23, 25
   wave/particle duality of, 26
Limulus, 270
lithium, 124, 135, 156, 160, 161
   formation in early universe, 124
lobefin, 265
Lorentz contraction, 60-62, 94
Lorentz, Hendrik, 60, 61, 62
Lorenz, Edward, 198
Los Alamos Laboratories, 204
Luca, 234
   Last Universal Common
      Ancestor, 234
lycine, 225
Lyell, Charles, 283, 284
magma, 276
Magnesium, 162
Mantell, Gideon, 278, 279
Marcel Grossman, 85
Mars, 42, 63, 143, 144
Mars Rover project, 63

Marsden, Brian, 348
mass and energy equivalence
   special relativity, 79
mass defect, 128
Massachussets Institute of
   Technology, MIT, 116, 301
Maxwell, James Clerk, 16, 17, 26,
   46–51, 355
   electromagnetism, 26, 46
McElwain, Jennifer, 275, 276
megaparsec, 103
Mesozoic era, 279
Metatheria, 263
meteor, 146, 147, 286, 287, 346, 347
   effect of size on impact, 146
meteorites, 348
methane, 171
   bonding with hydrogen, 170
   conversion of sewage to, 238
   methanogens and swamp gas,
      237
   tetrahedral configuration, 166,
      167
methanogens, 238
methionine, 224, 225
Michaelson, Edward, 60
Michelson, Albert, 48, 49, 50
Michio Kaku, 10
Migdol Temple, 337
migration, 295, 296, 341
   by rafting, 295
   Old World Monkeys, Africa to
      Asia, 295
Milky Way, 8, 13, 95, 99, 103–8, 351
Miller, Stanley L., 222
Miocene Epoch, 290, 292, 293
Mississippian era, 269
mitochondria, 243, 310-315
mitochondrial DNA, 4, 310, 311,
   315
mitochondrial Eve, 4, 311, 312, 314,
   322
modern humans, 306, 311, 336, 352
   and Neanderthal, 320
   brain size, 306

How IT Happened

How IT Happened

www.ingramcontent.com/pod-product-compliance
Lightning Source LLC
Chambersburg PA
CBHW021547210326
41599CB00010B/340